AF779927

New Control Schemes for Actuators

New Control Schemes for Actuators

Guest Editors

Oscar Barambones
Jose Antonio Cortajarena
Patxi Alkorta

MDPI

Basel • Beijing • Wuhan • Barcelona • Belgrade • Novi Sad • Cluj • Manchester

Guest Editors

Oscar Barambones
Department of System
Engineering and Automation
Faculty of Engineering of
Vitoria-Gasteiz
University of the Basque
Country
Leioa
Spain

Jose Antonio Cortajarena
Engineering School of
Gipuzkoa
University of the Basque
Country
Eibar
Spain

Patxi Alkorta
Department of Systems
Engineering and Automation
Faculty of Engineering,
Gipuzkoa (Eibar)
University of the Basque
Country
Eibar
Spain

Editorial Office

MDPI AG
Grosspeteranlage 5
4052 Basel, Switzerland

This is a reprint of the Special Issue, published open access by the journal *Actuators* (ISSN 2076-0825), freely accessible at: https://www.mdpi.com/journal/actuators/special_issues/control_schemes.

For citation purposes, cite each article independently as indicated on the article page online and as indicated below:

Lastname, A.A.; Lastname, B.B. Article Title. *Journal Name* **Year**, *Volume Number*, Page Range.

ISBN 978-3-7258-3509-6 (Hbk)
ISBN 978-3-7258-3510-2 (PDF)
https://doi.org/10.3390/books978-3-7258-3510-2

© 2025 by the authors. Articles in this book are Open Access and distributed under the Creative Commons Attribution (CC BY) license. The book as a whole is distributed by MDPI under the terms and conditions of the Creative Commons Attribution-NonCommercial-NoDerivs (CC BY-NC-ND) license (https://creativecommons.org/licenses/by-nc-nd/4.0/).

Contents

actuators

Editorial

New Control Schemes for Actuators

Oscar Barambones [1,*], José Antonio Cortajarena [2] and Patxi Alkorta [2]

1 Engineering School of Vitoria, University of the Basque Country, UPV/EHU, Nieves Cano 12, 1006 Vitoria, Spain
2 Engineering School of Gipuzkoa, University of the Basque Country, UPV/EHU, Avda Otaola N29, 20600 Eibar, Spain; josean.cortajarena@ehu.eus (J.A.C.); patxi.alkorta@ehu.eus (P.A.)
* Correspondence: oscar.barambones@ehu.eus

An actuator is a device that moves or controls a mechanism, by turning a control signal into mechanical action, such as in an electric motor. Actuators may be hydraulic, pneumatic, electric, thermal or mechanical, and they may be powered by electric current, hydraulic fluid or pneumatic pressure. However, increasingly, these systems are being driven by software, with the control signal originating from a microcontroller programmed by software. Therefore, an element key to increasing the reliability and performance of these actuators is the control system. The limitations of traditional control techniques when coping with real control problems have motivated the invention of new and advanced control schemes, in order to improve actuator performance and reliability and to reduce the non-linear dynamics and uncertainties usually present in actuators. Control schemes refer to the strategies and methods used to regulate and manipulate the behavior of a system or process. These schemes are crucial in various fields, including engineering, automation and robotics, to achieve desired outcomes, improve performance and maintain stability [1–3].

In this sense, the objective of the control scheme is to drive the system outputs toward a desired state, while minimizing any steady-state errors, overshoot and delays. Moreover, the control scheme should ensure the stability of the controls, as well as optimizing them as far as possible.

Several types of control systems date back to ancient times. However, a more formal analysis of the field began with a dynamic analysis of the centrifugal governor (used to regulate speed), which was conducted by physicist James Clerk Maxwell in 1868 and titled On Governors.

However, modern control theory was not developed until the 1960s, which heralded the introduction of new mathematical tools and techniques for analyzing and designing new control systems. This theory was based in the state space and could deal with multiple-input and multiple-output (MIMO) systems. This overcame the limitations of using classical control theory for more sophisticated design problems, such as fighter aircraft control; however, there was a limitation, in that no frequency domain analysis was possible. In modern control theory, a system is represented as a set of first order differential equations, defined using state space variables. Under this new approach, many new control schemes were developed, including non-linear, multivariable, adaptive and robust control schemes, among others [4,5].

More recently, in the mid-20th century, the advent of computers paved the way for digital control systems. Digital controllers offered greater flexibility and precision, as well as the ability to implement complex algorithms. Since that point, the field of control schemes has continued to evolve, with ongoing advancements in areas such as artificial intelligence, machine learning and quantum control [6–8].

The most common control scheme used in industry currently is the traditional PID control. However, advancements in technology continually drive the development of new control schemes for actuators, which play a crucial role in various industries, ranging from manufacturing to robotics, and are essential for the precise and efficient control of motion and force [9–11].

check for updates

Citation: Barambones, O.; Cortajarena, J.A.; Alkorta, P. New Control Schemes for Actuators. *Actuators* **2024**, *13*, 99. https://doi.org/10.3390/act13030099

Received: 19 February 2024
Accepted: 28 February 2024
Published: 1 March 2024

Copyright: © 2024 by the authors. Licensee MDPI, Basel, Switzerland. This article is an open access article distributed under the terms and conditions of the Creative Commons Attribution (CC BY) license (https://creativecommons.org/licenses/by/4.0/).

Below is a brief summary of some advanced control techniques currently under development, in order to improve actuator performance, as follows [12–15]:

Model Predictive Control (MPC): Uses a dynamic model of the system to predict future behavior and optimize control inputs over a specified prediction horizon. MPC is effective for systems with constraints and varying operating conditions.

Fuzzy Logic Control: Utilizes fuzzy sets and linguistic variables to create rules for decision-making. Fuzzy logic is especially useful when dealing with systems that deal with uncertain or imprecise information.

Model Reference Adaptive Control (MRAC): adjusts the controller parameters based on the difference between the system's actual and desired responses, adapting to changes in system dynamics.

Self-Tuning Adaptive Control: automatically adjusts controller parameters in real time, based on changes in the system or operating conditions.

Optimal Control: Utilizes optimization techniques to find the best control inputs that minimize a cost function, considering system dynamics and constraints. Model Predictive Control (MPC) is a type of optimal control.

Robust Control: Focuses on maintaining stable performance in the presence of uncertainties and variations in system parameters. Robust control techniques provide a margin of safety against disturbances.

Sliding Mode Control: Applies a discontinuous control law to drive the system along a sliding surface, ensuring rapid convergence to the desired state. Sliding mode control is robust against uncertainties and disturbances.

Neural Network Control: Integrates artificial neural networks in the control scheme, to learn and adapt to system dynamics. Neural network controllers are particularly useful for non-linear systems or those with complex and uncertain behaviors.

Reinforcement Learning (RL): RL algorithms can adapt and optimize control strategies based on feedback from the environment. This can be particularly useful in scenarios where the system dynamics are complex or change over time.

Redundancy-Based Control: actuators utilising this scheme are designed with redundant components, enabling them to adapt to component failures and thereby ensuring the system's continued operation.

Hierarchical Control: Organizes control tasks into a hierarchy of levels, with each level responsible for a specific aspect of the system. This structure simplifies complex systems and allows for a modular design.

Quantum Control: in emerging fields such as quantum computing, control schemes are designed to manipulate quantum states for specific applications, such as quantum information processing and quantum communication.

These control schemes represent a diverse range of approaches aimed at enhancing the efficiency, adaptability and reliability of actuators in different applications. The choice of a specific control scheme depends on the characteristics of the system, the desired performance of the actuator and the environmental conditions in which the actuator operates.

The Special Issue entitled "New Control Schemes for Actuators" was an opportunity to share knowledge, experience and information regarding the design and implementation of different control schemes. In this Special Issue, eleven original research papers were published. Within these papers, the authors analyzed different control schemes, such as Boundary Control, Disturbance Rejection Control, Fault-Tolerant Control, Adaptive Fuzzy Logic, Robust Control, Digital Twin Concept, Sliding Mode Control, Damping Passivity-Based Control and Adaptive Control. These works are summarized below.

In the first paper, Acho et al. [16] proposed a boundary controller based on a peak detector system, in order to reduce vibrations in the cable–tip–mass system. The control procedure was built upon a recent modification of the controller itself, incorporating a non-symmetric peak detector mechanism to enhance the robustness of the control design. The crucial element lay in the identification of peaks within the boundary input signal, which were then utilized to formulate the control scheme. Its mathematical representation relied on just two tunable parameters. Numerical experiments were conducted to assess the performance of this novel approach and to compare it to that of the boundary damper control; the results demonstrated that the novel approach showcased superior efficacy to the boundary damper control.

In the work of Zhang et al. [17], a friction feed-forward compensation method, based on an improved active disturbance rejection control (IADRC), was designed. A mathematical model of EMA was also developed, and the relationship between friction torque and torque current was derived. Furthermore, the compound ADRC method, utilizing a second-order speed loop and a position loop, was studied, and an IADRC method was proposed. A real EMA was developed, and the working principles of the EMA driving circuit and current sampling were analyzed. The three methods—PI, ADRC, and IADRC—were verified by conducting speed step experiments and sinusoidal tracking experiments. The integral values of time multiplied by the absolute error of the three control modes under the step speed mode were approximately 47.7, 32.1 and 15.5, respectively. Disregarding the inertia of the reducer and assuming that the torque during no-load operations equals the friction torque during constant motion, the findings indicate that, under a load purely driven by inertia, the IADRC method enhanced the tracking accuracy.

In the work of Wu et al. [18], a fuzzy linear active disturbance rejection control strategy (FLADRC) for absolute pressure piston manometers was proposed, to address the internal uncertainties and external disturbances of a pressure-measuring instrument. First, the characteristics of the main components were analyzed, according to the actual working principle of the system, to establish a theoretical model of the controlled system. Second, the corresponding linear active disturbance rejection controller (LADRC) was designed, according to the model. The principle of fuzzy control was introduced, in order to adaptively adjust the controller parameters of the LADRC in real time. The LADRC parameters have several disadvantages which are difficult to rectify, including a poor immunity to disturbances due to their fixed nature; adaptively adjusting these parameters subsequently demonstrated the stability of the control method. Finally, a simulation model was built in the Simulink environment in MATLAB, and three different pressure operating points were selected for the corresponding experiments, in order to comparatively analyze Kp, PID and LADRC. The results showed that FLADRC enabled the absolute pressure piston manometer, achieving better stability and a greater immunity to disturbances. This also verified the effectiveness and feasibility of the control strategy in practical engineering applications.

In the paper authored by Zhu et al. [19], the fault problem in distributed-four-wheel-drive electric vehicle drive systems was addressed. First, a fault-factor-based active fault diagnosis strategy was proposed. Second, a fault-tolerant controller was designed, to reconstruct motor drive torque based on vehicle stability. This controller ensured that the vehicle maintained stability by providing fault-free motor output torque based on the fault diagnosis results. To validate the effectiveness of the fault diagnosis and the fault-tolerant control, SIL simulations were conducted, using MATLAB/Simulink and CarSim. A hardware-in-the-loop (HIL) simulation platform, with the highest possible confidence level, was established, based on NI PXI and CarSim RT. Through the HIL simulation experiments, the proposed control strategy was shown to be able to accurately diagnose the operating state of the motor, rebuild the motor torque based on the stability of the system, and demonstrate robust stability when the drive system failed. Under various fault conditions, the maximum error in the vehicle lateral angular velocity was less than 0.017 rad/s, and the maximum deviation in the lateral direction was less than 0.7 m. These findings substantiated the highly robust stability of the proposed method.

The next paper, authored by Sun et al. [20], aimed to highlight the critical role of robot manipulators in industrial applications and elucidate the challenges associated with achieving high-precision control. In particular, the detrimental effects of non-linear friction on manipulators were discussed. To overcome this challenge, a novel friction compensation controller (FCC), combining time delay estimation (TDE) and an adaptive fuzzy logic system (AFLS), was proposed in this paper. The friction compensation controller was designed to take advantage of the time delay estimation algorithm's strengths in eliminating and estimating the unknown dynamic functions of the system, using information from the previous sampling period. Simultaneously, the adaptive fuzzy logic system compensated for the hard non-linearities in the system and suppressed the errors generated by time delay estimation, thus improving the tracking accuracy of the robotic arm. The numerical experimental results demonstrated that the proposed friction compensation controller significantly enhanced the tracking accuracy of the robotic arm, and that the addition of the adaptive fuzzy logic system improved the performance of the time delay estimation by an average of 90.59.

In the work of Hashim et al. [21], two new versions of modified active disturbance rejection controls (MADRCs) were proposed, which aimed to stabilize a non-linear quadruple tank system and to control the water levels of the lower two tanks in the presence of exogenous disturbances, parameter uncertainties and parallel varying input set-points. The first proposed scheme was configured from the combination of a modified tracking differentiator (TD), a modified super twisting sliding mode (STC-SM) and a modified non-linear extended state observer (NLESO). The second proposed scheme was obtained by aggregating another modified TD, a modified non-linear state error feedback (MNLSEF) and a fal-function-based ESO. The MADRC schemes, with a non-linear quadruple tank system, were investigated by running simulations in the MATLAB/SIMULINK environment, and several comparison experiments were conducted, to validate the effectiveness of the proposed control schemes. Furthermore, a genetic algorithm (GA) was used as a tuning algorithm to parametrize the proposed MADRC schemes, with the integral time absolute error (ITAE), the integral square of the control signal (ISU) and the integral absolute of the control signal (IAU) as an output performance index (OPI). Finally, the simulation results showed the robustness of the proposed schemes, with a noticeable reduction in the OPI.

Chaiprabha et al. [22] proposed an advanced trajectory controller, based on a digital twin framework into which artificial intelligence (AI) was incorporated, which could effectively control a precision linear stage. A precision linear stage is an electro-mechanical system that includes a motor, electronics, flexible coupling, gear, ball screw and precision linear bearing. In these kind of systems, a tight fit can provide better precision but also generates a difficult-to-model friction that is highly non-linear and asymmetrical. This framework offered the following advantages: the detection of abnormalities, an estimation of performance and selective control over any situation. The digital twin was developed via Matlab's Simscape and ran concurrently, using a real-time controller.

In the work of Shiravani et al. [23], an enhanced integral sliding mode control (ISMC) for the mechanical speed of an induction motor (IM) was presented and experimentally validated. The design of the proposed controller was created in the DQ synchronous reference frame with indirect field-oriented control (FOC). Global asymptotic speed tracking, in the presence of model uncertainties and load torque variations, was guaranteed using an enhanced ISMC surface. Moreover, this controller provided a faster speed convergence rate than the conventional ISMC and proportional integral methods, and it eliminated the steady-state error. Furthermore, the chattering phenomenon was reduced through the use of a switching sigmoid function. The stability of the proposed controller under parameter uncertainties and load disturbances was proven, using the Lyapunov stability theory. Finally, the performance of this control method was verified through numerical simulations and experimental tests, achieving fast dynamics and good robustness for IM drives.

Montoya et al. [24] presented a paper describing the output voltage regulation control for an interleaved microgrid connected to a direct current (DC), which considered bidirectional current flows. The proposed controller was based on an interconnection and damping passivity-based control (IDA–PBC) approach, with integral action that regulated the output voltage profile at its assigned reference. These authors also designed a control law, using non-linear feedback, that ensured asymptotic stability in a closed loop, according to Lyapunov. Moreover, the IDA–PBC design added an integral gain to eliminate the tracking errors possible in steady-state conditions. Numerical simulations, carried out in the piecewise linear electrical circuit simulation (PLECS) package for MATLAB/Simulink, enabled the assessment of the effectiveness of the proposed controller and its comparison with a conventional proportional integral controller under different scenarios, considering strong variations in the current injected/absorbed by the DC microgrid.

Zhang et al. [25] derived a mathematical model of asymmetric thrust magnetic bearings for a cold compressor, and analyzed the changes in the system characteristics owing to changes in the equilibrium position. By constructing PID controllers associated with the structural parameters of the magnetic bearing, they realized the adaptive adjustment of the control parameters under different balanced position commands. The simulation and experimental results proved that the gain-scheduled control method proposed in this paper could achieve robust stability of the rotor, in the range of 50 to 350 μm, and not at the cost of the response speed, adjustment time and overshoot. These research results have significance for the structural design of asymmetric thrust magnetic bearings and play an important role in the commissioning and performance improvement of cold compressors.

In the last paper of this Special Issue, Chen et al. [26] proposed a Witty control system, using a revised recurrent Jacobi polynomial neural network (RRJPNN) control and two remunerated controls with an altered bat search algorithm (ABSA) method, in order to control the electromagnetic actuator systems employed in a rice milling machine system. The Witty control system, with a finer learning capability, could fulfil the RRJPNN control, which involved an attunement law, two remunerated controls, which also have two evaluation laws, and a dominator control. The aforementioned attunement and evaluation laws were derived from the Lyapunov stability principle. Moreover, the ABSA method could acquire adjustable learning rates, to quicken the convergence of the weights. Finally, the proposed control method exhibited a finer control performance, which was confirmed by the experimental results.

The number and the quality of the papers presented in this Special Issue have shown that the design and implementation of new control schemes for different actuators is an active research area that attracts the interest of the scientific community.

Finally, as the Guest Editors, we would like to thank all of the authors who submitted papers and, therefore, contributed to the success of this Special Issue. All the papers submitted were reviewed by experts in the field and I would like to extend my thanks to these reviewers; without their input, the Special Issue would not have been a success. We would also like to thank the Editorial Board for their assistance in managing this Special Issue.

Author Contributions: O.B., J.A.C. and P.A.: writing, review and editing. All authors have read and agreed to the published version of the manuscript.

Funding: This research received no external funding.

Acknowledgments: The authors wish to express their gratitude to the Basque Government through the project EKOHEGAZ II (ELKARTEK KK-2023/00051), to the Diputación Foral de Álava (DFA) through the project CONAVANTER, to the UPV/EHU through the project GIU23/002, and to the MobilityLab Foundation (CONV23/14, CONV23/12), for supporting this work.

Conflicts of Interest: The authors declare no conflicts of interest.

References

1. Tan, C.; Lin, M.; Xia, H.; Zhao, B. Fault-tolerant tracking control for nonlinear systems with multiplicative actuator faults in view of zero-sum differential games. *Neurocomputing* **2024**, *567*, 127037. [CrossRef]
2. Li, A.; Qin, D.; Guo, Z.; Xia, Y.; Lv, C. Wet clutch pressure hysteresis compensation control under variable oil temperatures for electro-hydraulic actuators. *Control Eng. Pract.* **2023**, *141*, 105723. [CrossRef]

3. Li, J.; Lu, Y.; He, F.; Miao, L. Motion control of hydraulic actuators with nonlinear friction compensation: Applied to variable valve systems of diesel engine. *ISA Trans.* **2023**, *137*, 561–573. [CrossRef]
4. Liu, Y.; Chen, X.; Yu, Z.; Yu, H.; Meng, L.; Yokoi, H. High-precision dynamic torque control of high stiffness actuator for humanoids. *ISA Trans.* **2023**, *141*, 401–413. [CrossRef]
5. Sai, H.; Xu, Z.; He, S.; Zhang, E.; Zhu, L. Adaptive nonsingular fixed-time sliding mode control for uncertain robotic manipulators under actuator saturation. *ISA Trans.* **2022**, *123*, 46–60. [CrossRef]
6. Ding, S.; Zhao, T.; Zhang, W.; Gao, F.; Zhu, B.; Tang, Z. Motion-inhibition control of a multi-robot mooring system using an actuating force fuzzy control method. *Ocean Eng.* **2023**, *267*, 113236. [CrossRef]
7. Fan, Y.; Shi, Y.; Mu, D.; Liu, J.; Li, Z. Global fixed-time adaptive fuzzy path following control for unmanned surface vehicles subject to lumped uncertainties and actuator saturation. *Ocean Eng.* **2023**, *286*, 115533. [CrossRef]
8. Yan, S.; Qian, H.; Ding, P.; Chu, S.; Wang, H. Finite-time tolerant containment control for IT2 T-S fuzzy network multi-agent systems with actuator faults, packet dropouts and DoS attacks. *ISA Trans.* **2023**, *137*, 199–209. [CrossRef]
9. Ziaei, A.; Sinafar, B.; Kharrati, H.; Rahimi, A. Concurrent-learning-based event-triggered fault tolerant attitude control for spacecraft with actuator faults. *Adv. Space Res.* **2024**, *73*, 95–107. [CrossRef]
10. Lázaro, R.P.S.; Fuentes-Aguilar, R.; Chairez, I. Trajectory tracking control with state restricted gains for a magnetic pendulum using electromagnetic actuators. *ISA Trans.* **2023**, *139*, 475–483. [CrossRef]
11. Wang, W.; Wen, T.; He, X.; Xu, G. Robust trajectory tracking and control allocation of X-rudder AUV with actuator uncertainty. *Control Eng. Pract.* **2023**, *136*, 105535. [CrossRef]
12. Wan, Q.; Cao, L.; Pan, Y.; Lin, W. Improved prescribed performance fuzzy control for nonlinear multiagent systems with actuator faults and dual input saturations. *Eur. J. Control* **2024**, *75*, 100937. [CrossRef]
13. Liu, A.; Lai, G.; Xiao, H.; Liu, Z.; Zhang, Y.; Chen, C.P. Resilient adaptive trajectory tracking control for uncalibrated visual servoing systems with unknown actuator failures. *J. Frankl. Inst.* **2024**, *361*, 526–542. [CrossRef]
14. Tang, L.; Liu, J.; Wang, L.; Wang, Y. Robust fixed-time trajectory tracking control of the dynamic positioning ship with actuator saturation. *Ocean Eng.* **2023**, *284*, 115199. [CrossRef]
15. Khan, A.M.; Kim, Y.; Shin, B.; Moghadam, M.H.; Mansour, N.A. Modeling and control analysis of an arc-shaped SMA actuator using PID, sliding and integral sliding mode controllers. *Sens. Actuators A Phys.* **2022**, *340*, 113523. [CrossRef]
16. Acho, L.; Pujol-Vázquez, G. Enhancing Vibration Control in Cable-Tip-Mass Systems Using Asymmetric Peak Detector Boundary Control. *Actuators* **2023**, *12*, 463. [CrossRef]
17. Zhang, P.; Shi, Z.; Yu, B. Research on Friction Compensation Method of Electromechanical Actuator Based on Improved Active Disturbance Rejection Control. *Actuators* **2023**, *12*, 445. [CrossRef]
18. Wu, H.; Zhai, X.; Gao, T.; Wang, N.; Zhao, Z.; Pang, G. Balancing Control of an Absolute Pressure Piston Manometer Based on Fuzzy Linear Active Disturbance Rejection Control. *Actuators* **2023**, *12*, 275. [CrossRef]
19. Zhu, S.; Li, H.; Wang, G.; Kuang, C.; Chen, H.; Gao, J.; Xie, W. Research on Fault-Tolerant Control of Distributed-Drive Electric Vehicles Based on Fuzzy Fault Diagnosis. *Actuators* **2023**, *12*, 246. [CrossRef]
20. Sun, Y.; Liang, X.; Wan, Y. Tracking Control of Robot Manipulator with Friction Compensation Using Time-Delay Control and an Adaptive Fuzzy Logic System. *Actuators* **2023**, *12*, 184. [CrossRef]
21. Hashim, Z.S.; Khani, H.I.; Azar, A.T.; Khan, Z.I.; Smait, D.A.; Abdulwahab, A.; Zalzala, A.M.; Jawad, A.J.M.; Ahmed, S.; Ibraheem, I.K.; et al. Robust Liquid Level Control of Quadruple Tank System: A Nonlinear Model-Free Approach. *Actuators* **2023**, *12*, 119. [CrossRef]
22. Chaiprabha, K.; Chancharoen, R. A Deep Trajectory Controller for a Mechanical Linear Stage Using Digital Twin Concept. *Actuators* **2023**, *12*, 91. [CrossRef]
23. Shiravani, F.; Alkorta, P.; Cortajarena, J.A.; Barambones, O. An Enhanced Sliding Mode Speed Control for Induction Motor Drives. *Actuators* **2022**, *11*, 18. [CrossRef]
24. Montoya, O.D.; Serra, F.M.; Gil-González, W.; Asensio, E.M.; Bosso, J.E. An IDA-PBC Design with Integral Action for Output Voltage Regulation in an Interleaved Boost Converter for DC Microgrid Applications. *Actuators* **2022**, *11*, 5. [CrossRef]
25. Zhang, S.; Wu, J. Gain-Scheduled Control of Asymmetric Thrust Magnetic Bearing. *Actuators* **2021**, *10*, 329. [CrossRef]
26. Chen, D.F.; Chiu, S.P.C.; Cheng, A.B.; Ting, J.C. Electromagnetic Actuator System Using Witty Control System. *Actuators* **2021**, *10*, 65. [CrossRef]

Disclaimer/Publisher's Note: The statements, opinions and data contained in all publications are solely those of the individual author(s) and contributor(s) and not of MDPI and/or the editor(s). MDPI and/or the editor(s) disclaim responsibility for any injury to people or property resulting from any ideas, methods, instructions or products referred to in the content.

actuators

MDPI

Article

Electromagnetic Actuator System Using Witty Control System

Der-Fa Chen, Shen-Pao-Chi Chiu, An-Bang Cheng and Jung-Chu Ting *

Department of Industrial Education and Technology, National Changhua University of Education, Changhua 500, Taiwan; dfchen@cc.ncue.edu.tw (D.-F.C.); baochyi@ms47.hinet.net (S.-P.-C.C.); jhenganbang@gmail.com (A.-B.C.)
* Correspondence: d0331019@gm.ncue.edu.tw; Tel.: +886-4-723-2105

Abstract: Electromagnetic actuator systems composed of an induction servo motor (ISM) drive system and a rice milling machine system have widely been used in agricultural applications. In order to achieve a finer control performance, a witty control system using a revised recurrent Jacobi polynomial neural network (RRJPNN) control and two remunerated controls with an altered bat search algorithm (ABSA) method is proposed to control electromagnetic actuator systems. The witty control system with finer learning capability can fulfill the RRJPNN control, which involves an attunement law, two remunerated controls, which have two evaluation laws, and a dominator control. Based on the Lyapunov stability principle, the attunement law in the RRJPNN control and two evaluation laws in the two remunerated controls are derived. Moreover, the ABSA method can acquire the adjustable learning rates to quicken convergence of weights. Finally, the proposed control method exhibits a finer control performance that is confirmed by experimental results.

Keywords: bat search algorithm; Jacobi polynomial neural network; Lyapunov stability principle; three-phase induction servo motor; rice milling machine system

Citation: Chen, D.-F.; Chiu, S.-P.-C.; Cheng, A.-B.; Ting, J.-C. Electromagnetic Actuator System Using Witty Control System. *Actuators* **2021**, *10*, 65. https://doi.org/10.3390/act10030065

Academic Editor: Oscar Barambones

Received: 1 March 2021
Accepted: 19 March 2021
Published: 22 March 2021

Publisher's Note: MDPI stays neutral with regard to jurisdictional claims in published maps and institutional affiliations.

Copyright: © 2021 by the authors. Licensee MDPI, Basel, Switzerland. This article is an open access article distributed under the terms and conditions of the Creative Commons Attribution (CC BY) license (https://creativecommons.org/licenses/by/4.0/).

1. Introduction

Compared to other three-phase motors, three-phase induction motors (IMs) are widely used in many industrial and commerce applications due to their simple structures and easy maintenance. In order to achieve better control performance, IMs have served as induction servo motors (ISMs) via structural improvement and encoder installation. Therefore, ISMs have been broadly applied to various servo fields such as computer numerical control (CNC) machine tools and milling machines [1–4]. Li et al. [1] proposed a new intelligent adaptive CNC system design for a milling machine by using the neural network controller to achieve better control characteristics. Huang et al. [2] proposed an approach for cutting the force control of CNC machines. This approach with a state estimator was executed by using the observed variables and cutting force to achieve robust control. Recently, the developed approach was applied to a milling machine center. Gomes and Sousa [3] proposed the adaptive control of milling machine cutting force by using an artificial neural network to obtain good results. Mikolajczyk [4] proposed a system of a numerical control conventional milling machine with electromagnetic clutches by using VB6 special software to control the machine with G-code. However, these control systems took a long time to fulfill nonlinear disturbances so that they resulted in lower calculation efficiency. Thereby, the aim of the proposed witty control system using a revised recurrent Jacobi polynomial neural network (RRJPNN) control and two remunerated controls with altered bat search algorithm (ABSA) and progressive weight pruning approaches for the ISM driving the rice milling machine is to reduce computing time and to quicken convergence of weights. Meanwhile, the proposed witty control system can increase machining efficiency and control characteristics.

Due to their good learning capability, many neural networks (NNs) have been applied in many linear and nonlinear systems, such as optimal hysteresis modelling methods of magnetic controlled shape memory alloys (MSMAs) [5], control of ionic electroactive polymer actuators [6], predicting the driving force for a multicyclic driving experiment of silicone/ethanol soft composite material actuators [7], and design of a 2-DOF ankle exoskeleton with a bidirectional tendon-driving actuator and a polycentric structure [8]. These NNs take a long time to conduct training and learning of a system. In order to reduce calculating time, some polynomial function NNs have been proposed for applications in function approximation [9], dynamic control for continuously variable transmission [10], approximation and estimation of nonlinear functions [11] and backstepping control of synchronous reluctance motor (SynRM) drive systems [12]. As heuristic comprehension methods were applied to adjust learning rates of weights, these NNs that were applied in control systems resulted in slower convergence. Thereby, the proposed NN that is combined with Jacobi polynomials [13] is a novel NN structure that has not yet been applied to estimate, predict and control nonlinear systems. The feedforward Jacobi polynomial neural network (FJPNN) [14] may not effectively approximate nonlinear dynamics because there is no recurrent path. Therefore, many recurrent NNs have been applied in nonlinear system identification [15], model predictive control for continuous pharmaceutical manufacturing [16], estimation in the effective connectivity of electroencephalography [17] and visual field prediction [18] because of higher accuracy and better identification. Due to it having more advantages than the feedforward FJPNN, the RRJPNN control by adopting the ABSA method with progressive weight pruning approach has not yet been used to control the ISM driving the rice milling machine system in order to cut down computation complexity.

The customary optimization algorithm is applied when solving the classic problem of smaller dimensions that are not easily applicable in reality. The swarm intelligence optimization algorithms (SIOAs) were discovered by adopting simulated natural biological systems. Recently, many researchers have proposed the SIOAs, such as bat algorithm (BA) [19], particle swarm optimization (PSO) [20], social learning optimization algorithm (SLOA) [21], chicken swarm optimization algorithm (CSOA) [22] and ant colony algorithm (ACO) [23]. Yang [24] first proposed the bat algorithm (BA), which is based on the swarm intelligence heuristic search algorithm. Due to it having fewer parameters, powerful robustness, and a simple and easy implementation, BA has attracted more and more attention in the search for a global optimal solution. The BA has been applied in multiobjective function optimizations with a neural network model [19], economic operations [25] and numerical optimizations [26]. However, the local search method by adopting BA has the shortcoming of precocious convergence and does not perform well in the early search stage. Therefore, the proposed altered bat search algorithm (ABSA) method is a novel method to avoid precocious convergence. Moreover, a progressive weight pruning approach based on the ABSA method is proposed to deal with nonconvex optimization problems. The ABSA method with modified loudness and modified pulse rate is used to adjust two optimal learning rates. Thereby, a novel fast-convergence algorithm applied to RRJPNN with two learning rates by adopting the ABSA method with the progressive weight pruning approach that is used to adjust two optimal learning rates and to quicken convergence of weights is proposed in this paper. At last, some tested results show that the fine control performances are confirmed by the proposed control method.

The organization of this research is as follows. Section 2 introduces some explanations of the ISM driving the rice milling machine system. Section 3 advocates the witty control system by adopting the RRJPNN control and two remunerated controls with ABSA. Section 4 promotes the tests and experimental results for the ISM driving the rice milling machine system. Section 5 presents the discussions and explanations of some experimental results. Section 6 describes some conclusions.

2. Materials

The complete system consists of three subparts that are: the ISM and drive system, the digital signal processor (DSP) control system and the rice milling machine system, as below.

2.1. System Description

The ISM and drive system consist of a mounted encoder three-phase ISM with a rotor that adopts a high moment of inertia and low frictional coefficient, a rectifier, an LC filter, an isolated circuit, a lockout-time circuit, a current sensing circuit, an analog-digital conversion, and the voltage fed converter with 3-leg 6-switch transistor power devices. The DSP control system is composed of a speed control, a proportional-integral (PI) current control and an indirect field-oriented control (IFOC) [27–29]. The IFOC consists of a space-vector pulse-width-modulation control, $\sin\theta_t / \cos\theta_t$ creation, a coordinate transformation and an inverse coordinate transformation. The rice milling machine system is composed of a feed chopper, milling room, main idler, idler 1, idler 2, chaff suction fan, jet fan, chaff outlet, inlet chopper mouth, rice outlet, thickness adjustment lever, belt 1 and belt 2. The arrangement of the ISM driving the rice milling machine system is shown in Figure 1.

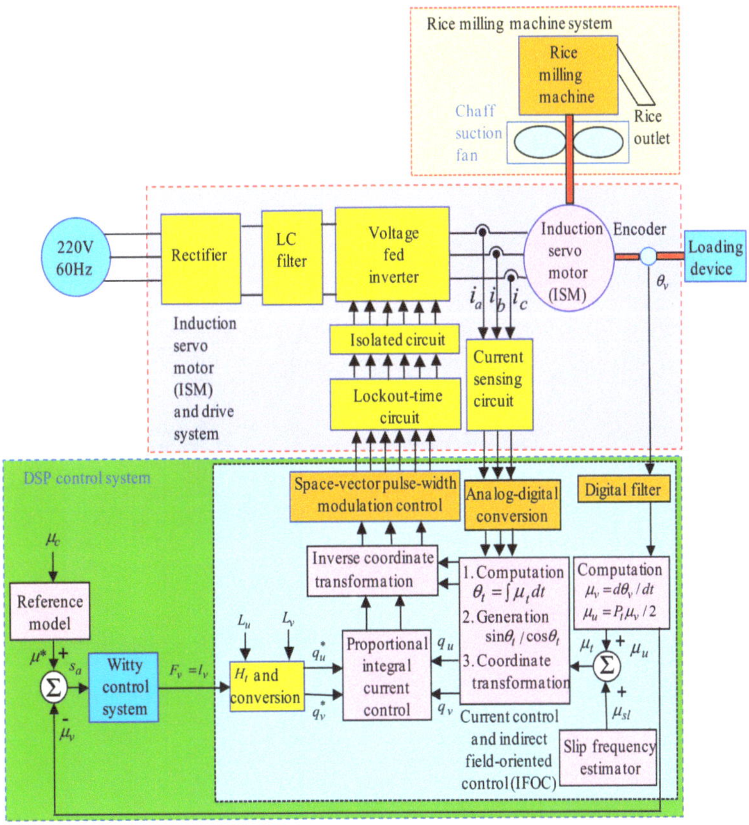

Figure 1. Arrangement of the induction servo motor (ISM) driving the rice milling machine system.

2.2. System Model

The voltage equations in the three-phase ISM by using the simplified coordinate frame transforms, from *a-b-c* to *u-v*, and the Clarke and Park transformations by using the IFOC [27–29] can be represented by

$$w_u = r_a q_u - \mu_t(L_v q_v + L_{uv} q_u) + \frac{d(L_u q_u + L_{uv} q_v)}{dt} \tag{1}$$

$$w_v = r_a q_v + \mu_t(L_u q_u + L_{uv} q_v) + \frac{d(L_v q_v + L_{uv} q_u)}{dt} \tag{2}$$

$$0 = r_b q_{ur} + (\mu_t - \mu_u)(L_u q_{ur} + L_{uv} q_{vr}) + \frac{d(L_v q_{vr} + L_{uv} q_{ur})}{dt} \tag{3}$$

$$0 = r_b q_{vr} + (\mu_t - \mu_u)(L_v q_{vr} + L_{uv} q_{ur}) + \frac{d(L_u q_{ur} + L_{uv} q_{vr})}{dt} \tag{4}$$

where w_u and w_v are the $u - v$ axis stator voltages. q_u and q_v are the $u - v$ axis stator currents. q_{ur} and q_{vr} are the $u - v$ axis rotor currents. L_u, L_v and L_{uv} are the $u - v$ axis self-inductances and mutual inductance, respectively. r_a and r_b are the stator and equalized rotor resistances. μ_v and μ_u are the mechanical and electrical angular speeds in the ISM. μ_t is the electrical angular speed of synchronous flux [27]. The electromagnetic torque F_a [28,29] in the ISM can be described as

$$F_v = \frac{3P_t[\varphi_v q_u - \varphi_u q_v]}{4} = \frac{3P_t[L_v q_v q_u - L_u q_u q_v]}{4} = \frac{3P_t[L_v - L_u]}{4} q_u q_v \tag{5}$$

where φ_u and φ_v are $u - v$ axis flux linkages. P_t is the number of poles.

The ISM driving the rice milling machine system led to a more sluggish performance of the system owing to a nonlinear uncertainty effect. The response of speed control for the ISM driving the rice milling machine system resulted in poor performance. The adumbration view of the ISM and the rice milling machine system is illustrated in Figure 2.

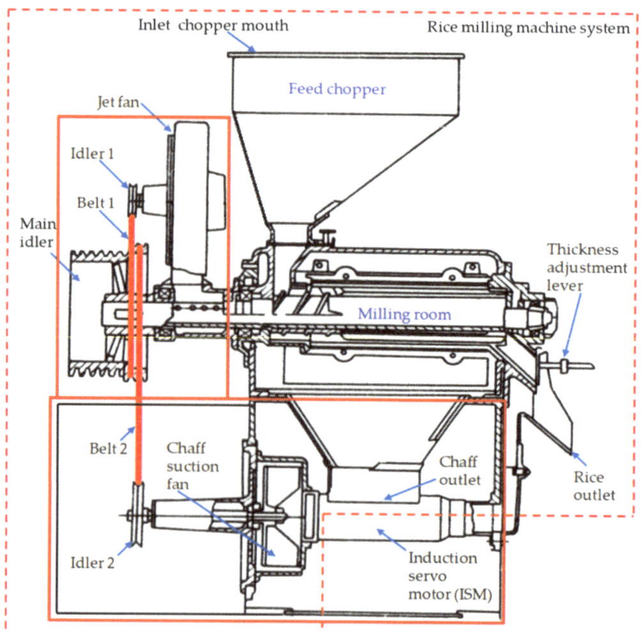

Figure 2. Adumbration view of the ISM and the rice milling machine system.

Considering that the power loss and sliding loss were insignificant, two dynamic equations with simplified kinematics of the rice milling machine system are described as [30–32]:

$$F_v = f_a \frac{d\mu_v}{dt} + g_a \mu_v + \phi_b \mu_b F_d / \mu_v + f_c \frac{d\mu_v}{dt} + g_c \mu_v \tag{6}$$

$$F_d = f_d \frac{d\mu_b}{dt} + g_d \mu_b + f_b \frac{d\mu_b}{dt} + g_b \mu_b + F_b^l(R_{vb}, F_{vb}, F_{vl}, g_b) \tag{7}$$

where F_v, F_d and F_{b1} are the electromagnetic torque of the ISM, the output torque of idler 2 and the output torque of the main idler, respectively. f_a, f_c, f_d and f_b are the four moments of inertia in the ISM, in idler 2, in the main idler and in idler 1, respectively. g_v, g_c, g_d and g_b are the four viscid frictional coefficients in the ISM, in idler 2, in the main idler and in idler 1, respectively. ϕ_b is the transposition ratios regarding idler 2 and the main idler for the rice milling machine system. $F_b^l(R_{vb}, F_{vb}, F_{vl}, g_b)$ is the nonlinear coalescence disturbances function including rolling force R_{vb}, wind force F_{vb}, and braking force F_{vl}. μ_v and μ_b are the speed in idler 2 and the speed in the main idler. Then, the torque equation can be transformed from the main idler to idler 2 by use of a transformed ratio. The modeling of the rice milling machine can be simplified by omitted sliding losses of two belts; thus, the dynamic equation in the ISM driving the rice milling machine system including the coalescence torque from Equations (6) and (7) can be expressed as

$$f_r \frac{d\mu_v}{dt} + g_r\mu_v + (\Delta F_a + F_{b1}) + F_t(F_{1c}, F_{2c}, F_{3c}) = F_v \tag{8}$$

$$F_{b1} = \frac{\phi_b\mu_b[(f_d + f_b)\frac{d\mu_b}{dt} + (g_d + g_b)\mu_b + F_b^l(u_{vb}, F_{vb}, F_{vl}, g_b)]}{\mu_v} \tag{9}$$

$$\Delta F_a = \Delta f_r \frac{d\mu_v}{dt} + \Delta g_r\mu_v \tag{10}$$

where $g_r = g_a + g_c$ and $f_r = f_a + f_c$ are the coalescence viscid friction coefficient and the coalescence moment of inertia including the main idler and the ISM. $\Delta F_a + F_{b1}$ is the huge comprehensive coalescence disturbances and parameter variations. $F_t(F_{1c}, F_{2c}, F_{3c})$ is the coalescence torque [27] including coulomb friction torque F_{3c}, Stribeck effect torque F_{2c} and adding load torque F_{1c}. F_{b1} represents the comprehensive coalescence disturbances. ΔF_a represents the comprehensive parameter variations.

The DSP control system with current control and IFOC can fulfill a speed control, an IFOC and a proportional-integral (PI) current control. The IFOC consists of a space-vector pulse-width-modulation control, $\sin\theta_t / \cos\theta_t$, creation, a coordinate transformation and an inverse coordinate transformation. The control gains of the PI current control are the proportional gain of 19.2 and the integral gain of 8.3 by using the heuristic method [33–35] to obtain a finer dynamic response. The drive system was operated under comprehensive coalescence disturbances and comprehensive parameter variations by adopting the DSP control system in this research.

3. Methods

In order to design the control structure, the dynamic equation of Equation (8) is modified as

$$\frac{d\mu_v}{dt} = h_v\mu_v + h_w(\Delta F_a + F_{b1} + F_e) + h_x l_v \tag{11}$$

where $\Delta F_a + F_{b1} + F_e$ represents the comprehensive coalescence disturbances. $h_v = -g_r f_r^{-1}$ is a friendly ratio constant and $|h_v\mu_v| \le R_v(\mu_v)$ is assumed to be bounded with functional-bounded value $R_v(\mu_v)$. $h_w = -f_r^{-1}$ is a friendly constant concerning the coalescence moment of inertia and $|h_w F_o| \le R_b$ is assumed to be bounded. $h_x = f_r^{-1}$ is a friendly constant concerning the coalescence moment of inertia and $h_x \le R_c$ is assumed to be bounded. R_b and R_c are two friendly values. $l_v = F_v$ is the electromagnetic torque of the ISM. The speed difference s_a is as follows.

$$s_a = \mu^* - \mu_v \tag{12}$$

If the comprehensive coalescence disturbances and the comprehensive parameter variations are favorable and affectionate, the excellent control law can be rewritten by

$$l_v^* = \frac{d\mu^*}{h_x dt} + \frac{c_v\, s_a}{h_x} - \frac{h_v\mu_v}{h_x} - \frac{h_w(\Delta F_a + F_{b1} + F_e)}{h_x} \tag{13}$$

where c_v is a positive control gain. Equation (12) and $l_v^* = l_v$ are substituted into Equation (11), so the error equation can be rewritten by

$$\frac{ds_a}{dt} + c_v s_a = 0 \tag{14}$$

The system will track the wished state value at $t \to \infty$ and $s_a(t) \to 0$. Nevertheless, the control system will exhibit a sluggish tracking response under the occurrence of uncertainty. Thereby, the proposed witty control system using an RRJPNN control and two remunerated controls with ABSA shown in Figure 3 were developed to control the ISM driving the rice milling machine system in order to enhance the speed of the tracking response.

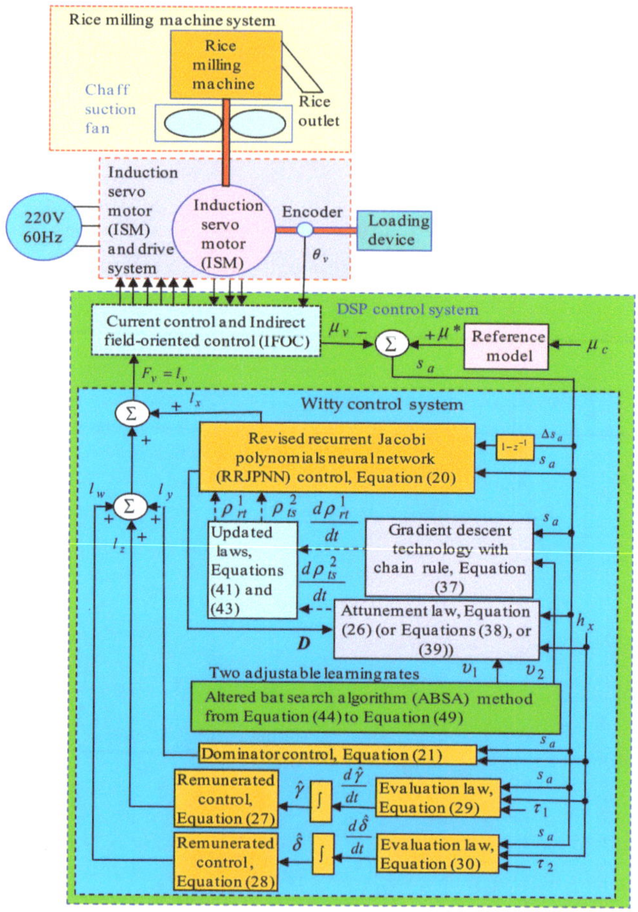

Figure 3. Control frame of witty control system.

The adopted control system is given by

$$l_v = l_x + l_y + l_z + l_w \tag{15}$$

We can use the differential equation of (11), then we can substitute Equations (8) and (12) into this equation. The error equation can be rewritten by

$$\frac{ds_a}{dt} = l_v^* h_x - l_x h_x - l_y h_x - l_z h_x - l_w h_x - c_v s_a \tag{16}$$

where l_x is the RRJPNN control that acts as the main tracking controller to impersonate the excellent control rule. l_y is the dominator control that will act in the appropriate region. l_z and l_w are two remunerated controls that acts as two remunerated controllers to repay the difference between the excellent control and the RRJPNN control. Then, the three-layer RRJPNN, which is shown in Figure 4, consists of the forehead, center and readward layers. All the informations of all layers are as follows.

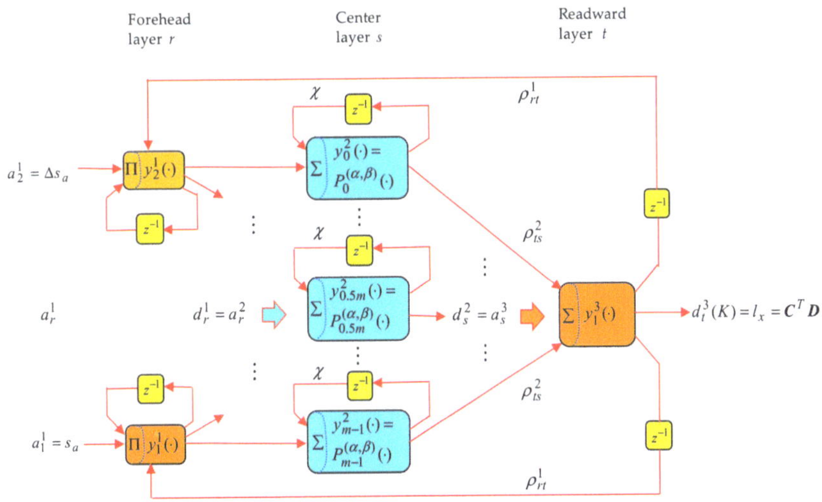

Figure 4. Constitution of the revised recurrent Jacobi polynomial neural network (RRJPNN).

The input and the output informations in the forehead layer are shown below.

$$b_r^1 = \prod_t a_r^1(K)\, \rho_{rt}^1(K)\, d_t^3(K-1)\, d_r^1(K-1), d_r^1(K) = y_r^1(b_r^1) = b_r^1,\ r = 1,\ 2 \tag{17}$$

The input and the output informations in the center layer are shown below.

$$b_s{}^2 = \sum_{r=1}^{2} d_r^1(K) + \chi d_s^2(K-1),\ d_s^2(K) = y_s^2(b_s^2) = P_s^{(\alpha,\beta)}(b_s^2), \quad s = 0,\ 1,\ \cdots,\ m-1 \tag{18}$$

The input and the output informations in the readward layer are shown below.

$$b_t{}^3 = \sum_{s=0}^{m-1} \rho_{ts}^2(K)\, d\,_s^2(K), d_t^3(K) = y_t^3(b_t^3) = b_t^3,\ t = 1 \tag{19}$$

where Π and $\mathbf{\Sigma}$ are the multiplication and summation symbols. $a_2^1 = s_a(1 - z^{-1}) = \Delta s_a$ and $a_1^1 = \mu * -\mu_v = s_a$ are the speed difference alteration and the speed difference. m, χ and K are the number of nodes of the center layer, the recurrent gain of the center layer and the iteration number, respectively. $\rho_{rt}^1(K)$ and $\rho_{ts}^2(K)$ are the recurrent weight between the readward layer and the forehead layer, and the conjoined weight between the center layer and the readward layer. y_r^1, y_s^2 and y_t^3 are the three linear activation functions in the forehead, center and readward layers, respectively. $d_r^1(K)$, $d_s^2(K)$ and $d_t^3(K)$ are the information of three outputs of nodes in the forehead, center and readward layers, respectively. $P_s^{(\alpha,\beta)}(b)$ is the Jacobi polynomial function [13,14] with $-1 < x < 1$ adopted as the activation function in the center layer—i.e., $y_s^2 = P_s^{(\alpha,\beta)}(b)$. $P_0^{(\alpha,\beta)}(b) = 1$, $P_1^{(\alpha,\beta)}(b) = (\alpha+1)(\alpha+\beta+2)(b-1)/2$ and $P_2^{(\alpha,\beta)}(b) = 0.5(\alpha+1)(\alpha+2) + 0.5(\alpha+2)(\alpha+\beta+3)(b-1) + 0.125(\alpha+\beta+3)(\alpha+\beta+4)(b-1)^2$ are the 0-, 1- and 2-order Jacobi polynomial functions, respectively. The Jacobi polynomial function with the recurrence relation [13,14] is as below $(2s+\alpha+\beta-1)\{(2s+\alpha+\beta)(2s+\alpha+\beta-2)+\alpha^2-\beta^2\}P_{s-1}^{(\alpha,\beta)}(b) - 2(s+\alpha-1)(s+\beta-1)(2s+\alpha+\beta)P_{s-2}^{(\alpha,\beta)}(b) = 2s(s+\alpha+\beta)(2s+\alpha+\beta-2)P_s^{(\alpha,\beta)}(b)$. The output information in the readward layer can be rewritten as $d_t^3(K) = l_x$. The RRJPNN control is thus described by

$$d_t^3(K) = l_x = \mathbf{C}^T\mathbf{D} \tag{20}$$

where $\mathbf{D} = \begin{bmatrix} d_0^2 & \cdots & d_{m-1}^2 \end{bmatrix}^T$ and $\mathbf{C} = \begin{bmatrix} \rho_{10}^2 & \cdots & \rho_{1,m-1}^2 \end{bmatrix}^T$ are the input information and weight vectors in the readward layer.

The dominator control l_y can be represented by

$$l_y = \frac{1}{h_x}\left[\left|\frac{d\mu_v}{dt}\right| + |c_v s_a| + R_v(\mu_v) + R_b\right]\mathrm{sgn}(s_a h_x) \tag{21}$$

The dominator control will act in the appropriate region if the RRJPNN control cannot be guaranteed.

To fulfill the remunerated mechanism, a minimum difference λ can be described as

$$\lambda = (l_x - l_x^*) - (l_x - l_v^*) \tag{22}$$

where $l_x^* = d_t^* = (\mathbf{C}^*)^T\mathbf{D}$ is the excellent control rule of the RRJPNN control; $\mathbf{C}^*$ is the excellent weight vector; $|\lambda| < \gamma < \delta$, $\delta + ((R_v(\mu_v) + R_b + |d\mu_v/dt| + |c_v s_a|)/h_x) > 0$ and δ are greater than zero. By using Equation (22), $l_x = d_t^3(K) = \mathbf{C}^T\mathbf{D}$ and $l_x^* = (\mathbf{C}^*)^T\mathbf{D}$, then Equation (16) can be described as

$$\begin{aligned} \tfrac{ds_a}{dt} &= h_x l_v^* - h_x l_x - h_x l_y - h_x l_z - h_x l_w - c_v s_a \\ &= h_x[(l_x - l_x^*) - (l_x - l_v^*)] + h_x(l_x^* - l_x) - h_x l_y - h_x l_z - h_x l_w - c_v s_a \\ &= h_x\lambda + h_x[(\mathbf{C}^*)^T\mathbf{D} - \mathbf{C}^T\mathbf{D}] - h_x l_y - h_x l_z - h_x l_w - c_v s_a \end{aligned} \tag{23}$$

To obtain two remunerated controls, the attunement law and the two evaluation laws, the Lyapunov function is described as

$$V_x = \frac{s_a^2}{2} + \frac{(\mathbf{C}^* - \mathbf{C})^T(\mathbf{C}^* - \mathbf{C})}{2v_1} + \frac{(\hat{\gamma} - \gamma)^2}{2\tau_1} + \frac{(\hat{\delta} - \delta)^2}{2\tau_2} \tag{24}$$

where v_1 is the learning rate of the conjoined weight; τ_1 and τ_2 are the two positive evaluation rates; $\hat{\gamma} - \gamma$ and $\hat{\delta} - \delta$ are the evaluation differences. By using Equations (22) and (23), then the differential equation of (24) can be described as

$$\begin{aligned} \tfrac{dV_x}{dt} &= s_a\tfrac{ds_a}{dt} - \tfrac{(\mathbf{C}^*-\mathbf{C})^T}{v_1}\tfrac{d\mathbf{C}}{dt} + (\hat{\gamma}-\gamma)\tfrac{d(\hat{\gamma}-\gamma)}{\tau_1 dt} + (\hat{\delta}-\delta)\tfrac{d(\hat{\delta}-\delta)}{\tau_2 dt} \\ &= s_a\{h_x\lambda + h_x[(\mathbf{C}^*-\mathbf{C})^T\mathbf{D}] - h_x l_z - h_x l_w - h_x l_x - c_v s_a\} - \tfrac{(\mathbf{C}^*-\mathbf{C})^T}{v_1}\tfrac{d\mathbf{C}}{dt} + (\hat{\gamma}-\gamma)\tfrac{d\hat{\gamma}}{\tau_1 dt} + (\hat{\delta}-\delta)\tfrac{d\hat{\delta}}{\tau_2 dt} \end{aligned} \tag{25}$$

The attunement law $\frac{d\mathbf{C}}{dt}$, two remunerated controls, l_z and l_w, and two evaluation laws, $d\hat{\gamma}/dt$ and $d\hat{\delta}/dt$ to fulfill $\frac{dV_x}{dt} \le 0$ can be described as

$$\frac{d\mathbf{C}}{dt} = v_1 s_a h_x \mathbf{D} \tag{26}$$

$$l_z = \hat{\gamma}\,\mathrm{sgn}(s_a h_x) \tag{27}$$

$$l_w = \hat{\delta}\,\mathrm{sgn}(s_a h_x) \tag{28}$$

$$\frac{d\hat{\gamma}}{dt} = \tau_1 |s_a h_x| \tag{29}$$

$$\frac{d\hat{\delta}}{dt} = \tau_2 |s_a h_x| \tag{30}$$

By using Equations (26)–(30) and (17), then Equation (25) can be rewritten by

$$\begin{aligned}
\tfrac{dV_x}{dt} &= s_a\{h_x\lambda + h_x[(\boldsymbol{C}^* - \boldsymbol{C})^T\boldsymbol{D}] - h_x l_y - h_x l_z - h_x l_w - c_v s_a\} - \tfrac{(\boldsymbol{C}^*-\boldsymbol{C})^T}{\upsilon_1}\tfrac{d\boldsymbol{C}}{dt} + (\hat{\gamma} - \gamma)\tfrac{d\hat{\gamma}}{\tau_1 dt} + (\hat{\delta} - \delta)\tfrac{d\hat{\delta}}{\tau_2 dt} \\
&= -c_a s_a^2 + s_a h_x\{\lambda + (\boldsymbol{C}^* - \boldsymbol{C})^T\boldsymbol{D} - (\hat{\gamma} + \hat{\delta})\mathrm{sgn}(s_a h_x) - \tfrac{1}{h_x}[\left|\tfrac{d\mu_v}{dt}\right| + |c_v s_a| + R_a(\mu_v) + R_b]\mathrm{sgn}(s_a h_x)\} \\
&\quad - \tfrac{(\boldsymbol{C}^*-\boldsymbol{C})^T}{\upsilon_1}\upsilon_1 s_a h_x \boldsymbol{D} + (\hat{\gamma} - \gamma)\tfrac{\tau_1|s_a h_x|}{\tau_1} + (\hat{\delta} - \delta)\tfrac{\tau_2|s_a h_x|}{\tau_2} \\
&= -c_a s_a^2 + (s_a h_x \lambda - \gamma|s_a h_x|) - \{\delta + \tfrac{1}{h_x}[\left|\tfrac{d\mu_v}{dt}\right| + |c_v s_a| + R_v(\mu_v) + R_b]\}|s_a h_x|
\end{aligned} \tag{31}$$

By using $|\lambda| < \gamma < \delta$ and $\delta + ((R_v(\mu_v) + R_b + |d\mu_v/dt| + |c_v, s_a|)/h_x) > 0$, then Equation (31) can be rewritten by

$$\begin{aligned}
\tfrac{dV_x}{dt} &\leq -c_v s_a^2 + (|\lambda| - \gamma)|s_a h_x|) \\
&\leq -c_v s_a^2 \\
&\leq 0
\end{aligned} \tag{32}$$

s_a and $(\boldsymbol{C}^* - \boldsymbol{C})$ are represented as bounded when $\frac{dV_x}{dt} \leq 0$, which is a negative semidefinite. Additionally, the uniformly continuous function $x_a(t)$ can be described by

$$x_a(t) = -\frac{dV_x}{dt} = c_v s_a^2 \tag{33}$$

The integral of $x_a(t)$ can be rewritten by

$$\int_0^t x_a(\varepsilon)d\varepsilon = \int_0^t [-\frac{dV_x}{d\varepsilon}]d\varepsilon = V_x(0) - V_x(t) \tag{34}$$

The differential of Equation (33) can be described as

$$\frac{dx_a(t)}{dt} = 2c_v s_a \frac{ds_a}{dt} \tag{35}$$

The limitation of Equation (34) when $V_x(0)$ and $V_x(t)$ are bounded can be described as

$$\lim_{t\to\infty}\int_0^t x_a(\varepsilon)d\varepsilon < \infty \tag{36}$$

As all variables on the right side of Equation (23) are bounded, which implies $\frac{ds_a}{dt}$ is also bounded and it can be shown that $\lim_{t\to\infty}\int_0^t x_a(\varepsilon)d\varepsilon = 0$, thus $s_a(t) \to 0$ as $t \to \infty$ by using Barbalat's lemma [36,37]. Therefore, the proposed witty control system is gradually stable from proof. Moreover, the tracking error $s_a(t)$ of the system will converge to zero.

Therefore, to describe the online training process of the RRJPNN, an objective function can be defined by

$$V_y = \frac{s_a^2}{2} \tag{37}$$

The conjoined weight by using the backpropagation technology and the gradient descent technology from the attunement law $\frac{dC}{dt}$ can be described as

$$\frac{d\rho_{ts}^2}{dt} = \upsilon_1 s_a h_x q_t^3 \tag{38}$$

By using the two above technologies, the conjoined weight from Equation (38) can be expressed by

$$\frac{d\rho_{ts}^2}{dt} = -v_1\frac{\partial F_2}{\partial \rho_{ts}^2} = -v_1\frac{\partial V_y}{\partial s_a}\frac{\partial s_a}{\partial d_t^3}\frac{\partial d_t^3}{\partial b_t^3}\frac{\partial b_t^3}{\partial \rho_{ts}^2} = -v_1\frac{\partial V_y}{\partial d_t^3}\frac{\partial d_t^3}{\partial b_t^3}\frac{\partial b_t^3}{\partial \rho_{ts}^2} = -v_1 q_t^3\frac{\partial V_y}{\partial d_t^3} \tag{39}$$

In comparison with Equations (38) and (39), this can be obtained as

$$\frac{\partial V_y}{\partial d_t^3} = -s_a h_x \tag{40}$$

The updated law of the conjoined weight can be denoted by [38]

$$\rho_{ts}^2(K+1) = \rho_{ts}^2(K) + \frac{d\rho_{ts}^2}{dt} \tag{41}$$

By using the two above technologies, the recurrent weight of the attunement law can be expressed by

$$\frac{d\rho_{rt}^1}{dt} = -v_2\frac{\partial V_y}{\partial d_t^3}\frac{\partial d_t^3}{\partial d_s^2}\frac{\partial d_s^2}{\partial b_s^2}\frac{\partial b_s^2}{\partial d_r^1}\frac{\partial d_r^1}{\partial b_r^1}\frac{\partial b_r^1}{\partial \rho_{rt}^1} = v_2 h_x s_a \rho_{ts}^2 P_s(\cdot) q_r^1(K) d_t^3(K-1) d_r^1(K) \tag{42}$$

where v_2 is the learning rate of the recurrent weight. The updated law of the recurrent weight can be denoted by [38]

$$\rho_{rt}^1(K+1) = \rho_{rt}^1(K) + \frac{d\rho_{rt}^1}{dt} \tag{43}$$

Moreover, the ABSA method was applied to search for two optimal adjustable learning rates and to improve convergent speed of the weights in the RRJPNN in this research. The existing algorithm with the excellent behavior of miniature bat echolocation is the important development of the BA method. By adopting a random technology, this algorithm produces a set of solutions. Then, the optimal solution is found by using the loop search. The local solution is generated by random flight and generates a global optimal solution. For all bats, the position of the bat i is $z_i(n-1)$, the flight velocity is $f_i(n-1)$ and the current global optimal position is $z*$ when their foraging space is part of the d-dimension at $n-1$ time. The flight velocity $f_i(n)$ and position $z_i(n)$ of bat i at n time can be calculated by

$$k_i = k_{\min} + (k_{\max} - k_{\min})n/N_{\max},\ i = 1,\ 2 \tag{44}$$

$$f_i(n) = f_i(n-1) + (z_i(n-1) - z^*)k_i,\ i = 1,\ 2,\ n = 1,\ldots,j \tag{45}$$

$$z_i(n) = z_i(n-1) + f_i(n),\ \ i = 1,\ 2,\ n = 1,\ldots,j \tag{46}$$

where $N_{\max}$ is the maximum number of iterations; $k_{\max}$ and $k_{\min}$ are the maximum and minimum frequencies of the soundwaves produced by the bat. In the initial process, the frequency of the bat's soundwaves is uniformly distributed between $k_{\min} = 0$ and $k_{\max} = 1$. The concerning frequency is obtained by adopting Equation (44). By adopting Equations (45) and (46), the local search is realized. The bat randomly goes along the optimal solution, and the new solution is updated by

$$z_{new}(n) = z_{old}(n-1) + \sigma \times \overline{d}_i(n),\ \ i = 1,\ 2,\ n = 1,\ldots,j \tag{47}$$

where σ is a random number at $[-1, 1]$. $z_{old}(n-1)$ is the solution selected from the current optimal solution by adopting a random skill. $\overline{d}_i(n)$ is the average loudness from the bat generation at n time. Additionally, it achieves a global search by controlling modified loudness $d_i(n+1)$ and modified pulse rate $e_i(n+1)$. The modified pulse rate $e_i(n+1)$ and modified loudness $d_i(n+1)$ of the bat launch pulse can be updated by

$$e_i(n+1) = [e_i(n) - e_i(0)][1 - \exp(-n\xi)], \quad i = 1, 2, \; n = 1, \ldots, j \tag{48}$$

$$e_i(n)^{d_i(n+1)} = \varsigma[d_i(n) - d_i(0)], \quad i = 1, 2, \; n = 1, \ldots, j \tag{49}$$

where $e_i(0)$ is an initial rate and $d_i(0)$ is an initial loudness. ς is a constant between 0 and 1 and ξ is a positive constant. When the bat is conscious of the presence of the prey, it will increase its pulse emission rate and reduce the response of its pulsed emission. At last, $z_i(n), \; i = 1, 2$ is the best solution concerning the learning rates $v_i(n), \; i = 1, 2$ of the two weights in the RRJPNN. Thereby, the two adjustable values may be optimized by adopting ABSA method to find the two learning rates of weights. Moreover, a progressive weight pruning approach is based on the ABSA method to quicken convergence of weight.

4. Tests and Results

The arrangement of the ISM driving the rice milling machine system by adopting a DSP control system is illustrated in Figure 1. In Figure 5, an experimental photo of the ISM and the rice milling machine system is illustrated.

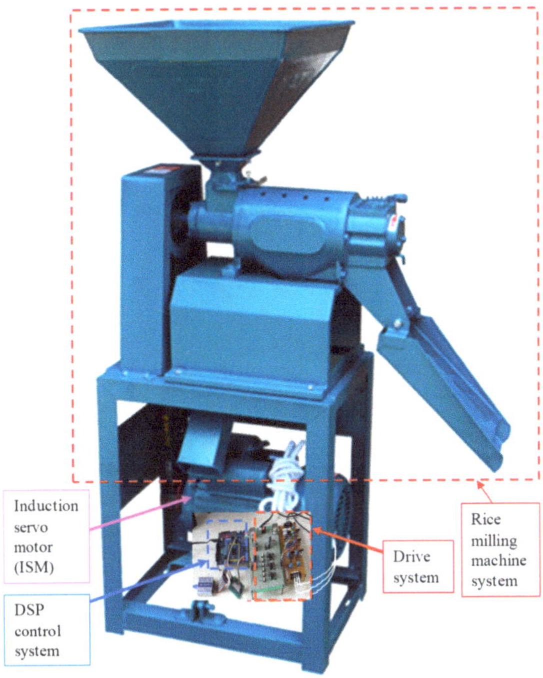

Figure 5. An experimental photo of the ISM and the rice milling machine system.

The conversion ratio for the rice milling machine system is 2.2. The profile formats are the belt 1 length is 42.2 mm, the belt 2 length is 52.2 mm, the main idler diameter is 92.6 mm, the diameter of idler 1 is 45.2 mm, and the diameter of idler 2 is 64.2 mm. The specification of the ISM is three-phase two-pole 220 V, 60 Hz, 3 kW, 3582 rpm. The position and speed conversion ratios are 1 V = 50 rad and 1 V = 50 rad/s. The internal parameters of the ISM are $r_a = 1.08 \; \Omega$, $r_b = 1.02 \; \Omega$, $L_u = 8.65$ mH, $L_v = 10.68$ mH, $L_{uv} = 8.89$ mH, $f_a = 16.22 \times 10^{-3}$ Nms$^2 = 0.811$ Nmsrad/V, and $g_a = 1.12 \times 10^{-3}$ Nms/rad $= 0.056$ Nm/V. Figure 6 illustrates the control flowchart of executive program by adopting the DSP control system.

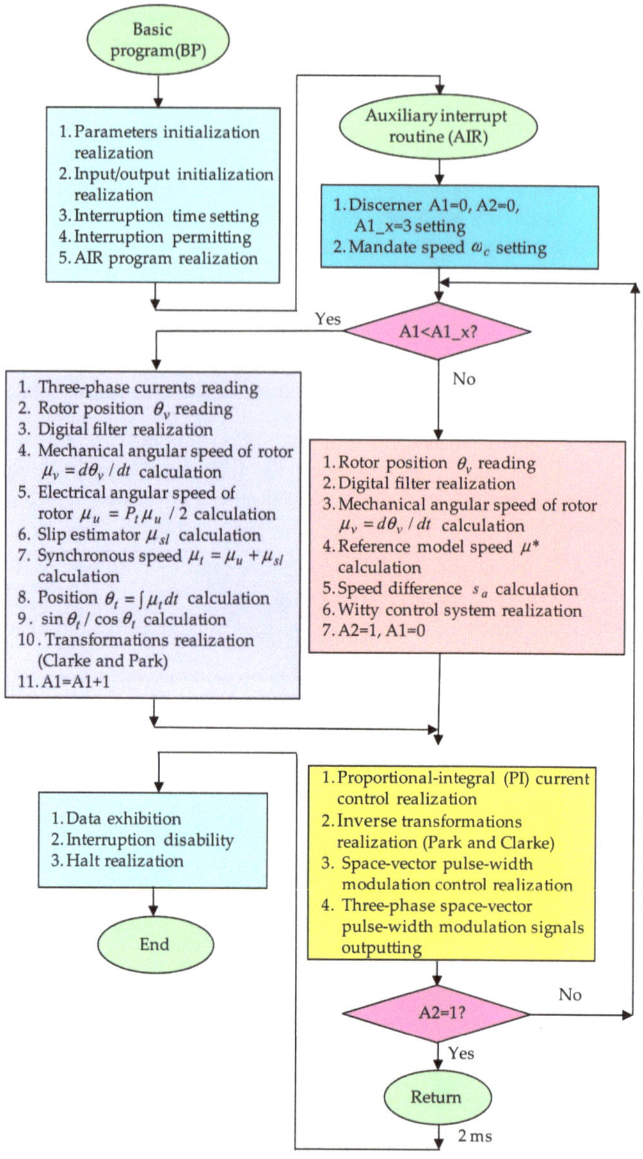

Figure 6. Control flowchart of executive program.

The program in the experimental tests consists of the basic program (BP) and the auxiliary interrupt routine (AIR). The BP conducts all initializations for the adopted parameters and all settings for the input/output interfaces. The AIR achieves the interrupt execution within 2 ms. The executed processes by AIR are as follows: three-phase currents read from analog–digital conversions, rotor position read from encoder interface circuit, rotor speed computation, speed difference computation, lookup table generation, coordinate transformations realization, PI current control realization, the proposed control system realization, and three-phase space-vector pulse-width-modulation outputs for switching the voltage fed converter. Three discerners A1_x, A1, A and 2 are set as 3, 0 and 0, respectively. The DSP control system with the IFOC applied the discerner A2 to act as the executing number of the proposed control method. If the IFOC is executed less than three times, i.e.,

A1 < A1_x, the IFOC needs to be enforced repeatedly. The proposed control method is executed one time and the IFOC is executed three times. Then, the AIR will back to the BP.

The experimental results with three test examples are shown to show some of the control performances. Firstly, test JA is the case with huge comprehensive coalescence disturbances and parameter variations $\Delta F_a + F_{b1}$ at 1.2 s, starting with a mandate speed of 1600 rpm (167.47 rad/s). Secondly, test JB is the case with double huge comprehensive coalescence disturbances and parameter variations $\Delta F_a + F_{b1}$ at 1.2 s, starting with a mandate speed of 3300 rpm (345.40 rad/s). Thirdly, test JC has a mandate speed of 2000 rpm (209.33 rad/s) starting at 2 s and a mandate speed of 3000 rpm (314.00 rad/s) at 10 s, with acceleration and added external load torque disturbance and parameter variations 8 $Nm(F_{1c}) + F_{b1}$ at 14 s, with a mandate speed of 3000 rpm (314.00 rad/s). The PI controller as the TA controller and the proposed witty control system as the TB controller are the two adopted controllers that are compared with control performances. Firstly, two gains of the PI control as the TA controller are the proportional gain of 24.1 and the integral gain of 10.2 by adopting the heuristic method [30–32] to obtain finer dynamic response under the requirement of stability consideration. The control gains by using the proposed witty control system as the TB controller are given as $c_v = 5.41$, $\chi = 0.092$, $\tau_1 = 0.12$ and $\tau_2 = 0.13$ to better measure transient performance under the demand of stability planning. In addition, the number of nodes of the RRJPNN, by adopting the progressive weight pruning approach based on ABSA method to quicken convergence of conjoined weight, are 2, 4 and 1 in the forehead, center and readward layers to better measure transient-state and steady-state control properties.

Firstly, Figure 7a,b display speed responses for measured speed μ_v, mandate speed μ_c and reference model speed $\mu*$ via experimental results of test JA by adopting the TA and TB controllers for the ISM driving the rice milling machine system.

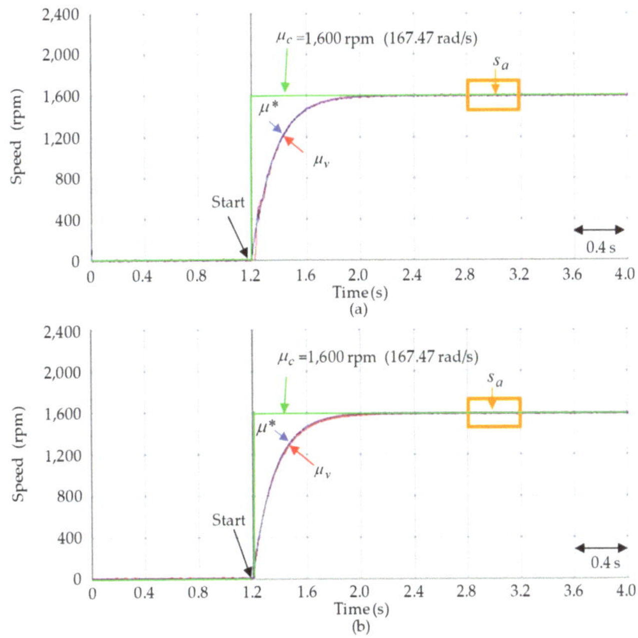

Figure 7. Speed responses via experimental results for the ISM driving the rice milling machine system at test JA by adopting the controllers: (**a**) TA; (**b**) TB.

Figure 8a,b display increase in speed difference s_a responses between 2.8 and 3.2 s via experimental results at test JA by adopting the TA and TB controllers for the ISM driving the rice milling machine system.

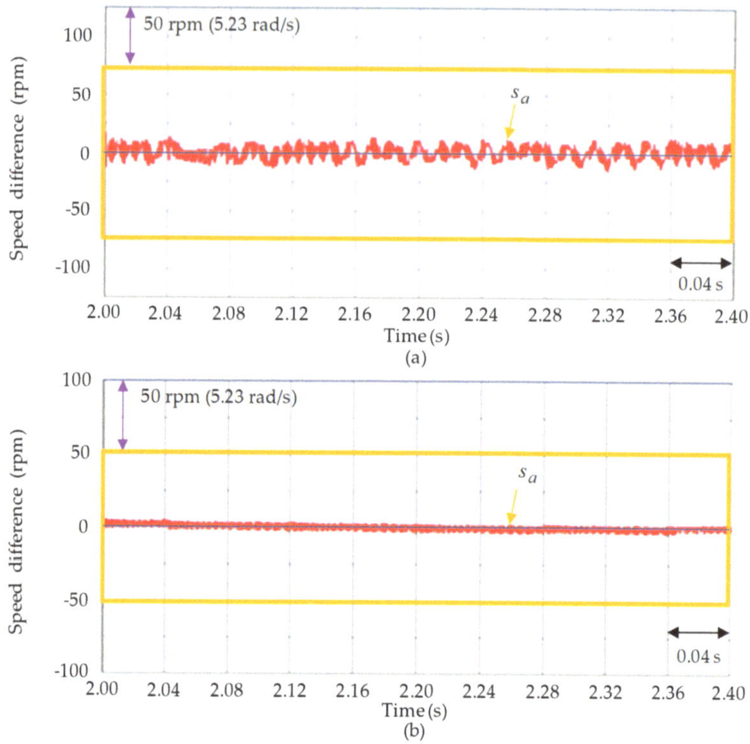

Figure 8. Increase in speed difference responses via experimental results for the ISM driving the rice milling machine system at test JA by adopting the controllers: (**a**) TA; (**b**) TB.

Figure 9a,b display the responses of three-phase currents via experimental results at test JA by adopting the TA and TB controllers for the ISM driving the rice milling machine system.

Figure 7a,b show that by adopting the TA and TB controllers in test JA, a better speed tracking performance was achieved because of smaller disturbance. However, the increase in tracking error when adopting the TA controller shown in Figure 8a is larger than the increase in tracking error when adopting the TB controller shown in Figure 8b. The response of three-phase currents when adopting the TA controller shown in Figure 9a generates a larger harmonic wave than when adopting the TB controller shown in Figure 9b.

Secondly, Figure 10a,b display speed responses for measured speed μ_v, mandate speed μ_c and reference model speed $\mu*$ via experimental results of test JB by adopting the TA and TB controllers for the ISM driving the rice milling machine system.

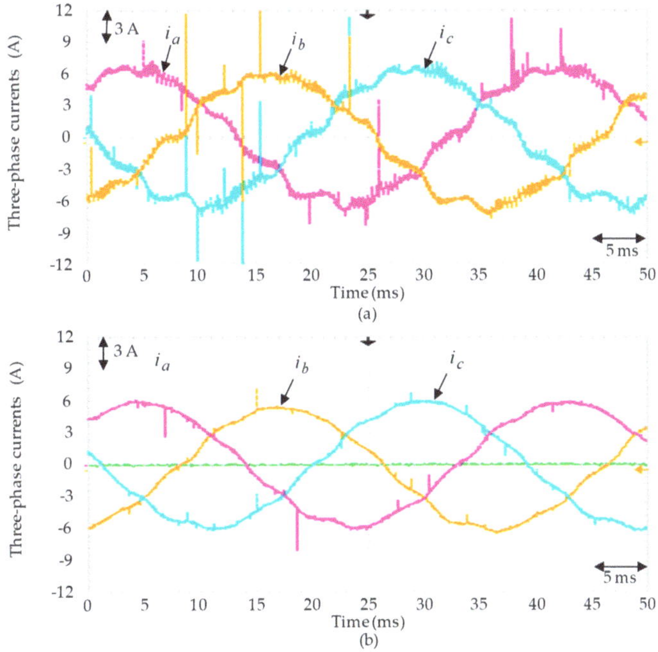

Figure 9. Responses of three-phase currents via experimental results for the ISM driving the rice milling machine system in test JA by adopting the controllers: (**a**) TA; (**b**) TB.

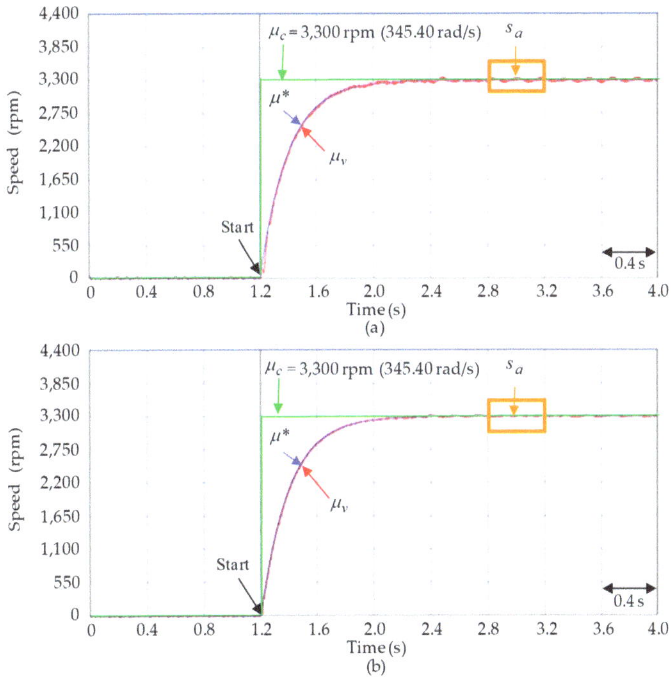

Figure 10. Speed responses via experimental results for the ISM driving the rice milling machine system in test JB by adopting the controllers: (**a**) TA; (**b**) TB.

Figure 11a,b display in increase speed difference s_a responses between 2.8 s and 3.2 s via experimental results at test JB by adopting the TA and TB controllers for the ISM driving the rice milling machine system.

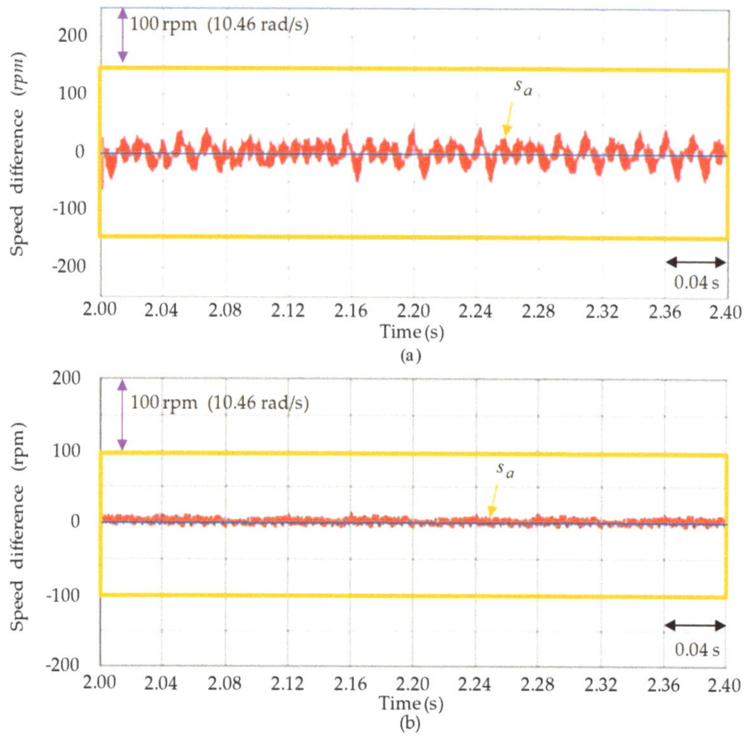

Figure 11. Increase in speed difference responses via experimental results for the ISM driving the rice milling machine system in test JB by adopting the controllers: (**a**) TA; (**b**) TB.

Figure 12a,b display responses of three-phase currents via experimental results of test JB by using the controllers TA, and TB for the ISM driving the rice milling machine system.

Figure 10a by using the TA controller in test JB appeared to show a dilatory speed response because of no good gain adjustment in the TA controller. In Figure 10b, by adopting the TB controller at test JB, a good speed response is demonstrated, owing to online adjustable method of RRJPNN control and two remunerated controls. However, the increase in tracking error when adopting the TA controller shown in Figure 11a is larger than the increase in tracking error when adopting the TB controller shown in Figure 11b. The response of three-phase currents when adopting the TA controller shown in Figure 12a results in larger harmonics than when adopting the TB controller shown in Figure 12b.

Thirdly, Figure 13a,b display two various speed-regulated responses when adding load torque via experimental results of test JC by using the TA and TB controllers for the ISM driving the rice milling machine system.

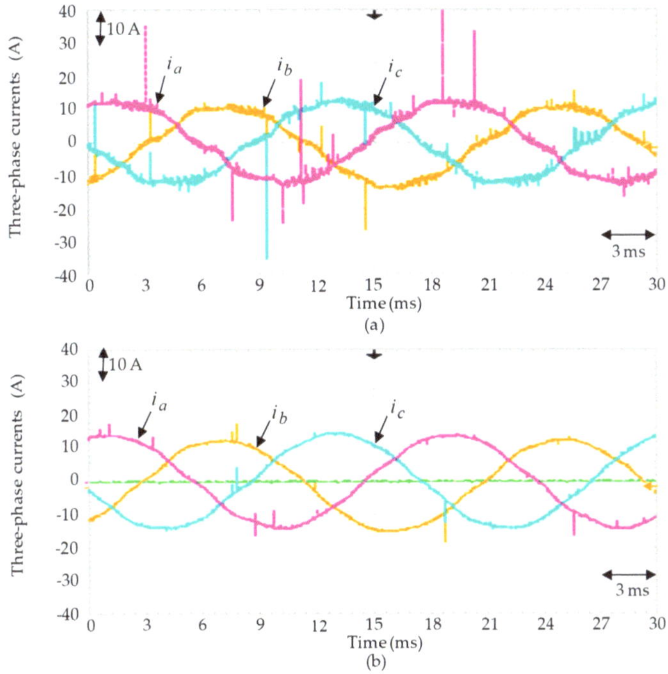

Figure 12. Responses of three-phase currents via experimental results for the ISM driving the rice milling machine system at test JA by adopting the controllers: (**a**) TA; (**b**) TB.

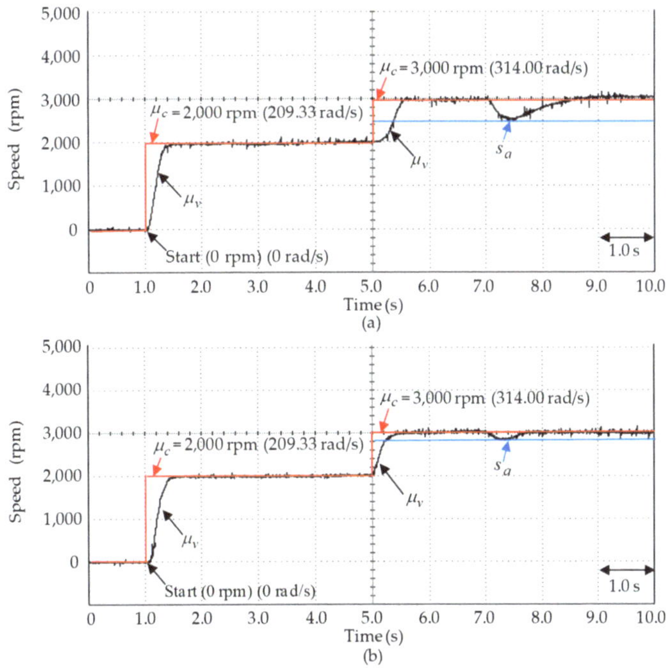

Figure 13. Two speed-regulated responses when adding load torque via experimental results of test JC by adopting the controllers: (**a**) TA; (**b**) TB.

Figure 14a,b display responses of three-phase currents with addition of load torque via experimental results of test JC by using the TA and TB controllers for the ISM driving the rice milling machine system.

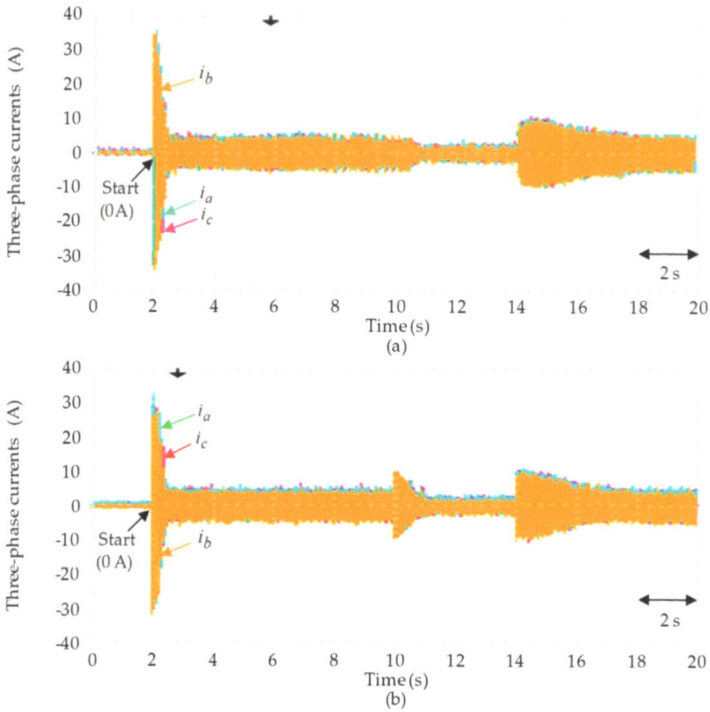

Figure 14. Responses of three-phase currents with adding load torque via experimental results of test JC by adopting the controllers: (**a**) TA; (**b**) TB.

The two speed-regulated responses when adding load torque via experimental results by adopting the TA controller, shown in Figure 13a, are worse than the two speed-regulated responses when adding load torque via experimental results by using the TB controller, also shown in Figure 13a. Responses of three-phase currents with adding load torque via experimental results by adopting the TA controller shown in Figure 14a has a greater harmonic than responses of three-phase currents when adding load torque via experimental results by adopting the TB controller shown in Figure 14b.

Moreover, responses of the two learning rates curves in test JB shown in Figure 15a,b using calculated learning rates according to the proposed ABSA method are compared to two learning rates of conjoined weight and recurrent weight by utilizing the PSO method [20] and ACO method [23] to demonstrate the usefulness of this novel technique. This study shows that convergence to optimal values can be achieved by using the proposed ABSA method. The proposed method also achieves faster convergence and less computational complexity.

Additionally, responses of two weights at test JB by using the PSO method [20] and the ACO method [23] and the proposed ABSA method are shown in Figure 16a,b. The convergences of conjoined weight and recurrent weight by using the proposed ABSA method with progressive weight pruning approach are superior to the ACO method and the PSO method. Thereby, the proposed method with progressive weight pruning approach can quicken convergence of weights.

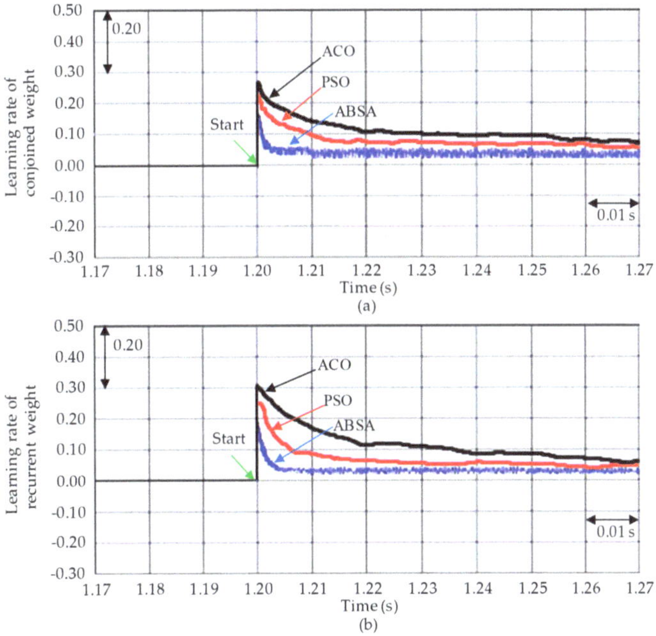

Figure 15. Responses of two learning rates via experimental results of test JB by adopting the ant colony algorithm (ACO), particle swarm optimization (PSO) and altered bat search algorithm (ABSA) methods for: (**a**) learning rate of conjoined weight, (**b**) learning rate of recurrent weight.

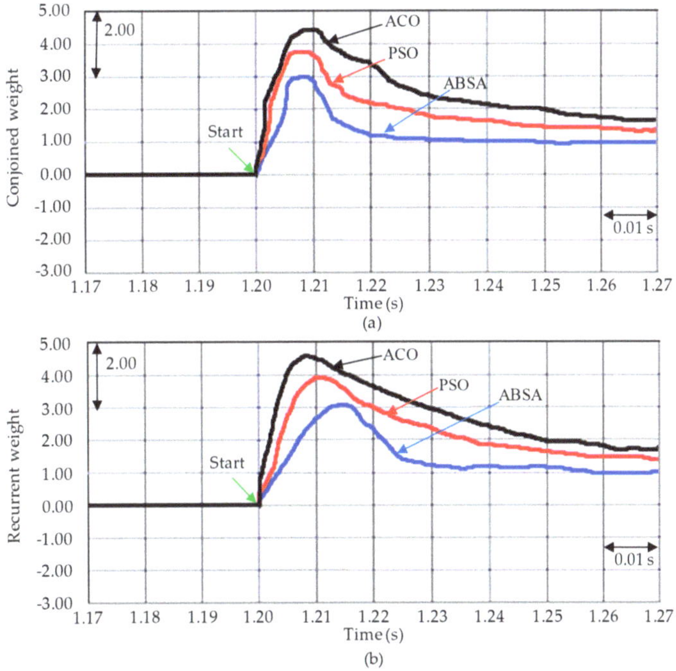

Figure 16. Responses of two weights via experimental results of test JB by adopting the ACO, PSO and ABSA methods for: (**a**) conjoined weight; (**b**) recurrent weight.

Furthermore, response of the numbers of conjoined weight at test JB by using the PSO method [20] and the ACO method [23] and the proposed ABSA method by adopting the progressive weight pruning approach is shown in Figure 17. The convergence of numbers of conjoined weight by using the proposed ABSA method and the progressive weight pruning approach is superior to the ACO method and the PSO method. Thereby, the proposed method can achieve faster convergence in conjoined weight.

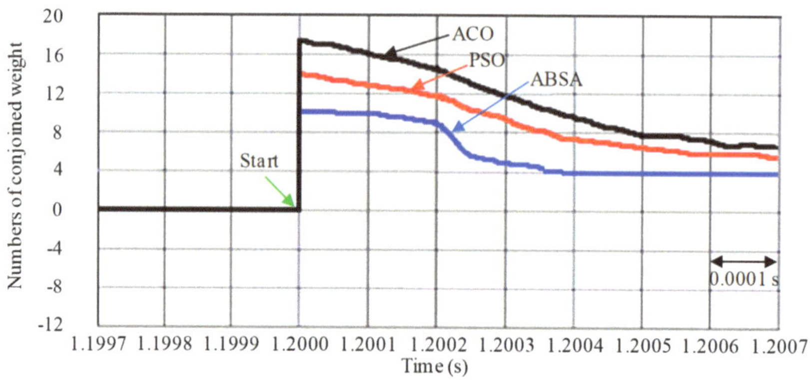

Figure 17. Responses of numbers of conjoined weight via experimental results of test JB by adopting the ACO, PSO and ABSA methods.

5. Analyses and Discussion

Dynamic responses for the PI controller as the TA controller and the proposed witty control system as the TB controller at three tested examples via experimental results that are listed in Table 1 are explained as below. For test JA, the maximum differences for the TA and TB controllers are 82 (8.58 rad/s) and 30 rpm (3.14 rad/s), and the quadratic mean differences for the TA and TB controllers are 48 (5.02 rad/s) and 17 rpm (1.78 rad/s). For test JB, the maximum differences for the TA and TB controllers are 128 (13.40 rad/s) and 35 rpm (3.66 rad/s), and the quadratic mean differences for the TA and TB controllers are 53 (5.55 rad/s) and 19 rpm (1.99 rad/s). For test JC, the maximum differences for the TA and TB controllers are 489 (51.18 rad/s) and 192 rpm (20.10 rad/s), and the quadratic mean differences for the TA and TB controllers are 188 (19.68 rad/s) and 46 rpm (4.81 rad/s). The TB controller has better dynamic responses than the TA controller according to the experimental results in tests JA, JB and JC.

Table 1. Dynamic responses for the two controllers.

Controllers / Three Test Examples / Performance	TA Controller		TB Controller	
	Maximum Differences of s_a	Quadratic Mean Differences of s_a	Maximum Differences of s_a	Quadratic Mean Differences of s_a
Test JA	82 rpm (8.58 rad/s)	48 rpm (5.02 rad/s)	30 rpm (3.14 rad/s)	17 rpm (1.78 rad/s)
Test JB	128 rpm (13.40 rad/s)	53 rpm (5.55 rad/s)	35 rpm (3.66 rad/s)	19 rpm (1.99 rad/s)
Test JC	489 rpm (51.18 rad/s)	188 rpm (19.68 rad/s)	192 rpm (20.10 rad/s)	46 rpm (4.81 rad/s)

Additionally, the peculiarity performances for the TA and TB controllers according to the three tested examples via experimental results that are listed in Table 2 are represented as below. The total harmonic distortion (THD) values in the three-phase currents for the TA and TB controllers in test JB are 21% and 5%. The responses of rising times for the TA and TB controllers in test JB are 0.92 s and 0.75 s. The regulation capabilities with adding load torque for the TA and TB controllers in test JC are 489 rpm (51.18 rad/s) in maximum difference and 192 rpm (20.10 rad/s) in maximum difference. The speed tracking differences for the TA and TB controllers in test JB are 128 rpm (13.40 rad/s) in maximum difference and 35 rpm (3.66 rad/s) in maximum difference. The denial potentialities of parameter disturbance for the TA and TB controllers in test JB are 128 rpm (13.40 rad/s) in maximum difference and 35 rpm (3.66 rad/s) in maximum difference. The above performances concerning the harmonic values in the three-phase currents, the dynamic responses, the regulation capabilities for adding load torque, the speed tracking differences and the denial potentialities of parameter disturbance in the TB controller are better than the TA controller. Thereby, the TB controller has better peculiarity performance than the TA controller from experimental results of tests JB and JC.

Table 2. Peculiarity performances for the two controllers.

Controllers / Peculiarity Performances	TA Controller	TB Controller
Total harmonic distortion (THD) values in the three-phase currents in test JB	21%	5%
Responses of rising times in test JB	0.92 s	0.75 s
Regulation capabilities with adding load torque in test JC	489 rpm (51.18 rad/s) in maximum difference	192 rpm (20.10 rad/s) in maximum difference
Speed tracking differences in test JB	128 rpm (13.40 rad/s) in maximum difference	35 rpm (3.66 rad/s) in maximum difference
Denial potentialities of parameter disturbance in test JB	128 rpm (13.40 rad/s) in maximum difference	35 rpm (3.66 rad/s) in maximum difference

6. Conclusions

The proposed witty control system has been applied to control the the ISM driving the rice milling machine system with better robustness. The proposed witty control system that can realize the RRJPNN control, which involves an attunement law, two remunerated controls, which have two evaluation laws, and a dominator control were proposed to obtain a fine control performance.

The contributions of this research are as below. (a) The dynamic models of the ISM driving the rice milling machine system have been developed. (b) The ISM driving the rice milling machine system under huge comprehensive nonlinear synthesized disturbances and parameter variations affect has been controlled by using the proposed witty control method. (c) On the basis of the Lyapunov stability principle, the attunement law in the RRJPNN control and the two evaluation laws in the two remunerated controls have been developed. (d) The ABSA method was utilized to find the learning rates of conjoined and recurrent weights in the RRJPNN to obtain optimal values and to quicken convergence of weights. (e) The proposed witty control system has better sinusoidal shapes than the PI control in terms of the harmonics values of three-phase currents.

Finally, all results show that the proposed witty control system is better than the PI control for the ISM driving the rice milling machine system from all experimental results and control behaviors.

Author Contributions: Conceptualization, D.-F.C., S.-P.-C.C., A.-B.C. and J.-C.T.; methodology, S.-P.-C.C.; software, S.-P.-C.C. and A.-B.C.; validation, S.-P.-C.C. and J.-C.T.; formal analysis, S.-P.-C.C. and A.-B.C.; investigation, S.-P.-C.C. and J.-C.T.; writing—original draft preparation, S.-P.-C.C. and J.-C.T.; writing—review and editing, S.-P.-C.C. and J.-C.T. All authors have read and agreed to the published version of the manuscript.

Funding: This research received no external funding.

Institutional Review Board Statement: Not applicable.

Informed Consent Statement: Not applicable.

Data Availability Statement: Data is contained within the article.

Conflicts of Interest: The authors declare no conflict of interest.

Abbreviations

w_u, w_v	$u-v$ axis stator voltages
q_u, q_v	$u-v$ axis stator currents
q_{ur}, q_{vr}	$u-v$ axis rotor currents
L_u, L_v, L_{uv}	$u-v$ axis self inductances, mutual inductance
r_a, r_b	stator and equalized rotor resistances.
μ_v, μ_u, μ_t	mechanical and electrical angular speeds, electrical angular speed of synchronous flux in the ISM
F_a	electromagnetic torque
φ_u, φ_v	$u-v$ axis flux linkages
P_t	number of pole
F_v, F_d, F_{b1}	electromagnetic torque of the ISM, the output torque of idler 2 and the output torque of the main idler
f_a, f_c, f_d, f_b	four moments of inertia in the ISM, in idler 2, in the main idler and in idler 1
g_v, g_c, g_d, g_b	four viscid frictional coefficients in the ISM, in idler 2, in the main idler and in idler 1
ϕ_b	transposition ratios regarding idler 2 and the main idler for the rice milling machine system
$F_b^l(R_{vb}, F_{vb}, F_{vl}, g_b)$	nonlinear coalescence disturbances function
R_{vb}, F_{vb}, F_{vl}	rolling force, wind force, braking force
μ_v, μ_b	speed in idler 2 and the speed in the main idler.
$g_r = g_a + g_c$	coalescence viscid friction coefficient including the main idler and the ISM
$f_r = f_a + f_c$	coalescence moment of inertia including the main idler and the ISM
$\Delta F_a + F_{b1}$	huge comprehensive coalescence disturbances and parameter variations
$F_t(F_{1c}, F_{2c}, F_{3c})$	coalescence torque
F_{3c}, F_{2c}, F_{1c}	coulomb friction torque, Stribeck effect torque, adding load torque
F_{b1}	comprehensive coalescence disturbances
ΔF_a	comprehensive parameter variations
$\Delta F_a + F_{b1} + F_e$	comprehensive coalescence disturbances
$h_v = -g_r f_r^{-1}$	friendly ratio constant
$R_v(\mu_v)$	bounded with functional-bounded value
$h_w = -f_r^{-1}$	friendly constant concerning the coalescence moment of inertia
$h_x = f_r^{-1}$	friendly constant concerning the coalescence moment of inertia
R_b, R_c	two friendly values with bound
$l_v = F_v$	electromagnetic torque of the ISM
s_a	speed difference
c_v	positive control gain
l_x, l_y, l_z, l_w	RRJPNN control, dominator control, two remunerated controls
$a_2^1 = s_a(1-z^{-1}) = \Delta s_a$, $a_1^1 = \mu * -\mu_v = s_a$	speed difference alteration, speed difference
m, χ and K	node number of the center layer, the recurrent gain of the center layer and the iteration number
$\rho_{rt}^1(K), \rho_{ts}^2(K)$	recurrent weight, conjoined weight
y_r^1, y_s^2, y_t^3	three linear activation functions in the forehead, center and readward layers

$d_r^1(K), d_s^2(K), d_t^3(K)$	information of three outputs of nodes in the forehead, center and readward layers
$P_s^{(\alpha,\beta)}(b)$	Jacobi polynomial function
$P_0^{(\alpha,\beta)}(b), P_1^{(\alpha,\beta)}(b), P_2^{(\alpha,\beta)}(b)$	0-, 1- and 2-order Jacobi polynomial functions
$d_t^3(K) = l_x$	output information in the readward layer
$D = \begin{bmatrix} d_0^2 & \cdots & d_{m-1}^2 \end{bmatrix}^T$, $C = \begin{bmatrix} \rho_{10}^2 & \cdots & \rho_{1,m-1}^2 \end{bmatrix}^T$	input information and weight vectors in the readward layer
λ	minimum difference
$l_x^* = d_t^* = (C^*)^T D$	excellent control rule of the RRJPNN control
C^*	excellent weight vector
δ	greater than zero real number
$x_a(t)$	uniformly continuous function
$V_x(0), V_x(t)$	two bounded
$\mathrm{sgn}(\cdot)$	sign function
V_y	objective function
$\frac{dC}{dt}$	attunement law
υ_1, υ_2	learning rate of the conjoined weight, learning rate of the recurrent weight
τ_1, τ_2	two positive evaluation rates
$\hat{\gamma} - \gamma, \hat{\delta} - \delta$	two evaluation differences
$d\hat{\gamma}/dt, d\hat{\delta}/dt$	two evaluation laws
$z_i(n-1), f_i(n-1)$	position of the bat i at $n-1$ time, flight velocity of the bat i at $n-1$ time
$z_i(n), f_i(n)$	position of the bat i at n time, flight velocity of the bat i at n time
$z*$	current global optimal position
$N_{\max}$	maximum number of iterations
$k_{\max}, k_{\min}$	maximum and minimum frequencies of the soundwaves produced by the bat
σ	random number at $[-1, 1]$
$z_{old}(n-1)$	solution selected from the current optimal solution at $n-1$ time
$\overline{d}_i(n)$	average loudness from the bat generation at n time
$d_i(n+1), e_i(n+1)$	modified loudness at n + 1 time, modified pulse rate at n + 1 time
$e_i(0), d_i(0)$	initial rate, initial loudness
ς, ξ	constant between 0 and 1, positive constant

References

1. Zuo, L.; Cheng, T.; Zuo, J.; Liu, Y.; Yang, S. A hierarchical intelligent control system for milling machine. In Proceedings of the IEEE International Conference on Intelligent Processing Systems, Beijing, China, 28–31 October 1997; pp. 1–6.
2. Huang, S.; Tan, K.K.; Hang, G.S.; Wang, Y.S. Cutting force control of milling machine. *Mechatronics* **2007**, *17*, 533–541. [CrossRef]
3. Lobinho, G.; Armando, S. Control of milling machine cutting force using artificial neural networks. In Proceedings of the 6th Iberian Conference on Information Systems and Technologies (CISTI 2011), Chaves, Portugal, 15–18 June 2011; pp. 1–6.
4. Tadeusz, M. Numerical control system of conventional milling machine. *Appl. Mech. Mater.* **2016**, *841*, 179–183.
5. Tu, F.; Hu, S.; Zhuang, Y.; Lv, J.; Wang, Y.; Sun, Z. Hysteresis curve fitting optimization of magnetic controlled shape memory alloy actuator. *Actuators* **2016**, *5*, 25. [CrossRef]
6. Nakshatharan, S.S.; Vunder, V.; Poldsalu, I.; Johanson, U.; Punning, A.; Aabloo, A. Modelling and control of ionic electroactive polymer actuators under varying humidity conditions. *Actuators* **2018**, *7*, 7. [CrossRef]
7. Xia, B.; Miriyev, A.; Trujillo, C.; Chen, N.; Cartolano, M.; Vartak, S.; Lipson, H. Improving the actuation speed and multi-cyclic actuation characteristics of silicone/ethanol soft actuators. *Actuators* **2020**, *9*, 62. [CrossRef]
8. Lee, T.; Kim, I. Design of a 2DOF ankle exoskeleton with a polycentric structure and a bi-directional tendon-driven actuator controlled using a PID neural network. *Actuators* **2021**, *10*, 9. [CrossRef]
9. Wang, L.; Duan, M.; Duan, S. Memristive Chebyshev neural network and its applications in function approximation. *Math. Probl. Eng.* **2013**, *2013*, 1–7. [CrossRef]
10. Ting, J.C.; Chen, D.F. Novel mingled reformed recurrent hermite polynomial neural network control system applied in continuously variable transmission system. *J. Mech. Sci. Technol.* **2018**, *32*, 4399–4412. [CrossRef]

11. Soltani, M.; Hegde, C. Fast and provable algorithms for learning two-layer polynomial neural networks. *IEEE Trans. Signal Process.* **2019**, *67*, 3361–3371. [CrossRef]
12. Ting, J.C.; Chen, D.F. Nonlinear backstepping control of SynRM drive systems using reformed recurrent Hermite polynomial neural networks with adaptive law and error estimated law. *J. Power Electron.* **2018**, *8*, 1380–1397.
13. Richard, A.; James, W. *Some Basic Hypergeometric Orthogonal Polynomials that Generalize Jacobi Polynomials*; American Mathematical Soc.: Providence, RI, USA, 1985; p. 54.
14. Guo, B.Y.; Shen, J.; Wang, L.L. Generalized Jacobi polynomials/functions and their applications. *Appl. Numer. Math.* **2009**, *59*, 1011–1028. [CrossRef]
15. Rubio, J.J.; Yu, W. Nonlinear system identification with recurrent neural networks and dead-zone Kalman filter algorithm. *Neurocomputing* **2007**, *70*, 2460–2466. [CrossRef]
16. Chen, D.F.; Shih, Y.C.; Li, S.C.; Chen, C.T.; Ting, J.C. Mixed modified recurring Rogers-Szego polynomials neural network control with mended grey wolf optimization applied in SIM expelling system. *Mathematics* **2020**, *8*, 754. [CrossRef]
17. Abbasvandi, Z.; Nasrabadi, A.M. A self-organized recurrent neural network for estimating the effective connectivity and its application to EEG data. *Comput. Biol. Med.* **2019**, *110*, 93–107. [CrossRef]
18. Park, K.; Kim, J.; Lee, J. Visual field prediction using recurrent neural network. *Sci. Rep.* **2019**, *9*, 8385. [CrossRef]
19. Jaddi, N.S.; Abdullah, S.; Hamdan, A. Multi-population cooperative bat algorithm-based optimization of artificial neural network model. *Inf. Sci.* **2015**, *294*, 628–644. [CrossRef]
20. Wang, L.; Geng, H.; Liu, P.; Lu, K.; Kolodziej, J.; Ranjan, R.; Zomaya, A.Y. Particle swarm optimization based dictionary learning for remote sensing big data. *Knowl. Based Syst.* **2015**, *79*, 43–50. [CrossRef]
21. Liu, Z.Z.; Chu, D.H.; Song, C.; Xue, X.; Lu, B.Y. Social learning optimization (SLO) algorithm paradigm and its application in QoS-aware cloud service composition. *Inf. Sci.* **2016**, *326*, 315–333. [CrossRef]
22. Wu, D.; Xu, S.; Kong, F. Convergence analysis and improvement of the chicken swarm optimization algorithm. *IEEE Access* **2016**, *4*, 9400–9412. [CrossRef]
23. Gheraibia, Y.; Djafri, K.; Krimou, H. Ant colony algorithm for automotive safety integrity level allocation. *Appl. Intell.* **2017**, *8*, 555–569. [CrossRef]
24. Yang, X.S. A new metaheuristic bat-inspired algorithm. In *Nature Inspired Cooperative Strategies for Optimization (NICSO 2010)*; Springer: Berlin, Heidelberg, 2010; Volume 284, pp. 65–74.
25. Adarsh, B.A.; Raghunathan, T.; Jayabarathi, T.; Yang, X.S. Economic dispatch using chaotic bat algorithm. *Energy* **2016**, *96*, 666–675. [CrossRef]
26. Wang, Y.; Wang, P.; Zhang, J.; Cui, H.; Cai, X.; Zhang, W.; Chen, J. A novel bat algorithm with multiple strategies coupling for numerical optimization. *Mathematics* **2019**, *7*, 135. [CrossRef]
27. Chen, D.F.; Cheng, A.B.; Chiu, S.P.C.; Ting, J.C. Electromagnetic torque control for six-phase induction motor expelled continuously variable transmission clumped system using sage dynamic control system. *Int. J. Appl. Electromagn. Mech.* **2021**, *65*, 579–608. [CrossRef]
28. Singh, G.K.; Nam, K.; Lim, S.K. A simple indirect field oriented control scheme for multiphase induction machine. *IEEE Trans. Ind. Electron.* **2005**, *52*, 1117–1184. [CrossRef]
29. Guo, Z.; Zhang, J.; Sun, Z.; Zheng, C. Indirect field oriented control of three-phase induction motor based on current-source inverter. *Procedia Eng.* **2017**, *174*, 588–594. [CrossRef]
30. Han, Y.; Jia, F.; Zeng, Y.; Cao, B. Effects of rotation speed and outlet opening on particle flow in a vertical rice mill. *Powder Technol.* **2016**, *297*, 153–164. [CrossRef]
31. Zareiforoush, H.; Minaee, S.; Alizadeh, M.R.; Samani, B.H. Design, development and performance evaluation of an automatic control system for rice whitening machine based on computer vision and fuzzy logic. *Comput. Electron. Agric.* **2016**, *124*, 14–22. [CrossRef]
32. Ruekkasaem, L.; Sasananan, M. Optimal parameter design of rice milling machine using design of experiment. *Mater. Sci. Forum* **2018**, *911*, 107–111. [CrossRef]
33. Astrom, K.J.; Hagglund, T. *PID Controller: Theory, Design, and Tuning*; Instrument Society of America: Research Triangle Park, NC, USA, 1995.
34. Hagglund, T.; Astrom, K.J. Revisiting the Ziegler-Nichols tuning rules for PI control. *Asian J. Control* **2002**, *4*, 364–380. [CrossRef]
35. Hagglund, T.; Astrom, K.J. Revisiting the Ziegler-Nichols tuning rules for PI control-part II: The frequency response method. *Asian J. Control* **2004**, *6*, 469–482. [CrossRef]
36. Astrom, K.J.; Wittenmark, B. *Adaptive Control*; Addison-Wesley: New York, NY, USA, 1995.
37. Slotine, J.J.E.; Li, W. *Applied Nonlinear Control*; Prentice-Hall: Englewood Cliffs, NJ, USA, 1991.
38. Chen, D.F.; Shih, Y.C.; Li, S.C.; Chen, C.T.; Ting, J.C. Permanent-magnet SLM drive system using AMRRSPNNB control system with DGWO. *Energies* **2020**, *13*, 2914. [CrossRef]

actuators

MDPI

Article

Gain-Scheduled Control of Asymmetric Thrust Magnetic Bearing

Shuyue Zhang [1,2] and Jihao Wu [1,2,*]

[1] State Key Laboratory of Technologies in Space Cryogenic Propellants, Technical Institute of Physics and Chemistry, Chinese Academy of Sciences, Beijing 100190, China; zhangshuyue@mail.ipc.ac.cn
[2] University of Chinese Academy of Sciences, Beijing 100049, China
* Correspondence: wujihao@mail.ipc.ac.cn; Tel.: +86-8254-3505

Abstract: The thrust position of the magnetic levitation rotor can be changed, bringing convenience to the practical application of cold compressors. This paper derives the mathematical model of asymmetric thrust magnetic bearings for a cold compressor and analyzes the changes in the system characteristics with the equilibrium position. By constructing PID controllers associated with the structural parameters of the magnetic bearing, the adaptive adjustment of the control parameters under different balanced position commands is realized. The simulation and experimental results prove that the gain-scheduled control method proposed in this paper can achieve a robust stability of the rotor in the range of 50 to 350 μm, and not at the cost of the response speed, adjustment time, and overshoot. The research results have reference significance for the structure design of asymmetric thrust magnetic bearings and play an important role in the commissioning and performance improvement of cold compressors.

Keywords: asymmetric; thrust magnetic bearings; gain-scheduled control; PID; changing balanced position

Citation: Zhang, S.; Wu, J. Gain-Scheduled Control of Asymmetric Thrust Magnetic Bearing. *Actuators* **2021**, *10*, 329. https://doi.org/10.3390/act10120329

Academic Editors: Oscar Barambones, Jose Antonio Cortajarena and Patxi Alkorta

Received: 31 October 2021
Accepted: 9 December 2021
Published: 13 December 2021

Publisher's Note: MDPI stays neutral with regard to jurisdictional claims in published maps and institutional affiliations.

Copyright: © 2021 by the authors. Licensee MDPI, Basel, Switzerland. This article is an open access article distributed under the terms and conditions of the Creative Commons Attribution (CC BY) license (https://creativecommons.org/licenses/by/4.0/).

1. Introduction

Nowadays, it is prevalent to apply cold centrifugal compressors to reduce the pressure of the sub-cooling tank in order to obtain 2 K superfluid helium on a large scale [1]. Grease-lubricated and gas bearings make it challenging to meet the low temperature, negative pressure, and high-speed operating conditions of cold compressors. The active magnetic bearing (AMB) uses electromagnetic force to suspend the rotor in space without friction and lubrication. Thus, it has universal advantages, such as no wear, high speed, and a long life [2]. Furthermore, AMB can improve the stability and efficiency of the compressor by adjusting its supporting characteristics. It is the best choice for the rotor support assembly of the cold compressor in the current superfluid helium refrigeration system [3].

The cold compressor is the critical component, and its efficiency is critical to the production of superfluid helium. In the case of the same rotation speed, the efficiency improvement of the magnetic levitation compressor can be achieved by actively controlling the axial position of the rotor to reduce the tip clearance [4]. However, most existing control methods [5–7] are based on a linear model. They have fixed control parameters, which only guarantee stability in a single equilibrium position or within a narrow range. In the past, we tested cold compressors with conventional control methods. When the balanced position was adjusted to be close to 30 μm, the response speed started to slow down; when the amplitude modulation was close to 100 μm, the rotor vibrated back and forth.

The unique working environment of the cold compressor determines that only one axial position is impossible to be perfect. The impeller is at an ultra-low temperature near 3 K, whereas the drive end is located in a room-temperature environment, so the rotor spans a temperature zone close to 300 K. Thermal expansion and contraction under the working conditions change the blade tip clearance, which has already been assembled at room temperature. The tip clearance can ensure the highest efficiency at a low temperature, resulting in very narrow operating conditions. For safety reasons, a large tip clearance is often reserved in reality, even though this is at the expense of efficiency. In our previous experience, the tip clearance has been set as large as 200 μm, and the efficiency is only close to 50%. Even so, there are still other problems. During the commissioning process of the superfluid helium system, the cold compressor needs to be shut down repeatedly. The impeller connected to the rotor at room temperature is quickly reheated, whereas the volute in the cold box can be kept low. A larger tip clearance than 200 μm is required to prevent the impeller and the volute from bonding and damage caused by thermal deformation, which will affect the usual start next time. In response to these problems, it is necessary to propose a control strategy to ensure that the rotor can quickly and robustly stabilize at any specified position within a large range, which is very important for the healthy and efficient operation of the compressor.

The balanced (tracking) position of the controlled object can be adjusted by modifying the reference command. Much research on this aspect comes from the positioning of lathe machining tools. Smirnov A. et al. [8] realize step, cone, and convex tracking in the range of 30 μm. Xudong Guan et al. [9] track different signals in the field of 70 μm. The literature [10] and [11] use advanced algorithms to achieve a higher tracking accuracy than ordinary algorithms. Overall, these studies focus more on improving the tracking accuracy, speed, and shape. On the other hand, they pay less attention to tracking a more extensive range, since the actuator can move synchronously with the tool it clamps.

Conversely, it is hoped that the adjustable range is extensive, since the tip clearance range of the cold compressor is as large as approximately 200 μm [4]. Apart from that, it also needs to have a strong anti-interference ability and have a minor overshoot when adjusting the balanced position in order to reduce the airflow impact. It may be possible to change the equilibrium position in a small range, but it is hard to achieve that within the entire clearance range. Suppose the balanced position is forcibly modified in a big step. In that case, the existing system will deviate significantly from the theoretical model, and the originally suitable controller may cause a poor response and even an unstable system. The literature [12] has studied how different balanced positions affect the tip clearance and ultimately influence the compression efficiency. Still, the clearance alteration in the research is not online, and the adjustment principle and methods of the control parameters under different clearances are not explained.

This paper derives the nonlinear model of the asymmetric thrust AMB for a vertical cold compressor, analyzes the system changes with the different balanced positions, and proposes a gain-scheduled controller whose control parameters automatically adjust with the balance position changing. The simulation and experimental results show that, compared with the parameter-fixed controller, the gain-scheduled control method ensures the stability and good performance of the AMB system, even if the balanced position changes in a wide range of 50 to 350 μm.

2. Model

2.1. Cold Compressor

The schematic diagram of the cold compressor in the superfluid helium refrigeration system is shown in Figure 1.

The working medium at the impeller is liquid helium of approximately 3 K. The rotor far away from the impeller is at room temperature to meet the working condition of the permanent magnet synchronous motor and the AMB. The motor stands in the middle position and it is used for driving the rotor rotating at high speed. Two radial AMBs lie in the upper and lower ends of the motor, and the thrust AMB is located at the upper position of the rotor. They are applied to support all of the degrees of freedom of the rotor. Two inductive displacement sensors are placed outside the two radial AMBs to measure the radial and axial displacement.

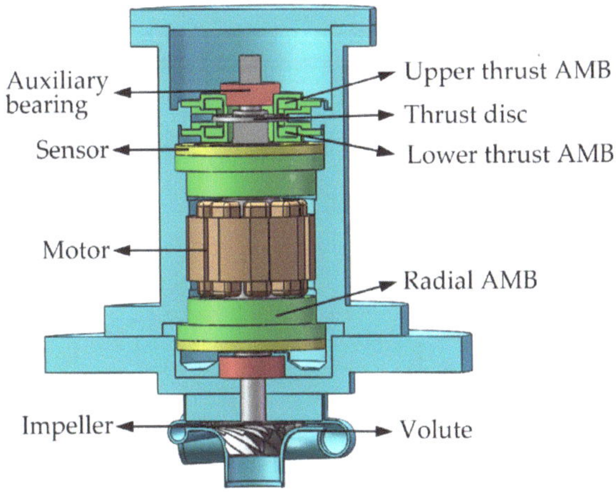

Figure 1. Sectional structure of magnetic suspension cold compressor.

The compressor is designed as a vertical structure to be fitted conveniently on the refrigeration system. Compared with the lower AMB, the upper AMB needs to bear the rotor gravity. In the regular operation, the balanced position of the compressor rotor is set in the center and remains unchanged. In this most common situation, if we expect that the currents of the upper and lower AMB are almost close, then we can only choose to make the axial AMB into a 'large upper and small lower' structure. The original intention of the asymmetric design is only to improve the conventional operating state of the compressor. The structure is not necessary for the application of the gain-scheduled control method proposed in this article.

The detailed design principle of the thrust AMB can be referred to [13]. In the actual assembly and testing, the air gap of the thrust AMB is further reduced based on the original design to facilitate the stability control of the rotor. As a result, the corresponding bias current and the heat generation of the thrust AMB can be decreased further. The structure diagram of the thrust AMB is shown in Figure 2, and its critical structural parameters are listed in Table 1. Moreover, the physical structure parameters of the rotor and radial AMB are listed in Table 2.

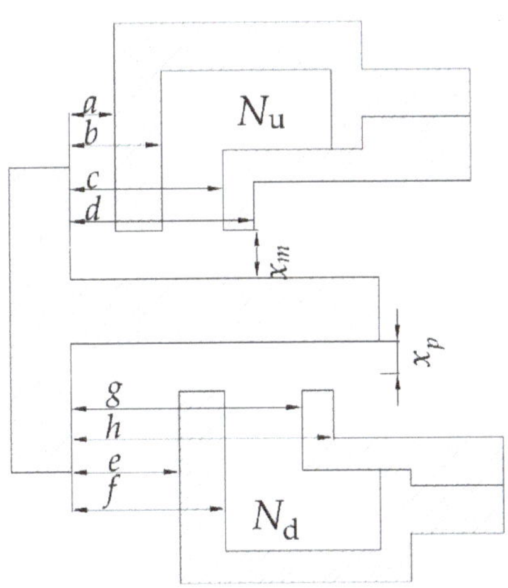

Figure 2. Structure diagram of thrust AMB model.

Table 1. Parameters of thrust AMB.

Name	Value
Bias current, i_0 (A)	0.85
Unilateral air gap of the thrust AMB, x_m (μm)	400
Unilateral air gap of the auxiliary bearing, x_p (μm)	300
Upper AMB inner magnetic pole inner/outer diameter, a/b (mm)	50/58.5
Upper AMB outer magnetic pole inner/outer diameter, c/d (mm)	74/80
Lower AMB inner magnetic pole inner/outer diameter, e/f (mm)	545/85
Lower AMB outer magnetic pole inner/outer diameter, g/h (mm)	76/80
Number of coil turns of upper AMB, N_u	225
Number of coil turns of lower AMB, N_d	175
Structure parameter of upper AMB, k_u	15.062
Structure parameter of lower AMB, k_d	13.662

Table 2. Parameters of radial AMB and rotor.

Name	Value
Bias current of radial AMB (A)	1.1
Air gap of radial AMB (μm)	350
Air gap of radial auxiliary bearing (μm)	150
Magnetic pole area of radial AMB (mm^2)	263
Number of coil turns of radial AMB	97
Number of magnetic poles of radial AMB	8
Total length of rotor (mm)	412
Rotor mass, m (kg)	6.2
Diameter of middle part of rotor (mm)	63

2.2. Non-Linear Electromagnetic Force

The electromagnetic force f_a generated by the single magnetic pole is related to the coil current i and the distance s between the stator and rotor [2]:

$$f_a = \underbrace{\frac{\mu_0 N^2 A}{4}}_{k} \frac{i^2}{s^2} \tag{1}$$

where μ_0 is the vacuum permeability, N denotes the coil turns, A is the magnetic pole area, and k is the AMB structure parameter, which does not change with the instantaneous state of the system.

Use the subscripts $_u$ and $_d$ to denote the upper and lower thrust AMB, respectively, and introduce the structural parameters k_u and k_d of the thrust AMB:

$$\begin{cases} k_u = \frac{\mu_0 A_u {N_u}^2}{4} \\ k_d = \frac{\mu_0 A_d {N_d}^2}{4} \end{cases} \tag{2}$$

Combining Formulas (1) and (2), the resultant electromagnetic force of the upper and lower thrust AMB can be expressed as [3]:

$$f_a = f_u - f_d = k_u \frac{(i_0 + i)^2}{(x_m + x_p - x)^2} - k_d \frac{(i_0 - i)^2}{(x_m - x_p + x)^2} \tag{3}$$

where x_m and x_p denote the unilateral air gap of the thrust AMB and the auxiliary bearing, respectively, the specified balanced position is labeled x_0 ($0 < x_0 < 2x_p$), the bias current is labeled i_0, the instantaneous distance of the rotor from the bottom end of the auxiliary bearing is indicated by x, and i indicates the instantaneous control current.

The first-order approximation of the Taylor series near the equilibrium ($x = x_0$, $i = 0$) for the Formula (3) is presented:

$$
\begin{aligned}
f_a = {} & k_u\left[\frac{i_0{}^2}{(x_m+x_p-x_0)^2}+\frac{2i_0}{(x_m+x_p-x_0)^2}i+\frac{2i_0^2}{(x_m+x_p-x_0)^3}(x-x_0)+\cdots\right] \\
& -k_d\left[\frac{i_0{}^2}{(x_m-x_p+x_0)^2}+\frac{-2i_0}{(x_m-x_p+x_0)^2}i+\frac{-2i_0^2}{(x_m-x_p+x_0)^3}(x-x_0)+\cdots\right] \\
\approx {} & \underbrace{i_0{}^2\left[\frac{k_u}{(x_m+x_p-x_0)^2}-\frac{k_d}{(x_m-x_p+x_0)^2}\right]}_{f_{a0}}+\underbrace{\left[\frac{2k_u i_0}{(x_m+x_p-x_0)^2}+\frac{2k_d i_0}{(x_m-x_p+x_0)^2}\right]}_{k_i(x_0)}i+\underbrace{\left[\frac{2k_u i_0{}^2}{(x_m+x_p-x_0)^3}+\frac{2k_d i_0{}^2}{(x_m-x_p+x_0)^3}\right]}_{k_x(x_0)}(x-x_0)
\end{aligned}
\tag{4}
$$

The resulting external force f on the rotor should be:

$$f = f_a - mg = f_{a0} + k_i i + k_x(x - x_0) - mg \tag{5}$$

In the above formula, k_i and k_x are the current stiffness and displacement stiffness, respectively. They are the two critical parameters related to the AMB geometry, the bias current, and the reference position.

According to Formula (4), the electromagnetic force f_a includes a constant force part f_{a0}, not varying with the instantaneous state of the system. Even when the rotor is balanced at the physical center, and even if the thrust AMB is asymmetric, the term is not zero. According to the design ideas mentioned above, when the rotor is suspended at the center, f_{a0} should be exactly the same as the gravity of the rotor.

It should be noted that the purpose of constructing the linearization model is to explain the variation regular of the model characteristics with the equilibrium position and thus put forward the automatic adjustment strategy of PID parameters. However, the simulation calculation will retain the nonlinear form of the model.

2.3. Model Analysis

When the equilibrium position is specified between 0 μm and 600 μm, the variation and their percentage changes in k_i and k_x characterizing the electromagnetic force model are illustrated in Figures 3 and 4, respectively. It should be noted that the magnitudes of k_i, k_x, and f_{a0} are also affected by the bias current, according to Formula (4). However, to highlight the variation characteristics with different positions, the bias current remains unchanged, with a value of 0.85 A, as shown in Table 1.

These figures show that the k_i and k_x are not symmetrical at the center of 300 μm. The change near the upper AMB is much more dramatic than that near the lower AMB due to the larger geometry of the upper AMB. These figures depict that the two parameters remain almost unchanged within the range of 100 μm, deviating downward from the center. However, k_i and k_x increase rapidly, and their slopes increase notably when the deviation increases. When the balanced position deviates from the scope of (200, 300) by 100 μm, the percentage change in k_i and k_x reaches 55% and 85%, respectively. It can be inferred from Figure 4 that the parameter-fixed controller may maintain the system as stable within the range of 200 to 300 μm. Still, beyond this range, the response performance is likely to deteriorate rapidly. The results are confirmed in the following simulation experiment.

The solid black line in Figure 5 shows the variation characteristics of f_{a0} with the balanced position. As shown in Formula (4), f_{a0} is a constant force generated by the asymmetrical AMB. When the rotor is balanced at the center, f_{a0} is not zero but is supposed to be equal to the rotor gravity, according to the previous design expectations. At this equilibrium point, the coil current of the upper and lower AMB should be close. When the balanced position deviates from the center, the resultant force of the electromagnetic force and the gravity is not zero. Thus, additional current is needed to offset the resultant force represented by the gray curve. The figure shows that the resultant force increases with the deviation. The increasing rate of the force close to the upper AMB is more significant than the lower.

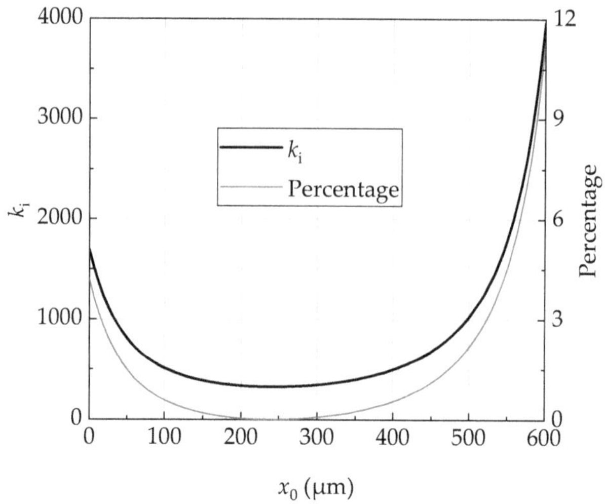

Figure 3. k_i and its percentage change with balanced position.

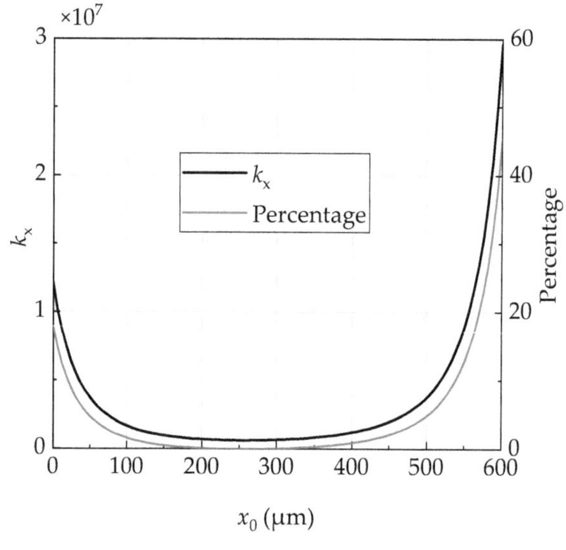

Figure 4. k_x and its percentage change with balanced position.

It should be mentioned that the feature analysis of these figures is based on Formula (1), ignoring magnetic flux leakage and core magnetization. The application of the Taylor expansion formula in Formula (4) makes the error tend to increase as x_0 moves away from the AMB center. However, it does not prevent us from analyzing the qualitative law of the problem.

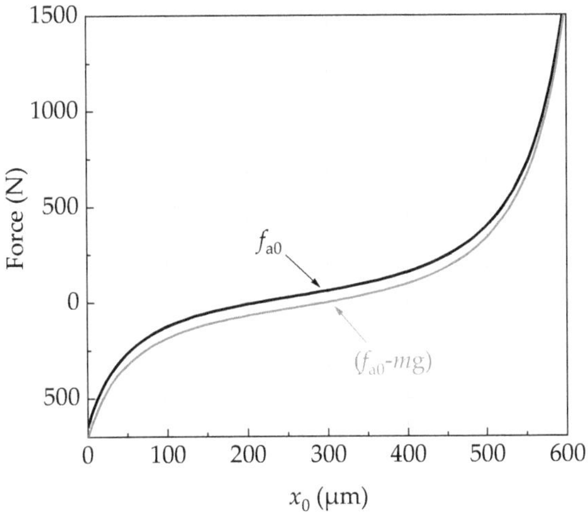

Figure 5. The constant force generated by AMB varies with the balanced position.

3. Controller Design

The research applies an incomplete differential PID controller G_c in the following form:

$$G_c(s) = K_p + \frac{K_i}{s} + \frac{K_d s}{T_d s + 1} \tag{6}$$

where K_p, K_i, and K_d are the proportional gain, integral gain, and derivative gain, respectively. The derivative time constant T_d is also introduced to prevent the controller from infinitely amplifying high-frequency noise signals.

The supporting characteristics of the magnetic suspension rotor in the closed-loop system depend on the AMB, rotor, controller, and other electronic control systems. In the following, the manuscript derives the physical relationship between the control parameters and the stiffness and damping of the closed-loop system and then uses it as the basis to recommend the value ranges of K_p, K_i, and K_d.

The rotor displacement signal x is captured by the sensor and then transmitted to the controller in the closed-loop feedback system. Afterward, it is calculated and converted into the control current i within an appropriate range through the power amplifier to drive the electromagnet coil. The relationship between x and i can be expressed as:

$$i = G_a G_c (1 - G_s (x - x_0)) \tag{7}$$

where G_a is the power amplifier's transfer functions and G_s is the sensor's. The sweep frequency test shown in Figure 6 displays that the gain of the power amplifier is 0.79 and that the bandwidth is approximately 2.3 kHz. The sensor gain is 20,000 V/m, and the bandwidth of the anti-aliasing filter in the signal processing circuit is 3.3 kHz. The bandwidth of both the power amplifier and the sensor is quite large, so their transfer functions can be simplified into a constant to highlight clearly the relationship between the control parameters and the system supporting performance.

According to Newton's second law, the force analysis in the axial direction of the rotor is presented as:

$$f = m\ddot{x} \tag{8}$$

Substituting Formulas (7) and (8) into Formula (5) and employing Fourier transform:

$$- mX\omega^2 = -(\mathrm{Re}\{G_a G_c G_s\} k_i - k_x) - \mathrm{j}[\mathrm{Im}\{G_a G_c G_s\} k_i] X \tag{9}$$

If k_e denotes the stiffness and d_e denotes the damping, the Fourier transform of the dynamic equation of the mass-spring-damping system is:

$$-mX\omega^2 = -k_e - jd_e\omega X \tag{10}$$

Using simultaneous equations from Formulas (6) and (9), and then comparing with the Formula (10), the equivalent stiffness and equivalent damping of the closed-loop system are:

$$k_e = k_i G_a G_s (K_p + K_i + \frac{K_d T_d \omega^2}{{T_d}^2\omega^2 + 1}) - k_x \tag{11}$$

$$d_e = k_i G_a G_s (\frac{K_d}{{T_d}^2\omega^2 + 1} - K_i) \tag{12}$$

The cut-off frequency of the differential term in Formula (6) is related to the operating frequency of the compressor, which is approximately 200 Hz. It is easy to prove that K_i and K_d in Formula (11) have little effect on the stiffness and thus can be ignored during the initial estimation stage when the working frequency is below 200 Hz,

$$k_e \approx k_i G_a G_s K_p - k_x \tag{13}$$

$$d_e \approx k_i G_a G_s K_d \tag{14}$$

Define the stiffness ratio $\beta = k_e/k_x$, and the stiffness coefficient should be greater than 1 to stabilize the rotor. However, excessive stiffness requires a high-bandwidth power amplifier and may cause magnetic field saturation. Therefore, according to (13):

$$K_p = \frac{\beta k_x(x_0)}{G_a G_s k_i(x_0)} \tag{15}$$

The design standards of compressors [14] and AMBs [15] require the system damping ratio within the operating range to be greater than 20% to obtain a better static and dynamic response performance. However, a large damping ratio will result in the amplification of the noise and the saturation of the control signal. Combining the damping ratio formula $\xi = d_e/(2\sqrt{k_e m})$ with the Formula (14) gives:

$$K_d = \frac{2\xi\sqrt{m k_x(x_0)}}{G_a G_s k_i(x_0)} \tag{16}$$

According to our debugging experience, the integral effect increases as the proportional coefficient increases. Thus,

$$K_i = \gamma K_p K_p \tag{17}$$

γ is assigned as 400 both in the simulation and experiment according to our experience.

3.1. Parameter-Fixed PID Controller

Let us set β 1.5 and ξ 0.75, and introduce these parameters in Table 1 into Formulas (15)–(17). Suppose the rotor is balanced at the AMB center; then, K_p, K_d, and K_i can be calculated. These parameters are listed in Table 3.

Under the parameter-fixed controller, the variation curve of the stiffness ratio and the relative damping coefficient with the balanced position is shown in Figure 7.

The figure indicates that β decreases rapidly as the balanced position deviates from the center. At approximately 170 and 380 μm, the system stiffness starts to be less than the displacement stiffness, and, as a consequence, the system may exhibit a poor response performance. If the deviation continues to increase, when the equilibrium position is less than 47 μm or greater than 545 μm, the stiffness k_e begins to decay to a negative value. Negative stiffness means that the characteristic root of the closed-loop system will move from the left half-plane to the real axis, and the rotor cannot remain stable anymore. When the balanced position moves from the center to the magnetic poles at both ends, the rotor is more likely to fall into the unstable area close to the larger upper axial magnetic bearing.

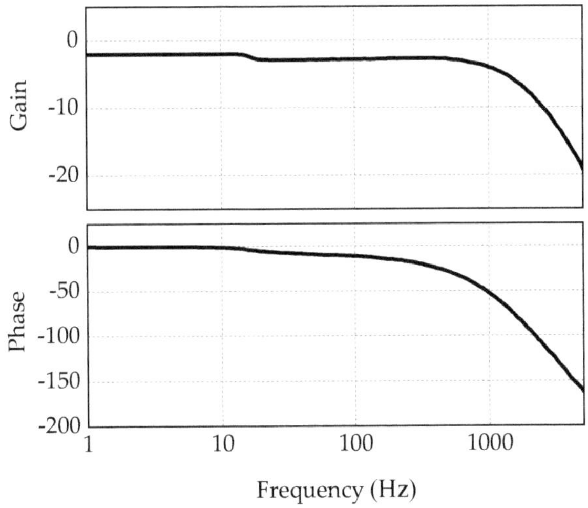

Figure 6. Sweep frequency test result of power amplifier.

Table 3. PID controller with fixed parameter.

Name	Value
Proportional gain K_{p0}	0.32
Integral gain K_{i0}	40
Differential gain K_{d0}	7.5×10^{-4}
Differential time constant T_d	$1/400\pi$

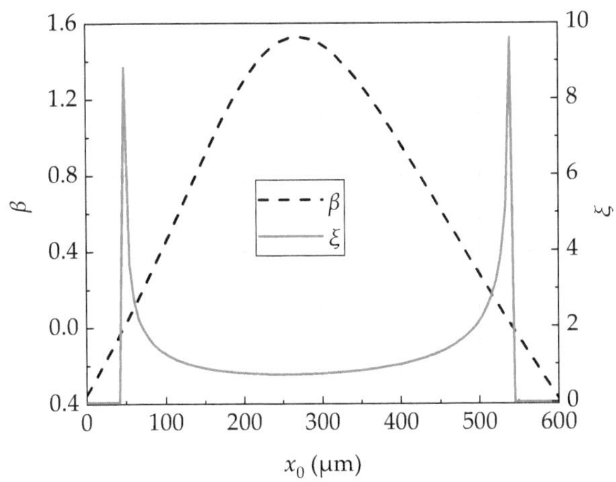

Figure 7. Variation curve of β and ξ with balanced position.

3.2. Gain-Scheduled PID Controller

According to Formulas (15) and (16), the variation curve of K_p and K_d with the balanced position is shown in Figure 8.

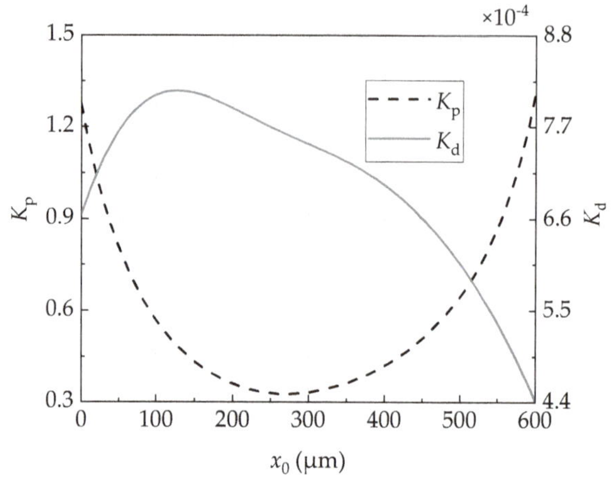

Figure 8. K_p and K_d vary with balanced position.

As the balanced position deviates from the physical center, K_p increases rapidly. K_p increases approximately three times at a distance of around 300 μm from the physical center and increases approximately two times at approximately 200 μm. In general, the variation of the balanced position in the asymmetric AMB causes K_p to vary greater, whereas K_d displays more obvious asymmetric features.

Let us select a group of gain-scheduled PID controllers with different balanced positions of 50 μm, 100 μm, 200 μm, 300 μm, 400 μm, and 500 μm, and draw their Bode diagrams shown in Figure 9.

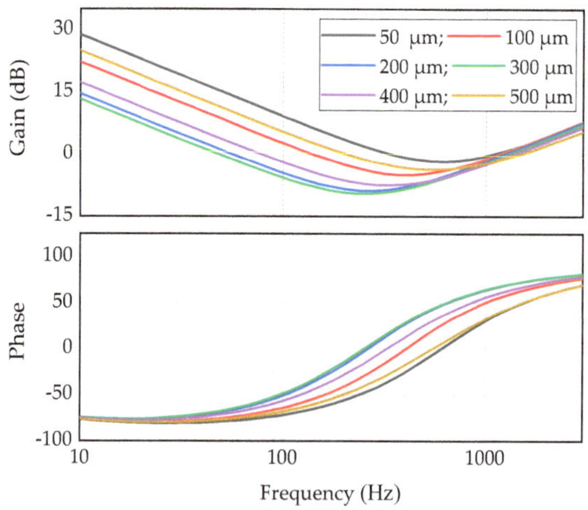

Figure 9. Bode plot of gain-scheduled PID controller.

4. Simulation and Experiment

4.1. Simulation Result Analysis

Based on MATLAB/SIMULINK, a simulation model is established, as shown in Figure 10. The deviation between the rotor displacement observed by the sensor and the set value is transmitted to the controller. Through the amplifier, the control voltage is converted into the current signal within a suitable range.

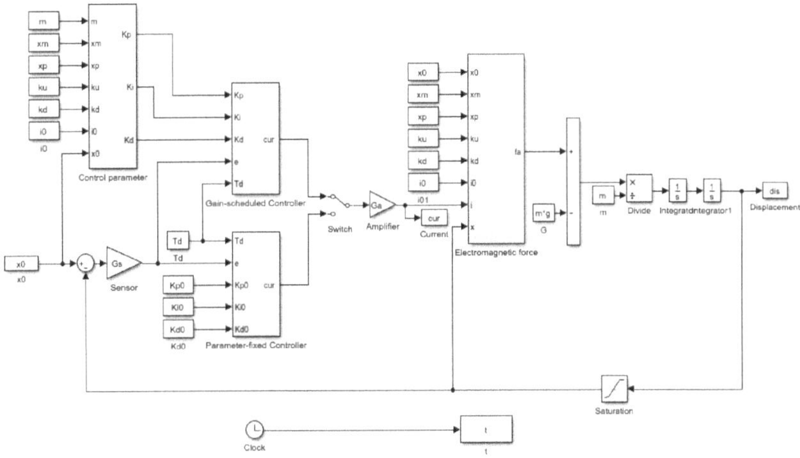

Figure 10. Simulation model based on SIMULINK.

In the simulation model, the user can freely set the balanced position of the rotor and independently select the controller as a parameter-fixed controller or a gain-scheduled controller. With the two different controllers, the simulation results of the step response are shown in Figures 11 and 12, respectively.

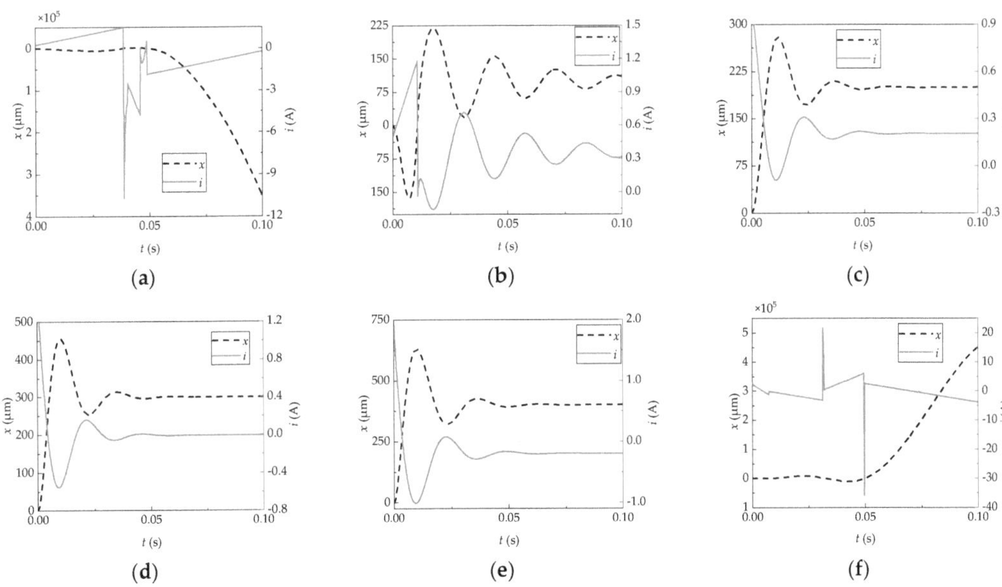

Figure 11. Step response simulation with parameter-fixed controller. (**a**) x_0 = 50 μm; (**b**) x_0 = 100 μm; (**c**) x_0 = 200 μm; (**d**) x_0 = 300 μm; (**e**) x_0 = 400 μm; (**f**) x_0 = 500 μm.

The lowest end of the air gap of the auxiliary bearing is defined as the initial position. When the parameter-fixed controller in Table 3 is utilized, the rotor is stepped from the initial appointment to 50, 100, 200, 300, 400, and 500 μm, and the corresponding step response curves are shown in Figure 11.

It can be found from Figure 11 that the system with the parameter-fixed controller can only keep just a satisfactory performance in a specific narrow area. For example, the response performance is good when the balanced position is from 200 to 300 μm, whereas it is poor in other areas. When it is less than 100 μm or more than 400 μm, the overshoot has exceeded the limit of the auxiliary bearing (0–600 μm). If it is less than 50 μm or more than 500 μm, the system will be unstable because the controller cannot provide adequate stiffness and damping. These phenomena are consistent with the stiffness and damping variation curves shown in Figure 7.

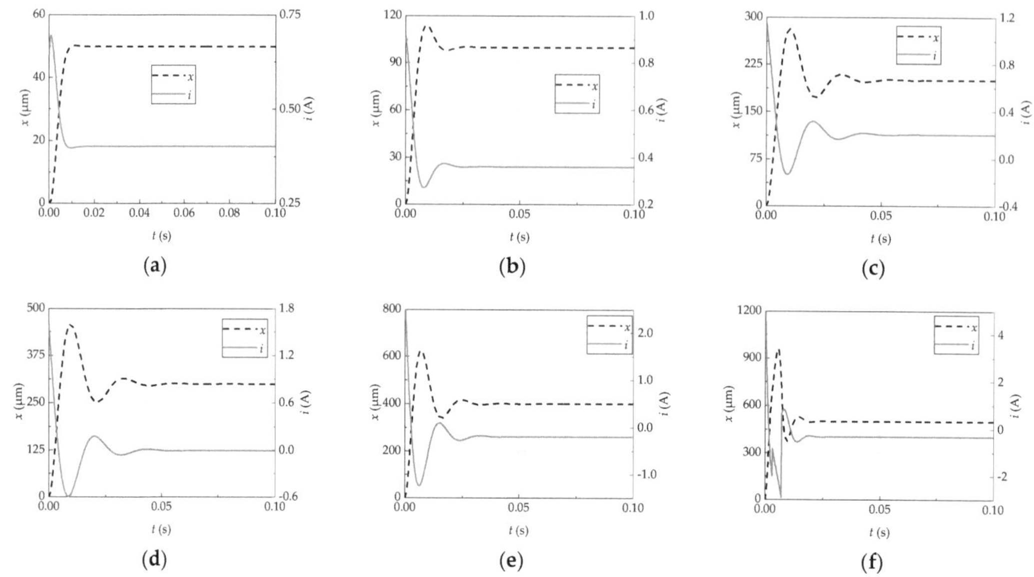

Figure 12. Step response simulation with gain-scheduled controller. (**a**) x_0 = 50 μm; (**b**) x_0 = 100 μm; (**c**) x_0 = 200 μm; (**d**) x_0 = 300 μm; (**e**) x_0 = 400 μm; (**f**) x_0 = 500 μm.

In contrast, when the abovementioned gain-scheduled controllers are applied, the floating characteristic curves at different balanced positions are shown in Figure 12.

These figures imply that the rotor with the gain-scheduled controller can remain stable at any balanced position within the range of 50 to 500 μm. As the balanced position moves upward, the overshoot of the current response increases, whereas the steady-state value of the current gradually decreases. In practice, it is difficult to achieve equilibrium at 500 μm when considering the current and displacement response. Furthermore, the displacement overshoot at 400 μm exceeds the upper limit of the auxiliary bearing. In the range of 50 to 400 μm, the response performance is good and does not change significantly with the balanced position. The above range can be recommended for the actual commissioning.

The above Figures 11 and 12 show that the stability effect of the rotor is not symmetrical, with a center point of approximately 300 μm. At the same symmetrical position, the control effect of the area close to the upper AMB is worse than that of the lower AMB.

4.2. Experiment Result Analysis

The main components of the experiment rig are shown in Figure 13. The thickness of the AMB shims and the auxiliary bearing was repeatedly adjusted, so that the physical centers of the two air gaps coincide. At the same time, the unilateral air gap between the thrust AMB and the auxiliary bearing is guaranteed to be 0.4 mm and 0.3 mm, respectively.

Formula (6) is discretized based on the bilinear transformation method, and then is written into the controller. The NI PXIe-8840 high-performance embedded product is employed as the control and monitoring system of the AMB. Its sampling frequency is 50 kHz, which can realize the real-time monitoring and control of the rotor state.

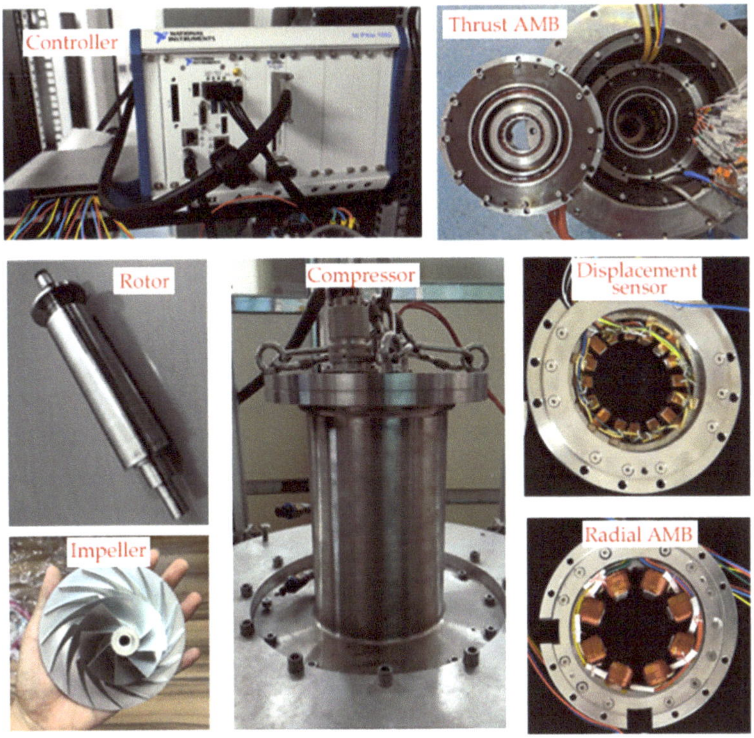

Figure 13. Photo of the cold compressor.

With the same gain-scheduled control strategy as the above simulation, the following step response characteristics of the rotor are obtained. Figure 14 is the step response with steps from 0 to 50, 100, 150, 250, and 350 μm, respectively. Figure 15 is the stair-like responses with a fixed step length of 50 μm. Experimental results demonstrate that the gain-scheduled control strategy proposed in this paper guarantees that the rotor remains stable in the range of 50 to 350 μm. When the rotor is balanced at 300 μm, the current of the upper and lower AMB is almost the same, and is equal to the bias current. Experimental phenomena match the previous design principle and verify the consistency between the actual and theoretical models at the design point.

It can be found from Figure 14 that, as the balanced position increases, the steady-state current decreases whereas the overshoot increases. When the balanced position is 50 μm, although the overshoot of the current response is the smallest, the stabilization time is the longest, and a lot of noise is introduced. The reason may be that the rotor is too far away from the upper AMB, which excites more prominent nonlinear characteristics of the electromagnetic force than those with a closer distance. When the rotor is balanced at 350 μm, both the response time and current overshoot are dramatic because of the large step size. It can be concluded that the equilibrium range is located below the physical center. In summary, the commissioning range of the compressor in actual operation is recommended to be in the field of 100 to 300 μm.

Figure 15 has a smaller step size, so the overall current overshoot is smaller than that in Figure 14. When the system needs a significant change in the balanced position and does not require a high response speed, it is recommended to use the stair-like strategy to obtain a better response effect.

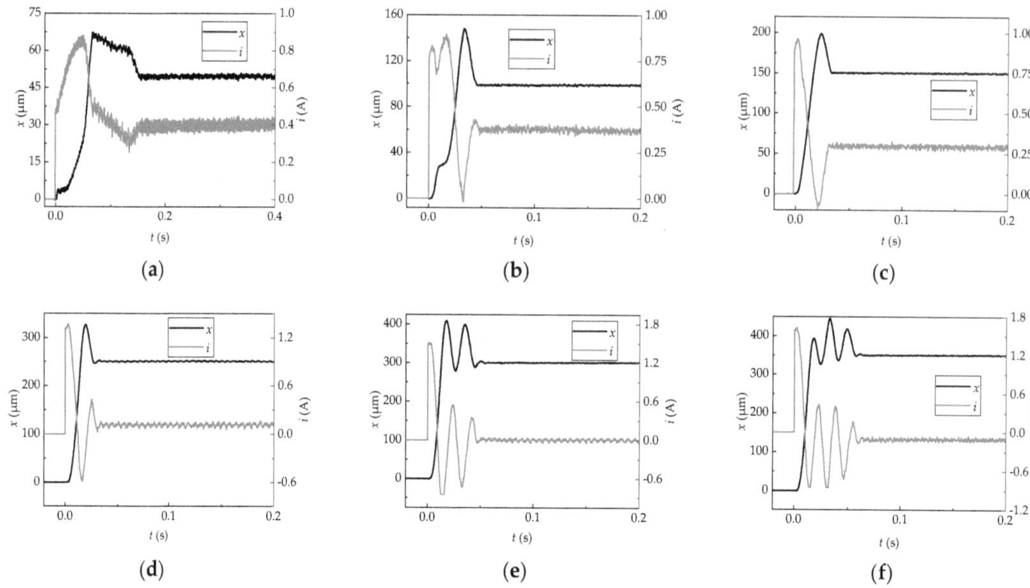

Figure 14. Experiment of step response from 0. (**a**) x_0 = 50 μm; (**b**) x_0 = 100 μm; (**c**) x_0 = 150 μm; (**d**) x_0 = 250 μm; (**e**) x_0 = 300 μm; (**f**) x_0 = 350 μm.

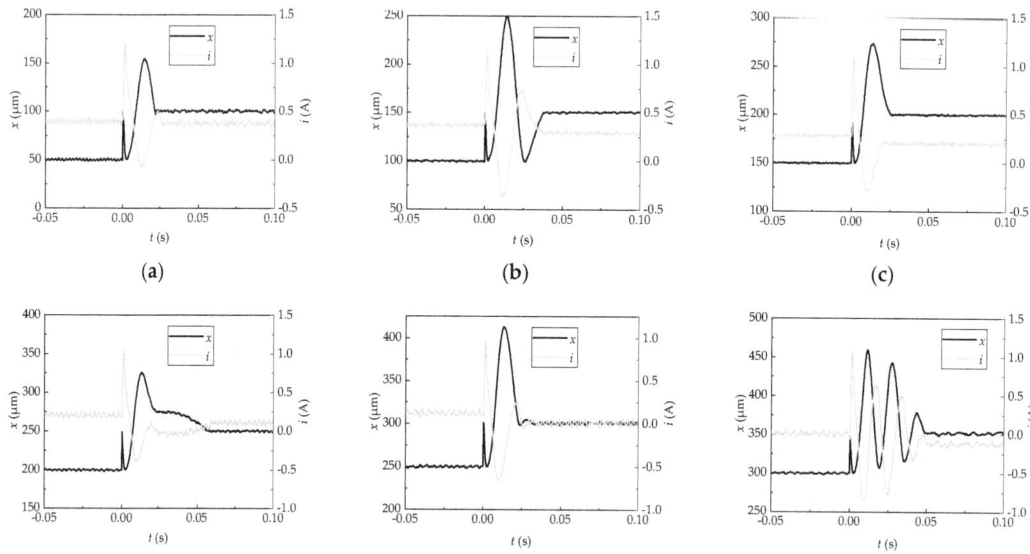

Figure 15. Experiment of stair-like response with a fixed step size of 50 μm. (**a**) from x_0 = 50 to 100 μm; (**b**) from x_0 =100 to 150 μm; (**c**) from x_0 = 150 to 200 μm; (**d**) from x_0 = 200 to 250 μm; (**e**) from x_0 = 250 to 300 μm; (**f**) from x_0 = 300 to 350 μm.

5. Conclusions

The paper studies the thrust AMB of the cold compressor, analyzes the change in the system stiffness and damping with the equilibrium position, and therefore proposes a gain-scheduled PID controller to cope with the variable equilibrium position. The research results show that, as the equilibrium position moves from the center to the magnetic poles on both sides, the system stiffness decreases whereas the damping increases. The asymmetry of the AMB exacerbates the unevenness of the variation. The control of the upper position is more complicated than that below the physical center because the upper AMB has a larger geometry. The simulation and experimental results indicate that gain-scheduled control ensures that the rotor remains stable in the range of 50 to 350 μm and possesses a breakneck response speed and slight overshoot in the field of 100 to 300 μm. This research has realized the automatic and real-time adjustment of the tip clearance of the cold compressor. The subsequent research will focus on the joint commissioning between the requirements of the superfluid helium system and the axial position of the magnetic suspension rotor.

Author Contributions: Conceptualization, S.Z. and J.W.; methodology, S.Z.; software, S.Z.; validation, S.Z. and J.W.; formal analysis, S.Z.; investigation, S.Z.; resources, J.W. data curation, S.Z.; writing—original draft preparation, S.Z.; writing—review and editing, S.Z. and J.W.; visualization, S.Z.; supervision, J.W.; project administration, J.W.; funding acquisition, S.Z. and J.W. All authors have read and agreed to the published version of the manuscript.

Funding: This research was funded by the fund of the State Key Laboratory of Technologies in Space Cryogenic Propellants, grant number SKLTSCPSKLTSCP202104.

Institutional Review Board Statement: Not applicable.

Informed Consent Statement: Not applicable.

Data Availability Statement: All of the data used in this paper can be obtained upon request to the corresponding author.

Conflicts of Interest: The authors declare no conflict of interest.

References

1. Dong, X.B.; Wu, J.H.; Li, J.J.; Xu, X.D.; Su, H.; Shang, J.; Li, Q.; Gong, L.H. Impact of external heat transfer on the performance of a cold compressor used in superfluid helium system. *Cryogenics* **2020**, *110*, 103141. [CrossRef]
2. Zhang, S.Y.; Pan, W.; Wei, C.B.; Wu, J.H. Structure design and simulation research of active magnetic bearing for helium centrifugal cold compressor. In Proceedings of the Materials Science and Engineering Conference, Madison, WI, USA, 9–13 July 2017; p. 268.
3. Larsonneur, R. Principle of Active Magnetic Suspension. In *Magnetic Bearings: Theory, Design, and Application to Rotating Machinery*, 2nd ed.; Schweitzer, G., Maslen, E.H., Eds.; Springer: Berlin/Heidelberg, Germany, 2009; pp. 27–68.
4. Liu, F.; Zhang, S.; Shang, J.; Wu, J. Influence of tip gap form on aerodynamic performance of cold compressor. *Cryo. Supercond.* **2021**, *49*, 49–57.
5. Pesch, A.; Smirnow, A.; Pyrhonen, O.; Sawicki, J.T. Magnetic bearing spindle tool tracking through μ-synthesis robust control. *IEEE/ASME Trans. Mech.* **2015**, *20*, 1448–1457. [CrossRef]
6. Yoon, S.Y.; Lin, Z.L.; Lim, K.T.; Goyne, C.; Allaire, P.E. Model Validation for an Active Magnetic Bearing Based Compressor Surge Control Test Rig. *J. Vib. Acoust.* **2010**, *132*, 71–81. [CrossRef]
7. Chen, L.L.; Zhu, C.S.; Zhong, Z.X.; Wang, C.K.; Li, Z.N. Radial position control for magnetically suspended high-speed flywheel energy storage system with inverse system method and extended 2-DOF PID controller. *IET Electr. Power Appl.* **2019**, *14*, 71–81. [CrossRef]
8. Smirnov, A.; Pesch, A.H.; Pyrhonen, O.; Sawicki, J.T. High-Precision Cutting Tool Tracking with a Magnetic Bearing Spindle. *J. Dyn. Syst. Meas. Control* **2015**, *137*, 051017. [CrossRef]
9. Guan, X.; Zhou, J.; Wu, H.; Zhang, Y. Trajectory Tracking of Rotating Shaft with Active Magnetic Bearings under Different Reference Signals. *Appl. Comput. Electron.* **2019**, *34*, 1435–1444.
10. Chen, S.Y.; Lin, F.J. Robust Nonsingular Terminal Sliding-Mode Control for Nonlinear Magnetic Bearing System. *IEEE Trans. Contr. Syst. Technol.* **2011**, *19*, 636–643. [CrossRef]
11. Lin, F.J.; Chen, S.Y.; Huang, M.S. Adaptive complementary sliding-mode control for thrust active magnetic bearing system. *Control Eng. Pract.* **2011**, *19*, 711–722. [CrossRef]
12. Han, B.C.; Cui, B.; Zheng, S.Q.; Zhang, Y.; Song, X.D. Model and experimental research on the relation between pressure and impeller clearance of magnetic levitation compressor. *J. Vib. Meas. Diagn.* **2019**, *39*, 286–291+441–442.
13. Zhang, S.Y.; Liu, F.; Qin, H.L. Structure design and numerical simulation of axial active magnetic bearing with an asymmetric form. *Cryo. Supercond.* **2020**, *48*, 12–37+45.
14. *API 617 Axial and Centrifugal Compressors and Expander-Compressors for Petroleum, Chemical and Gas Industry Services*; American Petroleum Institute API: Washington, DC, USA, 2002; pp. 180–184.
15. *ISO 14839-3 Mechanical Vibration—Vibration of Rotating Machinery Equipped with Active Magnetic Bearings—Part 3: Evaluation of Stability Margin*; International Organization for Standardization ISO: London, UK, 2006; pp. 21–23.

actuators

MDPI

Article

An IDA-PBC Design with Integral Action for Output Voltage Regulation in an Interleaved Boost Converter for DC Microgrid Applications

Oscar Danilo Montoya [1,2,*], Federico Martin Serra [3], Walter Gil-González [4], Eduardo Maximiliano Asensio [3] and Jonathan Emmanuel Bosso [3]

1 Facultad de Ingeniería, Universidad Distrital Francisco José de Caldas, Bogotá 110231, Colombia
2 Laboratorio Inteligente de Energía, Universidad Tecnológica de Bolívar, Cartagena 131001, Colombia
3 Laboratorio de Control Automático (LCA), Facultad de Ingeniería y Ciencias Agropecuarias, Universidad Nacional de San Luis—CONICET, Villa Mercedes, San Luis 5730, Argentina; fmserra@unsl.edu.ar (F.M.S.); maxiasensio@gmail.com (E.M.A.); jnthnbss@gmail.com (J.E.B.)
4 Facultad de Ingeniería, Institución Universitaria Pascual Bravo, Campus Robledo, Medellín 050036, Colombia; walter.gil@pascualbravo.edu.co
* Correspondence: odmontoyag@udistrital.edu.co

Abstract: This paper describes the output voltage regulation control for an interleaved connected to a direct current (DC) microgrid considering bidirectional current flows. The proposed controller is based on an interconnection and damping passivity-based control (IDA-PBC) approach with integral action that regulates the output voltage profile at its assigned reference. This approach designs a control law via nonlinear feedback that ensures asymptotic stability in a closed-loop in the sense of Lyapunov. Moreover, the IDA-PBC design adds an integral gain to eliminate the possible tracking errors in steady-state conditions. Numerical simulations in the Piecewise Linear Electrical Circuit Simulation (PLECS) package for MATLAB/Simulink demonstrate that the effectiveness of the proposed controller is assessed and compared with a conventional proportional-integral controller under different scenarios considering strong variations in the current injected/absorbed by the DC microgrid.

Keywords: nonlinear passivity-based control design; interleaved boost converter; voltage regulation; direct current microgrids; classic PI design

Citation: Montoya, O.D.; Serra, F.M.; Gil-González, W.; Asensio, E.M.; Bosso, J.E. An IDA-PBC Design with Integral Action for Output Voltage Regulation in an Interleaved Boost Converter for DC Microgrid Applications. *Actuators* **2022**, *11*, 5. https://doi.org/10.3390/act11010005

Academic Editors: Oscar Barambones, Jose Antonio Cortajarena and Patxi Alkorta

Received: 22 November 2021
Accepted: 26 December 2021
Published: 29 December 2021

Publisher's Note: MDPI stays neutral with regard to jurisdictional claims in published maps and institutional affiliations.

Copyright: © 2022 by the authors. Licensee MDPI, Basel, Switzerland. This article is an open access article distributed under the terms and conditions of the Creative Commons Attribution (CC BY) license (https://creativecommons.org/licenses/by/4.0/).

1. Introduction

Recent advances in power electronics from high- to low-power voltages for direct current (DC) applications make owning DC distribution networks in medium and low voltage levels with high levels of efficiency possible [1,2]. Since DC distribution does not require reactive power and frequency concepts to operate, energy losses are inferior in these systems when compared with its traditional alternating current (AC) counterparts [3]. Generally, a DC microgrid can be composed of multiple devices interconnected to the main regulated bus [4,5], which include renewable generation, batteries, linear loads (i.e., resistances), and constant power loads, requiring specialized controllers to ensure a stable operation in a closed-loop [6,7]. The interconnection of these devices is executed with power electronic converters that allow the controlling of each device into their operative range, ensuring secure working. Some of the most conventional converters for DC microgrid applications include buck [8], boost [6], buck-boost [9], non-inverting buck-boost [10], and Cuk converters [11]. The main characteristic of these devices is that, usually, due to the presence of commutated devices inside of their electrical circuits, they generate nonlinear dynamic models that complicate the usage of classical linear controllers such as PI (Proportional-Integral) and feedback designs [10,12].

The importance of the power electronic converters is illustrated in the development of the electricity service using DC microgrids (see Figure 1) [13]. In this paper, we explore the possibility of employing an interleaved boost converter for output voltage regulation in DC microgrid applications, considering that the grid current can be positive or negative as a function of the amount of renewable energy available and the total power consumption [14].

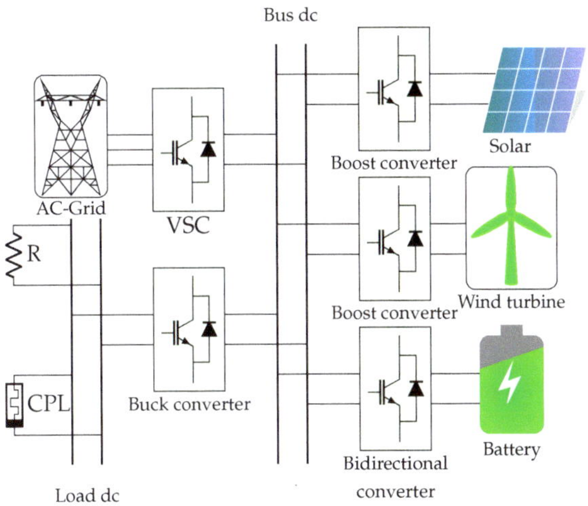

Figure 1. A typical DC microgrid with various sources and loads.

The interleaved boost converter is selected to support the voltage profile due to its simple structure and high capacity for transporting current owing to each parallel connection of inductor branches, allowing bidirectional current flow. Hence, this converter topology is ideal for applications that involve battery charging/discharging operating conditions, maintaining the voltage profile in the DC microgrid as consistently as possible [15].

Multiple control approaches have been applied to the interleaved boost converter for DC microgrid applications, some of which are discussed below. The authors of [14] present the design of a conventional PI controller in its digital version to control the current injected/absorbed by a battery pack to ensure reliable operation. Numerical and experimental validations of the proposed PI controller reveal that the objective of control is fulfilled by the PI design; however, the authors do not compare their control design with other control techniques, which do not permit measuring the effectiveness of their approach in terms of settling times and signal overshoots. The authors of [16] propose the application of the high-speed version of the model's predicted control approach to support voltage in DC microgrids using an interleaved boost converter with four inductive branches. Numerical simulations and experimental validations depicted the effectiveness of the proposed controller when compared with the classical model predictive and PI controllers. Cervantes et al. in [17] presented a time-varying switching-based controller for an interleaved boost converter composed of two inductor branches for electric vehicle applications. The proposed controller selects the switching surfaces distinctively for each interleaved converter cell to guarantee both maximum current and voltage ripple. Numerical simulations and experimental validations confirm its effectiveness in regulating the output voltage for a resistive load; however, the main flaw of this research was that the authors did not provide comparisons with other control methodologies. The authors of [18] present the application of the discrete-time inverse optimal control to an interleaved boost converter with two branches. They used the Euler approximation and the bilinear Tustin discretization to obtain a linear equivalent discrete model of the network, facilitating the control design via the inverse optimal control approach. Numerical results demonstrate the effectiveness of the proposed controller in supporting constant voltage to resistive loads independent of their variations; however, no comparisons with classical or nonlinear methodologies were provided in this study. Additional control methodologies include sliding control [19,20], exact state-feedback control [21,22], Hamiltonian-based controllers [23,24], neural and fuzzy controller designs [25,26], and [27], among others.

Based on the previous revision of the state-of-the-art regarding control applications for interleaved boost converters in this research, we present the following contributions:

i. The application of the IDA-PBC controller to the interleaved boost converter for output voltage regulation in DC microgrids with variable injected/demanded current: to improve the IDA-PBC design's performance, an integral action is added using the passive output of the system that does not affect the stability properties in closed-loop and allows eliminating the steady-state errors introduced by possible unmodeled dynamics;
ii. The proposed controller owns the advantage of not depending on the parameters of the interleaved boost converter which makes its robustness parametric variations.
iii. Select the current references through the inductor to maintain a balanced operation at each branch for positive and negative references.

Note that numerical validations in the Piecewise Linear Electrical Circuit Simulation (PLECS) simulation environment demonstrate the effectiveness and robustness of the proposed control design to maintain the output voltage on its reference with minimum overshoots when compared with the classical PI design that presents higher oscillations in the voltage output, which are also transferred to the inductor currents.

The remainder of this paper is structured as follows: Section 2 presents the dynamical modeling of the interleaved boost converter using averaging modeling theory; Section 3 describes the general IDA-PBC design for power electronic converters with port-Hamiltonian representation as well as the extension of this controller to include integral actions. Section 4 specifies all the numerical validations of the proposed controller with its corresponding comparisons with the conventional PI design for multiple operative conditions in the DC microgrid terminals that include positive and negative current inputs and voltage reference variations. Finally, Section 5 lists the main findings of this research as well as some possible future works.

2. Average Modeling for an Interleaved Boost Converter

An interlaced boost converter is featured by having two converters operate in parallel, which switch at the same frequency with a phase-shift between the control inputs [28]. The phase shift allows a decrease in the ripple of the input and output waveforms and a lower harmonic content in the converter [16]. The interleaved boost converter is widely used for battery charge/discharge because of its simplicity and high conversion rate.

Figure 2 illustrates the interleaved boost converter, where v_b and i_b are the battery voltage and current, i_{L_1} and i_{L_2} are the currents through the inductances, and v_{dc} is the capacitor voltage. The parameters L_1, L_2, and C are the two inductances and capacitance, respectively. The duty cycles $d_{1,2} \in [0,1]$ control the states of the converter and $\bar{d}_{1,2}$ are their negative values. The control objective in this paper corresponds with the voltage support in v_{dc} terminals for positive and negative variations of the input current (charging/neutral/discharging operative conditions on the battery). The currents through the inductors must have the same values to balance the operation of the converter and, thus, lead to a decrease in the ripple of current waveforms through the converter.

Applying Kirchhoff's laws over the interleaved boost converter presented in Figure 2 and defining the state variables and control signals as: $x = [x_1,\, x_2,\, x_3]^\top = [i_{L_1},\, i_{L_2},\, v_{dc}]^\top$, $u = [u_1,\, u_2]^\top = [d_1,\, d_2]^\top$, the following dynamic model is yielded:

$$\begin{aligned} L_1\dot{x}_1 &= v_b - x_3(1-u_1), \\ L_2\dot{x}_2 &= v_b - x_3(1-u_2), \\ C\dot{x}_3 &= x_1(1-u_1) + x_2(1-u_2) - i_{\text{bus}}. \end{aligned} \tag{1}$$

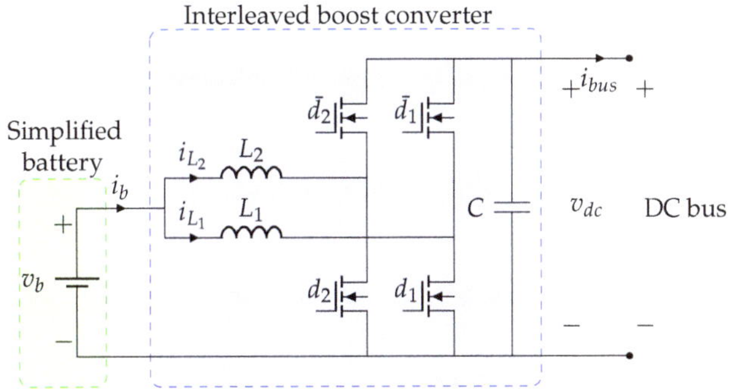

Figure 2. Electrical connection of an interleaved boost converter.

The dynamic model (1) can be rewritten in port-Hamiltonian (pH) structure as follows:

$$Q\dot{x} = (J - R)\frac{\partial H(x)}{\partial x} + G(x)u + \zeta, \tag{2}$$

where $\zeta = [v_b,\ v_b,\ -i_{\text{bus}}]^\top$ is the external inputs; $Q = Q^\top = \text{diag}(L_1,\ L_2,\ C) \succ 0$ is a positive definite matrix with the elements that store energy in the converter; R is the dissipation matrix, which takes a null form when no resistive effects on inductors are considered, that is, $R = 0_{3\times 3}$; $H(x)$ is a storage function (with a similar form to an energy storage function with normalized structure) of the system which is widely known as the Hamiltonian function, note that $\frac{\partial H(x)}{\partial x} = x$. $J = -J^\top$ is a skew-symmetry matrix; and $G(x)$ is the input matrix. The form of these functions and matrices is presented below:

$$H(x) = \frac{1}{2}\left(x_1^2 + x_2^2 + x_3^2\right),\ J = \begin{bmatrix} 0 & 0 & -1 \\ 0 & 0 & -1 \\ 1 & 1 & 0 \end{bmatrix},\ G(x) = \begin{bmatrix} x_3 & 0 \\ 0 & x_3 \\ -x_1 & -x_2 \end{bmatrix}.$$

Remark 1. *The main characteristic of the dynamical model (2) is that it is possible to obtain a nonlinear feedback controller via passivity-based control that ensures the closed-loop stability in the sense of Lyapunov. This can be made by selecting an adequate desired dynamics to sustain its pH properties as presented in [29].*

The design of the passivity-based controller with integral action is detailed in the next section.

3. IDA-PBC Design

3.1. Assignable Equilibrium Point

To stabilize any dynamical system via control around an equilibrium point, it is necessary to ensure its existence is independent of its stability properties [29,30]. In the case of DC microgrids in steady-state operating conditions, this equilibrium must be constant [10], that is, $\dot{x} = 0$, which implies that the dynamical system (2) must fulfill that:

$$(J - R)\frac{\partial H(x^\star)}{\partial x^\star} + G(x^\star)u^\star + \zeta = 0, \tag{3}$$

from (3), the following steady state operative conditions are reached:

$$u_1^\star = 1 - \frac{v_b}{x_3^\star}, \tag{4}$$

$$u_2^\star = 1 - \frac{v_b}{x_3^\star}, \tag{5}$$

$$x_1^\star + x_2^\star = \frac{x_3^\star}{v_b} i_{\text{bus}}. \tag{6}$$

In the case of the current reference for each inductor, the main challenge in the literature is to identify an adequate current balance in interleaved converters [14]; for this reason, we have chosen $x_1^\star = x_2^\star$, which from (5) implies that the reference for each inductor current is defined below:

$$x_1^\star = x_2^\star = \frac{1}{2}\frac{x_3^\star}{v_b} i_{\text{bus}}. \tag{7}$$

Remark 2. *Observe that the control inputs $u_1^\star$ and $u_2^\star$ are defined as a function of the desired variable $x_3^\star$ (see Equations (4) and (5)), since this is the desired control objective, that is, to maintain the voltage value of the DC microgrid in its reference value independent of the current variations [14].*

3.2. Classical IDA-PBC Design

The main idea of the IDA-PBC design is to redefine the closed-loop dynamics of a physical system by exploiting its passivity properties [31]. For this purpose, the closed-loop dynamics of the system can be selected with the following form:

$$Q\dot{x} = (J_d - R_d)\frac{\partial H_d(\tilde{x})}{\partial \tilde{x}}, \tag{8}$$

where J_d and R_d are the desired interconnection and damping matrices which are skew-symmetry and positive definite, respectively. Note that these matrices are adjusted to cancel some undesired interconnections in the open loop and inject sufficient damping to stabilize the system [29]. $H_d(\tilde{x})$ is the desired Hamiltonian function, which must be positive definite to ensure asymptotic stability in the sense of Lyapunov for the error state variables $\tilde{x} = x - x^\star$.

To determine the closed-loop controller, we equate the open loop dynamics (2) with the desired dynamics (8), which produces the following partial differential equation:

$$(J - R)\frac{\partial H(x)}{\partial x} + G(x)u + \zeta = (J_d - R_d)\frac{\partial H_d(\tilde{x})}{\partial \tilde{x}}, \tag{9}$$

which is easily solvable if we adequately select the interconnection matrix J_d, desired damping matrix R_d, and Hamiltonian function $H_d(\tilde{x})$. The selection of these matrices and the Hamiltonian function are the following:

$$J_d = J_a + J,\ R_d = R_a + R,\ H_d(\tilde{x}) = \frac{1}{2}\tilde{x}^\top \tilde{x},$$

where

$$J_a = \begin{bmatrix} 0 & 0 & 0 \\ 0 & 0 & 0 \\ 0 & 0 & 0 \end{bmatrix},\ R_a = \begin{bmatrix} R_1 & 0 & 0 \\ 0 & R_2 & 0 \\ 0 & 0 & R_3 \end{bmatrix}.$$

With the aforementioned definitions, it is possible to obtain the general control laws u_1 and u_2 from the first two equations of (9), which generates the following results:

$$u_1 = \frac{x_3^\star - v_b - R_1(x_1 - x_1^\star)}{x_3}, \tag{10}$$

$$u_2 = \frac{x_3^\star - v_b - R_2(x_2 - x_2^\star)}{x_3}. \tag{11}$$

Additionally, the third equation of (9) generates the following algebraic equation relating all the state variables and references:

$$\begin{aligned} (v_b - x_3^\star)(x_1 + x_2) + R_1 x_1(x_1 - x_1^\star) + R_2 x_2(x_2 - x_2^\star) \\ x_3 R_3(x_3 - x_3^\star) + x_3(x_1^\star + x_2^\star) = x_3 i_{\text{bus}}. \end{aligned} \tag{12}$$

Remark 3. *Once the system dynamical system (2) has been stabilized with the IDA-PBC using control laws (10) and (11), we can observe that in steady state conditions, Equations (10)–(12) take the form defined in Equations (4)–(7)—which confirms that the system have reached the desired operative point.*

To ascertain that the proposed controller has asymptotic stability in closed-loop, let us consider the candidate Lyapunov function as the desired Hamiltonian function, that is, $V(\tilde{x}) = \frac{1}{2}\tilde{x}^\top Q\tilde{x}$, which is a positive definite and $V(0) = 0$. If we consider the time derivative of this function, then, the following result yields:

$$\begin{aligned} \dot{V}(\tilde{x}) &= \tilde{x}^\top Q\dot{x}, \\ &= \tilde{x}^\top (J_d - R_d)\tilde{x}, \\ &= -\tilde{x}^\top R_d \tilde{x} < 0, \end{aligned} \tag{13}$$

which confirms that the desired dynamical system (9) is asymptotically stable in the sense of Lyapunov, that is, $x \to x^\star$ as $t \to \infty$.

3.3. IDA-PBC Redesign with Integral Action

The usage of an IDA-PBC design with integral action is necessary to eliminate possible steady-state errors introduced by unmodeled dynamics in the original systems, such as resistive effects in inductors, parasite resistances in capacitors, or energy losses in forced-commutated switches, among others [6,29].

To formulate the general IDA-PBC with integral action, let us consider the following augmented dynamical system:

$$\begin{bmatrix} Q\dot{x} \\ \dot{z} \end{bmatrix} = \begin{bmatrix} (J_d - R_d) & -G(x) \\ G^\top(x) & 0 \end{bmatrix} \begin{bmatrix} \frac{\partial H_d(\tilde{x})}{\partial \tilde{x}} \\ \frac{\partial H_z(z)}{\partial z} \end{bmatrix}, \tag{14}$$

where $G^\top(x)\frac{\partial H_d(\tilde{x})}{\partial \tilde{x}}$ is the passive output of the desired dynamical system [32]; K_i is a diagonal positive definite matrix that contains the integral grains, z is the vector of auxiliary state variables, and $H_z(z)$ is the Hamiltonian function defined in the set of auxiliary variables, which is defined as $H_z(z) = \frac{1}{2}z^\top K_i z$.

To confirm that the augmented dynamical system (14) is stable in the sense of Lyapunov, let us consider the following candidate Lyapunov function:

$$W(\tilde{x}, z) = V(\tilde{x}) + H_z(z). \tag{15}$$

Note that $W(\tilde{x}, z)$ is a positive definite function and $W(0,0) = 0$ only for the point in the origin of coordinates; now, taking the time derivative of (15), the following result is produced:

$$\begin{aligned} \dot{W}(\tilde{x}, z) &= \tilde{x}^\top Q\dot{x} + z^\top K_i \dot{z}, \\ &= \tilde{x}^\top (J_d - R_d)\tilde{x} - \tilde{x}^\top G(x) K_i z + z^\top K_i G^\top (x)\tilde{x}, \\ &= -\tilde{x}^\top R_d \tilde{x} < 0. \end{aligned} \tag{16}$$

Note that (16) ensures that the augmented dynamical system (14) is stable in the sense of Lyapunov.

Now, observe that the closed-loop dynamics will be defined by equaling the first row of (14) with the open-loop dynamics (2), which produces the following partial differential equation.

$$(J - R)\frac{\partial H(x)}{\partial x} + G(x)u + \zeta = (J_d - R_d)\frac{\partial H_d(\tilde{x})}{\partial \tilde{x}} - G(x)K_i\frac{\partial H_z(z)}{\partial z}. \tag{17}$$

If we solve (17) for the control laws u_1 and u_2, then, the following results are reached:

$$u_1 = \frac{x_3^\star - v_b - R_1(x_1 - x_1^\star)}{x_3} - K_1 z_1, \tag{18}$$

$$u_2 = \frac{x_3^\star - v_b - R_2(x_2 - x_2^\star)}{x_3} - K_2 z_2, \tag{19}$$

where z_1 and z_2 defines the integral components of the controller. These components are calculated with the second row of (14) as presented below:

$$z_1 = \int x_3(x_1 - x_1^\star)dt - \int x_1(x_3 - x_3^\star)dt, \tag{20}$$

$$z_2 = \int x_3(x_2 - x_2^\star)dt - \int x_2(x_3 - x_3^\star)dt. \tag{21}$$

Remark 4. *The control inputs u_1 and u_2 in (18) and (19) with the integral components (20) and (21) define the general IDA-PBC with integral action to stabilize the interleaved boost converter in the desired reference to support voltage in DC microgrid applications independent of the variations of the external input, that is, i_{bus}.*

4. Numerical Validation

This section exhibits the performance of the IDA-PBC with integral action implemented in an interleaved boost converter to regulate the output voltage under different load conditions. The proposed controller is validated in the test system displayed in Figure 2, and its parameters are outlined in Table 1. Moreover, the IDA-PBC is also compared to the conventional PI controller [14]. The simulations are conducted in the PLECS software, and the sample time for the proposed controller is configured in 10 µs.

Table 1. Interleaved boost converter parameters.

Element	Variable	Value	Element	Variable	Value	Element	Variable	Value
Battery Voltage	v_b	24 V	Bus Voltage	v_{dc}	48 V	Inductor	L_1, L_2	330 mH
Switching frequency	f_q	2 kHz	Capacitor	C	44 µF			

It is worth mentioning that the PLECS software is a complete interface to simulate electrical grids predominantly composed of power electronic converters that allow designing controllers using block diagrams or functions [33]. The main advantage of this simulation tool is that it works under the Simulink solution environment and can be easily implemented in real-time simulators for Hardware in the Loop applications [34].

The dynamic response of the proposed control is evaluated against different disturbances generated with current steps in the DC bus to emulate the charging/discharging processes of the converter. These current steps are 1 A, 1.5 A, and 2 A, both positives, and negatives with 30 ms intervals, as depicted in Figure 3.

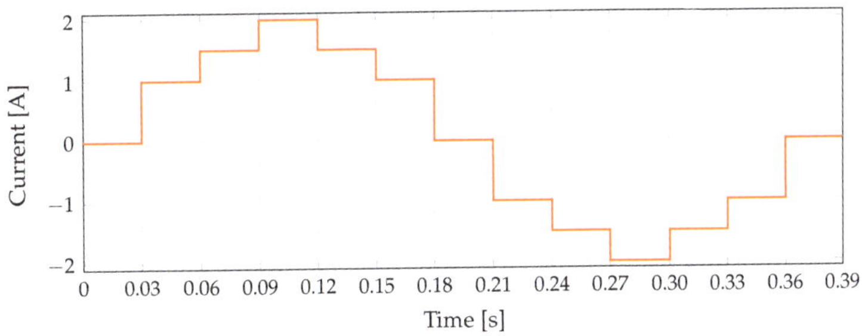

Figure 3. Variations in the input signal i_{bus}.

Figure 4 illustrates the output voltage of the interleaved boost converter under different DC current steps, in Figure 4a for the conventional PI controller and in Figure 4b for the proposed IDA-PBC.

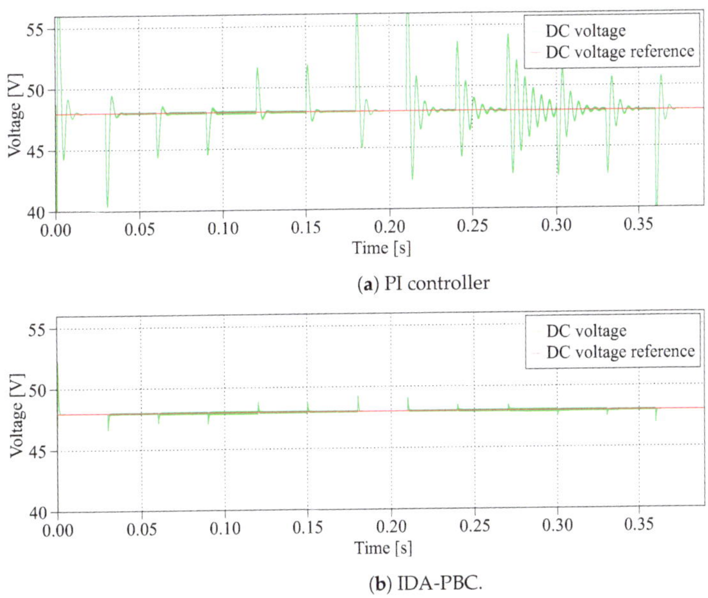

Figure 4. The dynamic response of output voltage of the interleaved boost converter under different DC current steps.

Comparing Figure 4a,b, it can be noted that the IDA-PBC adequately regulates the output voltage of the interleaved boost converter under different DC current steps, while conventional PI has higher oscillations, higher voltage overshoots, and it is slower than the proposed controller. This behavior is exacerbated by the the DC bus current's negative values. This analysis is supported by comparing the voltage overshoot and settling time. The voltage overshoot in the worst case for the proposed controller is 1.3 V, while the conventional PI controller is 8 V. For the settling time, the proposed controller requires 3 ms to stabilize in the worst case while the conventional PI controller needs 40 ms.

Figure 5 unveils the currents in the interleaved boost converter, DC current bus, and battery current. Figure 5a,b present the currents in the interleaved boost converter, DC current bus, and battery current when the conventional PI controller and IDA-PBC are used, respectively.

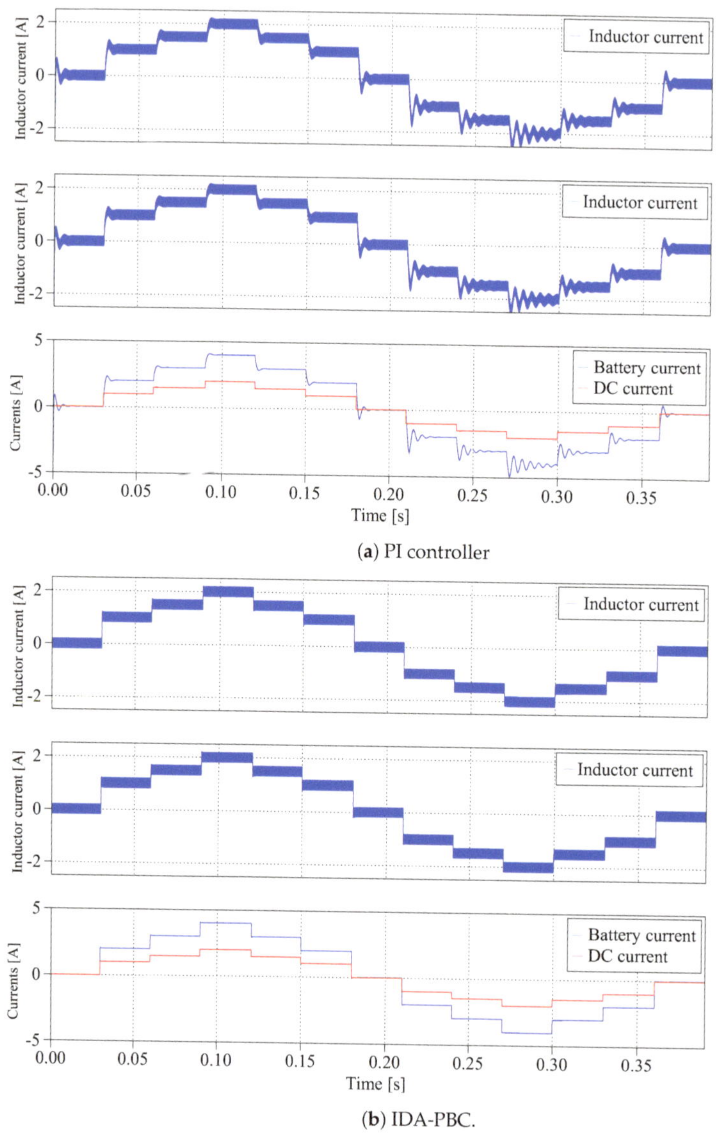

Figure 5. The dynamic response of currents associated with the interleaved boost converter under different DC current steps.

Comparing the behavior of the currents associated with the interleaved boost converter in Figure 5a,b, it is evident that the inductor currents stabilize faster when the IDA-PBC is implemented. Further, the conventional PI controller presents oscillation when the DC current bus has negative references, indicating that the interleaved boost may become unstable for large negative reference values on the DC current bus.

Figure 6 depicts control inputs for the proposed controller and PI control. For both controllers, the control inputs do not overshot their limits. Additionally, comparing the dynamic response of the voltage output with control inputs, it is observed that both exhibit the same behavior, that is, they are directly proportional.

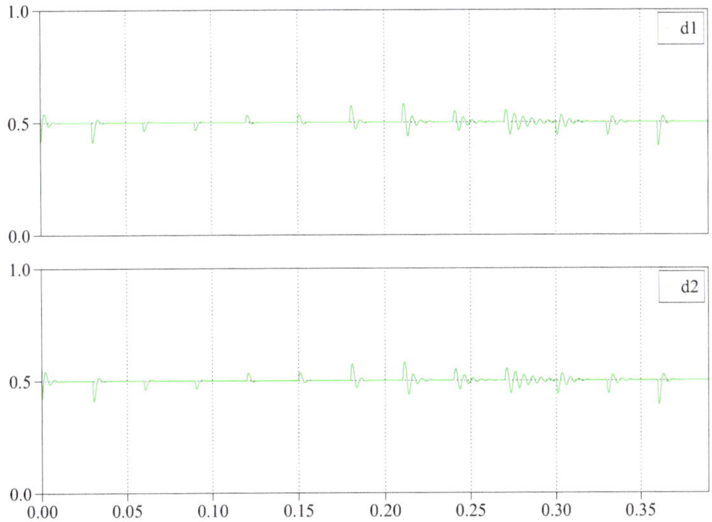

(**a**) PI controller.

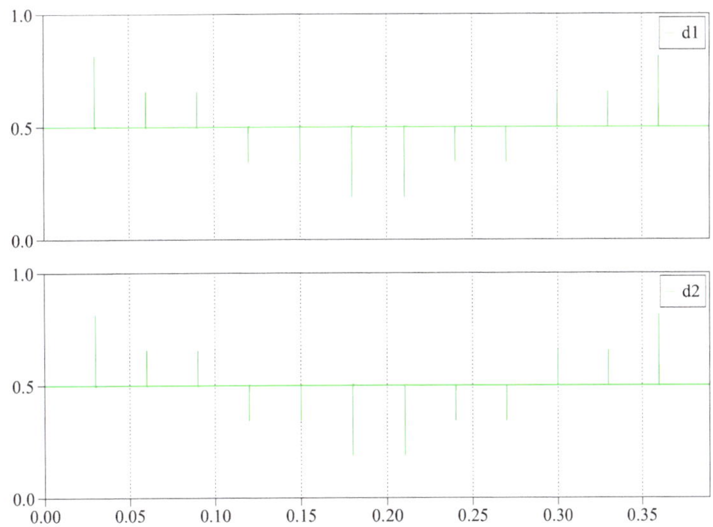

(**b**) IDA-PBC.

Figure 6. The dynamic response of control inputs with the interleaved boost converter under different DC current steps.

Now, the performance of the proposed control for different voltage value references is examined as displayed in Figure 7.

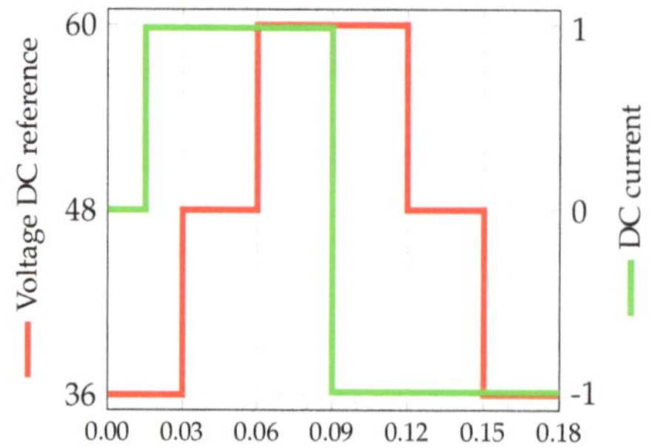

Figure 7. Different voltage DC references.

Figure 8 depicts the output voltage of the interleaved boost converter under different voltage DC references. Figure 8a,b show the output voltages of the interleaved boost converter when the conventional PI and IDA-PBC controllers are implemented.

Comparing Figure 8a,b, it can be perceived that the output voltage of the interleaved boost converter is correctly followed when the IDA-PBC is employed, whereas the output voltage has oscillations when the conventional PI controller is used. Therefore, the proposed controller continues presenting a better performance than the conventional PI controller. This is supported by comparing the voltage overshoot and settling time. The maximum voltage overshoot occurs when the current changes from 1 to -1 (time 0.9 s), while for the proposed controller, it is 0.775 V and 23.674 V for the conventional PI controller. For the settling time, the proposed controller requires 12 ms to stabilize in the worst case while the conventional PI controller needs 40 ms.

Figure 9 illustrates the currents associated with the interleaved boost converter. The currents in both inductors, DC bus and battery, are displayed in Figure 9a,b when the conventional PI and the IDA-PBC controller are implemented.

The difference in the behavior of the currents associated with the interleaved boost converter continues to be maintained as seen in Figure 5, establishing that the proposed IDA-PBC controller performs better than the conventional PI.

Figure 10 depicts the control inputs for the proposed controller and PI control when the interleaved boost converter is under different DC voltage and current steps. For both controllers, the control inputs do not overshoot their limits. The voltage output dynamics depict the same behavior as that of the control inputs, implying that the voltage output response is directly proportional to the control inputs.

It is worth emphasizing that, during the physical implementation of the proposed IDA-PBC controller, an adequate tuning process of the proportional and integral gains is mandatory to obtain the expected dynamical behavior. However, classical tuning techniques are not applicable due to the interleaved boost converter and IDA-PBC design generating a nonlinear feedback model [35]. For this reason, it is recommended to generate a mesh with multiple combinations of these gains and use an optimization technique to tune these using a performance index, which can be the mean square error or another [10].

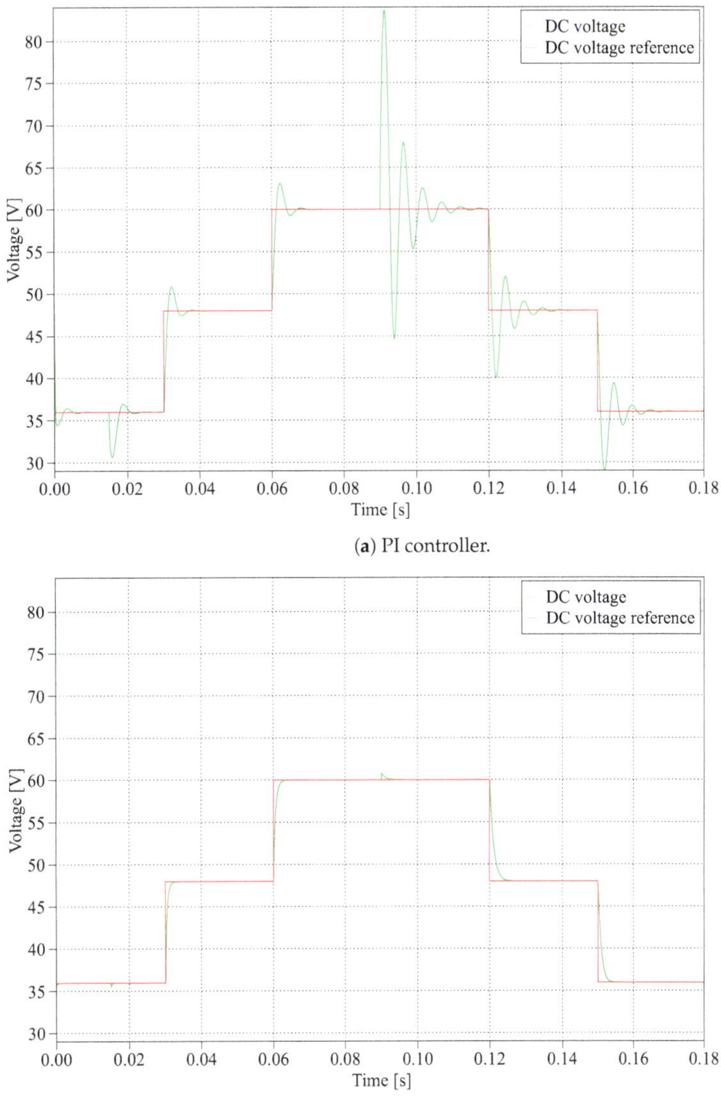

(**a**) PI controller.

(**b**) IDA-PBC.

Figure 8. The dynamic response of output voltage of the interleaved boost converter under different DC voltage and current steps.

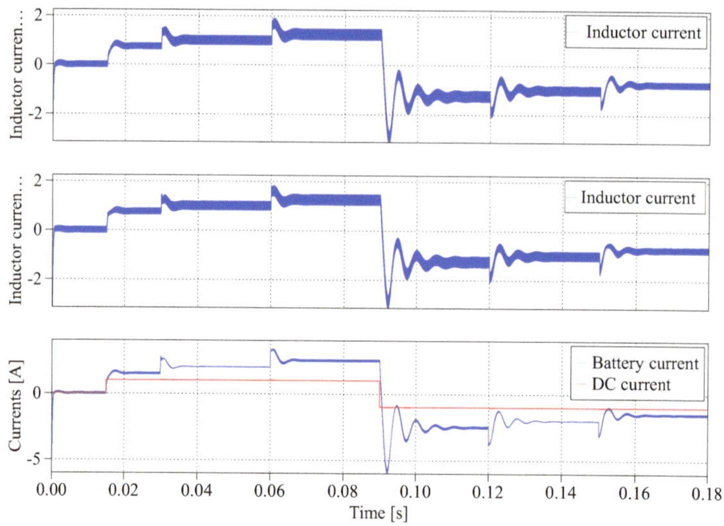

(**a**) PI controller.

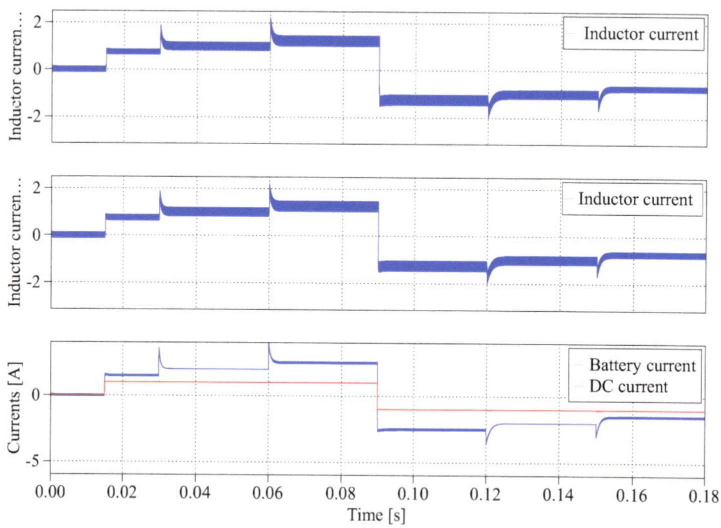

(**b**) IDA-PBC

Figure 9. The dynamic response of currents associated with the interleaved boost converter under different DC voltage and current steps.

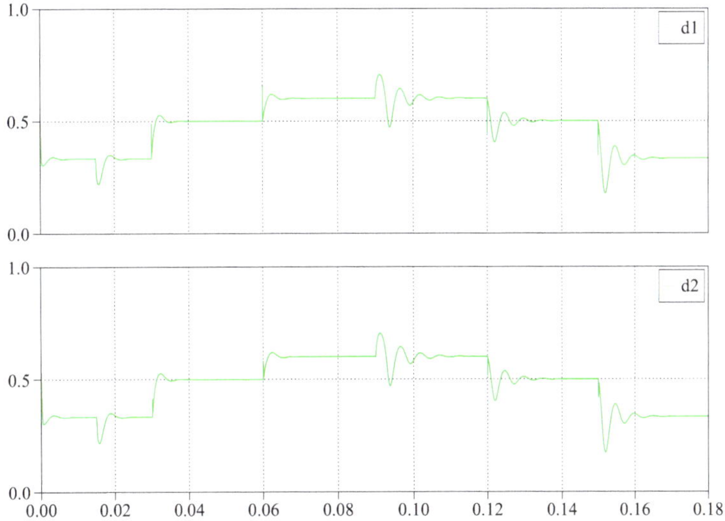

(**a**) PI controller.

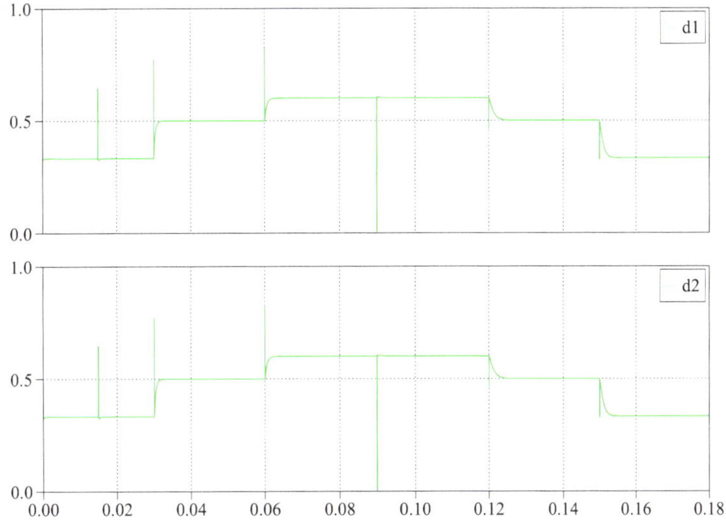

(**b**) IDA-PBC.

Figure 10. The dynamic response of control inputs with the interleaved boost converter under different DC voltage and current steps.

5. Conclusions and Future Works

A passivity-based control applied to the interleaved boost converter or output voltage regulation in DC microgrids with variable injected/demanded current was proposed in this paper. The passivity-based control is performed using the interconnection and damping assignment, which takes advantage of the system's dynamic in open-loop to design a control law guaranteeing the system's stability in closed-loop. Additionally, an integral action was included in the IDA-PBC design to enhance the performance of the proposed controller, thus, eliminating the steady-state errors introduced by possible unmodeled dynamics. This integral action did not affect the stability properties since it maintains the passive output of the system. The proposed controller was assessed under different simulation scenarios and compared with a conventional PI controller, exhibiting better performance than the PI controller. This was supported by comparing the integral of the voltage overshoot and settling time.

Future works could be developed from the following studies: (i) the application of the IDA-PBC with integral action to DC-DC converters, such as quadratic buck and boost converters; (ii) develop a complete experimental comparison among nonlinear control designs applied to the interleaved boost converter with a special focus on exact feedback linearization, inverse optimal control, and Lyapunov-based approaches; and (iii) extend the IDA-PBC design to the interleaved boost converter under the presence of nonlinear constant power loads.

Author Contributions: Conceptualization, methodology, software, and writing—review and editing, O.D.M., F.M.S., W.G.-G., E.M.A. and J.E.B. All authors have read and agreed to the published version of the manuscript.

Funding: This work was supported in part by the Centro de Investigación y Desarrollo Científico de la Universidad Distrital Francisco José de Caldas under grant 1643-12-2020 associated with the project: "Desarrollo de una metodología de optimización para la gestión óptima de recursos energéticos distribuidos en redes de distribución de energía eléctrica".

Institutional Review Board Statement: Not applicable.

Informed Consent Statement: Not applicable.

Data Availability Statement: No new data were created or analyzed in this study. Data sharing is not applicable to this article.

Acknowledgments: This work has been derived from the doctoral research project: "Análisis, operación y control de convertidores para la integración de recursos energéticos distribuidos en redes eléctricas aisladas" presented by the student Oscar Danilo Montoya to the Doctoral program of the Engineering Faculty at Universidad Nacional de Rio Cuarto as a partial requirement for the PhD in Engineering Sciences.

Conflicts of Interest: The authors declare no conflict of interest.

References

1. Lana, A.; Mattsson, A.; Nuutinen, P.; Peltoniemi, P.; Kaipia, T.; Kosonen, A.; Aarniovuori, L.; Partanen, J. On Low-Voltage DC Network Customer-End Inverter Energy Efficiency. *IEEE Trans. Smart Grid* **2014**, *5*, 2709–2717. [CrossRef]
2. Montoya, O.D.; Gil-González, W.; Serra, F.M.; Angelo, C.H.D.; Hernández, J.C. Global Optimal Stabilization of MT-HVDC Systems: Inverse Optimal Control Approach. *Electronics* **2021**, *10*, 2819. [CrossRef]
3. Garces, A. Uniqueness of the power flow solutions in low voltage direct current grids. *Electr. Power Syst. Res.* **2017**, *151*, 149–153. [CrossRef]
4. Justo, J.J.; Mwasilu, F.; Lee, J.; Jung, J.W. AC-microgrids versus DC-microgrids with distributed energy resources: A review. *Renew. Sustain. Energy Rev.* **2013**, *24*, 387–405. [CrossRef]
5. Lotfi, H.; Khodaei, A. AC Versus DC Microgrid Planning. *IEEE Trans. Smart Grid* **2017**, *8*, 296–304. [CrossRef]
6. Magaldi, G.L.; Serra, F.M.; de Angelo, C.H.; Montoya, O.D.; Giral-Ramírez, D.A. Voltage Regulation of an Isolated DC Microgrid with a Constant Power Load: A Passivity-based Control Design. *Electronics* **2021**, *10*, 2085. [CrossRef]
7. Serra, F.M.; Angelo, C.H.D. Control of a battery charger for electric vehicles with unity power factor. *Trans. Energy Syst. Eng. Appl.* **2021**, *2*, 32–44. [CrossRef]
8. Solsona, J.A.; Jorge, S.G.; Busada, C.A. Nonlinear Control of a Buck Converter Which Feeds a Constant Power Load. *IEEE Trans. Power Electron.* **2015**, *30*, 7193–7201. [CrossRef]
9. Salimi, M.; Siami, S. Cascade nonlinear control of DC-DC buck/boost converter using exact feedback linearization. In Proceedings of the 2015 4th International Conference on Electric Power and Energy Conversion Systems (EPECS), Sharjah, United Arab Emirates, 24–26 November 2015; IEEE: Piscataway, NJ, USA, 2015. [CrossRef]
10. Gil-González, W.; Montoya, O.D.; Restrepo, C.; Hernández, J.C. Sensorless Adaptive Voltage Control for Classical DC-DC Converters Feeding Unknown Loads: A Generalized PI Passivity-Based Approach. *Sensors* **2021**, *21*, 6367. [CrossRef]
11. Ramos-Paja, C.A.; Gonzalez-Motoya, D.; Villegas-Seballos, J.P.; Serna-Garces, S.I.; Giral, R. Sliding-mode controller for a photovoltaic system based on a Cuk converter. *Int. J. Electr. Comput. Eng. (IJECE)* **2021**, *11*, 2027. [CrossRef]
12. Jin, P.; Li, Y.; Li, G.; Chen, Z.; Zhai, X. Optimized hierarchical power oscillations control for distributed generation under unbalanced conditions. *Appl. Energy* **2017**, *194*, 343–352. [CrossRef]

13. Iskender, I.; Genc, N. Power Electronic Converters in DC Microgrid. In *Power Systems*; Springer International Publishing: Berlin/Heidelberg, Germany, 2019; pp. 115–137. [CrossRef]
14. Quintero, C.E.; Pérez, S.A.; Ceballos, J.P.V.; González-Montoya, D.; Garcés, S.S. Design and Digital Control of an Interleaved Boost Converter for Battery Charge/Discharge. *Tecnológicas* **2021**, *24*, e1556. (In Spanish) [CrossRef]
15. He, L.; Lin, Z.; Tan, Q.; Lu, F.; Zeng, T. Interleaved High Step-Up Current Sharing Converter with Coupled Inductors. *Electronics* **2021**, *10*, 436. [CrossRef]
16. Hausberger, T.; Kugi, A.; Eder, A.; Kemmetmüller, W. High-speed nonlinear model predictive control of an interleaved switching DC/DC-converter. *Control. Eng. Pract.* **2020**, *103*, 104576. [CrossRef]
17. Cervantes, I.; Mendoza-Torres, A.; Garcia-Cuevas, A.; Perez-Pinal, F. Switched control of interleaved converters. In Proceedings of the 2009 IEEE Vehicle Power and Propulsion Conference, Dearborn, MI, USA, 7–11 September 2009; IEEE: Piscataway, NJ, USA, 2009. [CrossRef]
18. Kumar, S.S.; Kanimozhi, G. A nonlinear control technique for interleaved boost converter. In Proceedings of the 2016 10th International Conference on Intelligent Systems and Control (ISCO), Coimbatore, India, 7–8 January 2016; IEEE: Piscataway, NJ, USA, 2016. [CrossRef]
19. Cid-Pastor, A.; Giral, R.; Calvente, J.; Utkin, V.I.; Martinez-Salamero, L. Interleaved Converters Based on Sliding-Mode Control in a Ring Configuration. *IEEE Trans. Circuits Syst. I Regul. Pap.* **2011**, *58*, 2566–2577. [CrossRef]
20. Tiwari, A.; Jaga, O.; Soni, S.S. Sliding mode controller based interleaved boost converter for fuel cell system. In Proceedings of the 2017 Recent Developments in Control, Automation & Power Engineering (RDCAPE), Noida, India, 26–27 October 2017; IEEE: Piscataway, NJ, USA, 2017. [CrossRef]
21. Gkizas, G.; Amanatidis, C.; Yfoulis, C.; Stergiopoulos, F.; Giaouris, D.; Ziogou, C.; Voutetakis, S.; Papadopoulou, S. State-feedback control of an interleaved DC-DC boost converter. In Proceedings of the 2016 24th Mediterranean Conference on Control and Automation (MED), Athens, Greece, 21–24 June 2016; IEEE: Piscataway, NJ, USA 2016. [CrossRef]
22. González, A.; López-Erauskin, R.; Gyselinck, J. Analysis, modeling, control and operation of an interleaved three-port boost converter for DMPPT systems including PV and storage at module level. *Heliyon* **2019**, *5*, e01402. [CrossRef] [PubMed]
23. Olmos-Lopez, A.; Guerrero, G.; Arau, J.; Aguilar, C.; Yris, J.C. Passivity-based control for current sharing in PFC interleaved boost converters. In Proceedings of the 2011 Twenty-Sixth Annual IEEE Applied Power Electronics Conference and Exposition (APEC), Fort Worth, TX, USA, 6–11 March 2011; IEEE: Piscataway, NJ, USA, 2011. [CrossRef]
24. Zhou, H.; Khambadkone, A.M.; Kong, X. A Passivity Based Control with Augmented Integration for an Interleaved Current Fed Full Bridge Converter as a Front End for Fuel Cell Source. In Proceedings of the 2007 IEEE Industry Applications Annual Meeting, New Orleans, LA, USA, 23–27 September2007; IEEE: Piscataway, NJ, USA, 2007. [CrossRef]
25. Bharathi, M.; Kirubakaran, D. Solar powered closed-loop controlled fuzzy logic-based three-stage interleaved DC-DC boost converter with an inverter. *Int. J. Adv. Intell. Paradig.* **2016**, *8*, 140. [CrossRef]
26. Sunarno, E.; Sudiharto, I.; Nugraha, S.D.; Qudsi, O.A.; Eviningsih, R.P.; Raharja, L.P.S.; Arifin, I.F. A Simple And Implementation of Interleaved Boost Converter For Renewable Energy. In Proceedings of the 2018 International Conference on Sustainable Energy Engineering and Application (ICSEEA), Tangerang, Indonesia, 1–2 November 2018; IEEE: Piscataway, NJ, USA, 2018. [CrossRef]
27. Barhoumi, E.; Belgacem, I.B.; Khiareddine, A.; Zghaibeh, M.; Tlili, I. A Neural Network-Based Four Phases Interleaved Boost Converter for Fuel Cell System Applications. *Energies* **2018**, *11*, 3423. [CrossRef]
28. Gonzalez, W.J.G.; Bocanegra, S.Y.; Serra, F.M.; Bueno-López, M.; Magaldi, G.L. Control Methods for Single-phase Voltage Supply with VSCs to Feed Nonlinear Loads in Rural Areas. *Trans. Energy Syst. Eng. Appl.* **2020**, *1*, 33–47. [CrossRef]
29. Serra, F.M.; Angelo, C.H.D.; Forchetti, D.G. Interconnection and damping assignment control of a three-phase front end converter. *Int. J. Electr. Power Energy Syst.* **2014**, *60*, 317–324. [CrossRef]
30. Herrera-Pérez, J.J.; Garcés-Ruiz, A. Análisis de estabilidad de convertidores de segundo orden con la metodología de optimización de suma de polinomios cuadráticos. *Trans. Energy Syst. Eng. Appl.* **2020**, *1*, 49–58. (In Spanish) [CrossRef]
31. Serra, F.M.; Angelo, C.H.D. IDA-PBC controller design for grid connected Front End Converters under non-ideal grid conditions. *Electr. Power Syst. Res.* **2017**, *142*, 12–19. [CrossRef]
32. Donaire, A.; Junco, S. On the addition of integral action to port-controlled Hamiltonian systems. *Automatica* **2009**, *45*, 1910–1916. [CrossRef]
33. Asadi, F.; Eguchi, K. *Simulation of Power Electronics Converters Using PLECS®*; Elsevier: Amsterdam, The Netherlands, 2020. [CrossRef]
34. Frivaldsky, M.; Morgos, J.; Prazenica, M.; Takacs, K. System Level Simulation of Microgrid Power Electronic Systems. *Electronics* **2021**, *10*, 644. [CrossRef]
35. Morales, J.A.; Castro, M.A.; Garcia, D.; Higuera, C.; Sandoval, J. IDA-PBC Controller Tuning Using Steepest Descent. In *Numerical and Evolutionary Optimization—NEO 2017*; Springer International Publishing: Berlin/Heidelberg, Germany, 2018; pp. 158–170. [CrossRef]

actuators

MDPI

Article

An Enhanced Sliding Mode Speed Control for Induction Motor Drives

Fahimeh Shiravani [1,*], Patxi Alkorta [1], Jose Antonio Cortajarena [1] and Oscar Barambones [2,*]

1 Engineering School of Gipuzkoa, University of the Basque Country, Otaola Hirib. 29, 20600 Eibar, Spain; patxi.alkorta@ehu.eus (P.A.); josean.cortajarena@ehu.eus (J.A.C.)
2 Engineering School of Vitoria, University of the Basque Country, Nieves Cano 12, 01006 Vitoria, Spain
* Correspondence: fahimeh.shiravani@ehu.eus (F.S.); oscar.barambones@ehu.eus (O.B.)

Abstract: In this paper, an enhanced Integral Sliding Mode Control (ISMC) for mechanical speed of an Induction Motor (IM) is presented and experimentally validated. The design of the proposed controller has been done in the d-q synchronous reference frame and indirect Field Oriented Control (FOC). Global asymptotic speed tracking in the presence of model uncertainties and load torque variations has been guaranteed by using an enhanced ISMC surface. Moreover, this controller provides a faster speed convergence rate compared to the conventional ISMC and the Proportional Integral methods, and it eliminates the steady-state error. Furthermore, the chattering phenomenon is reduced by using a switching sigmoid function. The stability of the proposed controller under parameter uncertainties and load disturbances has been provided by using the Lyapunov stability theory. Finally, the performance of this control method is verified through numerical simulations and experimental tests, getting fast dynamics and good robustness for IM drives.

Keywords: experimental validation; Induction Motor; Integral Sliding Mode Control; robustness; speed control

Citation: Shiravani, F.; Alkorta, P.; Cortajarena, J.A.; Barambones, O. An Enhanced Sliding Mode Speed Control for Induction Motor Drives. *Actuators* **2022**, *11*, 18. https://doi.org/10.3390/act11010018

Academic Editor: Shuxiang Dong

Received: 9 December 2021
Accepted: 3 January 2022
Published: 10 January 2022

Publisher's Note: MDPI stays neutral with regard to jurisdictional claims in published maps and institutional affiliations.

Copyright: © 2022 by the authors. Licensee MDPI, Basel, Switzerland. This article is an open access article distributed under the terms and conditions of the Creative Commons Attribution (CC BY) license (https://creativecommons.org/licenses/by/4.0/).

1. Introduction

Three-phase machines, such as motors and generators, are used extensively in industry and in civil engineering. Renewable energy sources, machine tools, servo drives, and robots are just a few examples. Because of its low cost, minimal maintenance, low moment of inertia, robust architecture, and operational reliability, IM has become widely employed in various applications as power electronics technology has advanced. In the last two decades, the FOC technique has been the most widely used method for regulating IM in high-performance applications, such as speed and position control of three-phase motors. The torque and flux control current commands for the IM are decoupled by using the FOC method. As a result, the machine is controlled as if it were an independent DC machine. However, uncertainties such as unexpected parameter variations, external load disturbances, and nonlinear dynamics continue to impact the IM's control performance. The Proportional Integral (PI) regulator, due to its simplicity, clear functionality, and effectiveness, is one of the most popular control techniques which has been used in electrical machines [1]. Nonetheless, because the IM is a nonlinear framework, a well-designed nonlinear regulator can improve action in the presence of disturbances and uncertainty [2]. Many authors have taken use of different advanced control approaches to govern the power electronics and drives area in this regard. For instance, adaptive control method [3], Backstepping algorithm [4,5], predictive control method [6] and Sliding Mode Control (SMC) [7–9]. Among the nonlinear control method, the SMC technique has become a fascinating nonlinear control method with a particular dynamic performance for IM, such as strong robustness, quick response, and simple software and hardware implementation [10].

Almost ever since sliding mode ideas imply, the considerable noise which some of the sliding mode controllers expose is not pleasant for control engineers and sometimes has led to resentments and even rejection of the technique. The phenomenon is best known as chattering. Chattering is a high-frequency oscillation around the equilibrium point, which arises because of the discontinuous nature of the control action. Due to this, the well-designed control action stands unsuitable for many practical applications. This behavior creates a problem of wear and tear within the mechanical parts, vibrations within the machines, or flapping of wing vanes in aerospace and hitting effect. Hence, it is unwanted in light of implementation [11]. Regarding chattering suppression, various types of chattering prevention schemes have been suggested [12–14]. For instance, many authors have been designing regulators for IM based on combining SMC theory and other advanced control methods such as backstepping algorithm, adaptive method, and fuzzy technique [15–17]. It should be noted that using SMC by employing other sorts of nonlinear control strategies increments the controller complexity and computational endeavors, which is incompatible with the ease of SMC. Another solution could be applying higher order sliding mode control to IM. For instance, in [18], authors have taken advantage of the super-twisting sliding mode method to eliminate the chattering phenomenon. Furthermore, discrete-time sliding mode control has been applied to the machine in this regard [19,20].

Another way to reduce the chattering phenomenon is to apply some changes to the traditional sliding surface, which is basically based on error and its derivative signal [21]. In [21–24], an integral sliding surface has been applied to the IM to eliminate steady-state error, which has been named ISMC. A sensorless adaptive ISMC for IM has been discussed in [21]. In [22], a speed estimator based on the ISMC method for IM has been designed where stator current controllers consist of a PI and an ISMC controller, which means more complexity. Besides, this speed estimator is not universal and can only be implemented in conjunction with the above current controller. Furthermore, the authors in [23] designed an ISMC anti-windup in the speed control loop of the IM. Furthermore, an ISMC method has been proposed in [24] to start the sensorless IM in the rotating condition.

In this paper, the considered IM is under load torque and it is perturbed by model uncertainties. The common objective in nearly all industrial control design procedures is to provide a fast and accurate response by employing a smooth and effective electromagnetic torque. This goal will be achieved by designing an efficient control law. The idea is to regulate the mechanical rotor speed of IM by using ISMC method to achieve asymptotic speed tracking and disturbance rejection. In the controller design process, the FOC theory has been applied to get fast dynamic performance. Furthermore, to tackle the chattering problem and to eliminate the steady-state error, the proposed robust controller is designed based on the ISMC while using a continuous switching function $arctan()$. Besides, the proposed controller has a faster speed convergence rate compared to the conventional ISMC method, due to the surface design difference. Compared with the conventional ISMC, in this paper, the $arctan()$ function of the error of the mechanical speed has been considered in the surface design. By employing this function, the control action becomes smoother and provides a faster dynamics. Additionally, the presented proposal offers good robustness under parametric uncertainties. Then, by using a Lyapunov-based approach, robust output tracking of rotor speed is achieved. Since most of the electric drives in the industry are controlled by PI regulators, this proposal also has been compared with the PI controller to demonstrate the ability of the proposed controller in fast convergence of the speed of the IM. Furthermore, worth noting that, in this work, the experiments have been done by use of a 7.5 kW commercial IM, which shows the applicability of the projected methodology for the real applications within the industry.

The paper has been organized as follows: the speed controller of IM is designed in Section 2. In Section 3, the experiment platform and the effectiveness of the proposed controller by employing several simulations and experimental tests are shown. Finally, in Section 4, the conclusion has been presented.

2. Robust Speed ISMC Design

2.1. Model of the Mechanical Loop of IM

The block diagram of the proposed control scheme is shown in Figure 1. In this figure by starting from the right side of this block set, by measuring the three stator phase currents, and by using the Clarke's and Park's transformations (ABC$\rightarrow$ dq block), the i_{sd} and i_{sq} components in the rotational reference frame are obtained. The resulting transformed currents i_{sd} and i_{sq} will be responsible for magnetizing rotor flux and electromagnetic torque, respectively. The block Calc θ_s calculates the rotor flux angle, θ_s , by using the indirect FOC control technique. The speed error between the reference and real speed is fed to the $ISMC_{\omega_m}$ block (speed regulator). The output of the block will be i_{sq}^*, which is responsible for the electromagnetic torque reference generation. Torque and flux currents are also controlled by the mean of the two $PI_{i_{sd}}$ and $PI_{i_{sq}}$ current regulators, which are providing the correspondent v_{sd}^* and v_{sq}^* voltage references in the rotating reference frame. They are transformed to the stationary reference frame by using the inverse park transformation block (dq$\rightarrow$ $\alpha\beta$ block), giving v_α^* and v_β^* voltage references. These two voltages are applied to the modulator Space Vector Pulse Width Modulation (SVPWM), which transforms stationary reference frame voltages to control signals (pulses) to drive the power three-phase inverter IGBTs.

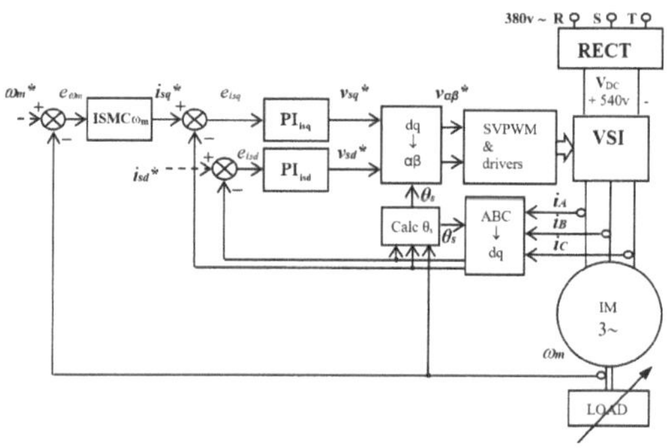

Figure 1. Block diagram of IM.

ISMC speed controller is designed in the d-q synchronous reference frame by using the indirect FOC, where it is assumed that $\psi_{rq} = 0$ and consequently $\psi_r = \psi_{rd}$. Thus, the equation of the inductions electromagnetic torque of the motor has the following expression,

$$T_e = \frac{3pL_m}{4L_r}\psi_r i_{sq} = K_T i_{sq} \tag{1}$$

where K_T is the torque constant:

$$K_T = \frac{3pL_m}{4L_r}\psi i_r \tag{2}$$

Taking the mechanical equation:

$$J\dot{\omega}_m + B_v\omega_m + T_L = T_e \tag{3}$$

it can be written as:

$$\dot{\omega}_m + a_\omega\omega_m + f_\omega = b_\omega i_{sq} \tag{4}$$

where the following parameters can be defined as: $a_\omega = B_v/J, b_\omega = K_T/J, f_\omega = T_L/J$

2.2. Basic Principles of ISMC

There are two phases to the ISMC regulator's design. The first step is to choose the adequate integral sliding surface to meet the control goals. The second step is to design the control law, which ensures that the system's trajectories reach and remain on the sliding surfaces in a finite amount of time (reaching phase). In this approach, the ISMC project may be categorized into two sections: defining an appropriate sliding surface $S(x)$ and developing a control law. To reach the sliding regime, the conventional ISMC requires an error as well as its integral signal [25].

$$S(x) = (\mu_f + d/dt)^{r-1} \int e d\tau \tag{5}$$

where $e = (x^* - x)$ is the error, x is the system state space, x^* is system state space reference, r is the degree of the sliding mode, and μ_f is the weighting factor. The sliding mode control was used to assess the generic system (6) in [23], and the design process was thoroughly described.

$$\begin{aligned} \dot{x} = f(x) + g(x)U_c \\ y = h(x) \end{aligned} \tag{6}$$

where $x \in \mathbb{R}^n$ is the state space vector, $U_c \in \mathbb{R}^m$ is the input control action, and $y(t) \in \mathbb{R}^p$ is the system output. U_c can be obtained by using equivalent control method [23]:

$$U_C = U_{equ} + U_n \tag{7}$$

where U_{equ} is the equivalent control action that ensures the system's convergence. It is calculated off-line with the use of a model that precisely represents the plant. Furthermore, U_n is a switching control action that assures the attractiveness of the surface in the system state space.

$$U_n = \beta sgn(S(x)) \tag{8}$$

The positive gain β in the above equation will be designed to ensure the Lyapunov stability criterion.

2.3. Conventional ISMC for IM (D1 Design)

To continue, the mechanical (4) is considered under parameter uncertainty terms of the a_ω, f_ω and b_ω as (Δa_ω, Δf_ω, Δb_ω),

$$\dot{\omega}_m = -(a_\omega + \Delta_\omega)\omega_m - (f_\omega + \Delta f_\omega) + (b_\omega + \Delta b_\omega)i_{sq} \tag{9}$$

now the speed tracking error is defined as:

$$e_\omega = \omega_m - \omega_m^* \tag{10}$$

which ω_m^* is the mechanical rotor speed reference, and by taking its derivative:

$$\dot{e}_\omega = \dot{\omega}_m - \dot{\omega}_m^* = -a_\omega e_\omega + u_\omega + d_\omega \tag{11}$$

where the control law is defined as:

$$u_\omega = -a_\omega \omega_m^* + b_\omega i_{sq} - f_\omega - \dot{\omega}_m^* \tag{12}$$

and compiling the uncertainty terms in d_ω term, the following expression is obtained:

$$d_\omega = -\Delta a_\omega \omega_m - \Delta f_\omega + \Delta b_\omega i_{sq} \tag{13}$$

Now the sliding variable $s_\omega(t)$ is defined with an integral component as:

$$s_\omega = e_\omega + \int_0^t K_\omega e_\omega dt \tag{14}$$

The following assumptions are formulated in order to obtain speed tracking: **(A1)** The constant K_ω should be chosen such that $K_\omega > 0$ for all time. The law u_ω should be designed in a way that guarantees convergence to the sliding surface in a finite time. Therefore:

$$u_\omega = a_\omega e_\omega - K_\omega e_\omega - \beta_\omega sign(s_\omega) \tag{15}$$

(A2) The gain β_ω must be chosen so that $|d_\omega| < \beta_\omega$, for all times.

Finally, the torque current reference, i_{sq}^* , is obtained directly by substituting (15) in (12),

$$i_{sq}^* = 1/b(a_\omega e_\omega - K_\omega e_\omega - \beta_\omega sign(s_\omega) + a_\omega \omega_m^* + f_\omega + \dot{\omega}_m^*) \tag{16}$$

Theorem 1. *According to (1), the torque current command (16) will control the T_e. Consequently, based on (3), the rotor speed will be regulated so that speed tracking error (10) tends to zero asymptotically, as the time tends to infinity.*

Proof. Taking the derivatives of sliding surface s_ω gives,

$$\begin{aligned} \dot{s}_\omega &= \dot{e}_\omega + d/dt \int_0^t K_\omega e_\omega d\tau \\ &= -a_\omega e_\omega + u_\omega + d_\omega + K_\omega e_\omega \end{aligned} \tag{17}$$

Substituting the control law (15) into (17) yields:

$$\dot{s}_\omega = d_\omega - \beta_\omega sgn(s_\omega) \tag{18}$$

now, by considering the Lyapunov function as $v_\omega = \frac{1}{2}s_\omega^2$, then:

$$\dot{v}_\omega = s_\omega \dot{s}_\omega = s_\omega(d_\omega - \beta_\omega sgn(s_\omega)) \tag{19}$$

based on (A2),

$$\dot{v}_\omega \leqslant -\varepsilon_\omega |s_\omega| \leq -\varepsilon_\omega |v_\omega|^{1/2} \tag{20}$$

where ε_ω is a positive constant. Based on the Lyapunov's direct method, since v_ω is positive, $\dot{v}_\omega$ is negative definite and v_ω tends to infinity as s_ω tends to infinity. Therefore, $s_\omega = 0$ is globally asymptotically stable which means s_ω tends to zero as time tends to infinity (sliding phase). Furthermore, all trajectories must reach to sliding surface in the finite time (reaching phase). When the sliding phase occurs, $s_\omega = \dot{s}_\omega = 0$, and as a result, the dynamic behavior of the tracking problem (11) is equivalently governed by the following equation:

$$\dot{s}_\omega = 0 \Rightarrow \dot{e}_\omega = -K_\omega e_\omega \tag{21}$$

□

The reduced order model (21) represents the system error. It can be said that based on (A1) the speed error tends to zero exponentially. Besides, in (21), K_ω is the rate of error convergence to zero. However, based on (16) it may be deduced that a high value of K_ω may produce a high control signal that could saturate the actuator. As it can be seen in the control law (19), thef_ω term needs to be calculated and it is dependent on the load torque. Therefore, T_L is estimated by using mechanical Equation (3).

$$\hat{T}_L = T_e - J\dot{\omega}_m - B_v\omega_m \tag{22}$$

2.4. Enhanced ISMC for IM (D2 Design)

This subsection designs the ISMC for IM mechanical rotor speed enhancing the surface, by integrating the *arctan*() of the mechanical speed error. In this regard by considering the generic sliding surface as:

$$S(x) = (\mu_f + d/dt)^{r-1} \int arctan(e)d\tau \tag{23}$$

for rotor speed of the IM could be obtained:

$$s_\omega = e_\omega + \int_0^\tau K_\omega arctan(e_\omega)d\tau \tag{24}$$

Taking and following the same steps as the previous section, the mechanical rotor speed control law is designed.

Remark 1. *The control law (15) is a discontinuous function of the sliding surface which may cause to undesirable chattering problem as it contains a hard switch. Therefore, to alleviate ISMC chattering phenomena and to have a smooth transition, sign() function is replaced by arctan() function which is a continous approximation of this function [26]. Consequently, the control law will be rewritten as:*

$$u_\omega = a_\omega e_\omega - K_\omega arctan(e_\omega) - \beta_\omega arctan(s_\omega) \tag{25}$$

and as a result, the torque current reference, i^*_{sq} , is obtained as,

$$i^*_{sq} = 1/b(a_\omega e_\omega - K_\omega arctan(e_\omega) - \beta_\omega arctan(s_\omega) + a_\omega\omega^*_m + f_\omega + \dot{\omega}^*_m) \tag{26}$$

where arctan() function is the inverse function of tan() function.

It should be noted that the stability analysis is as same as in Section 2.3.

3. Simulation and Experimental Design

In this section, the experimental platform and the performance of the proposed speed regulation has been verified by means of MatLab/Simulink simulation and real tests using a commercial induction motor.

The experimental validation of proposed ISMC regulators has been carried out by means of the platform shown in Figure 2. This platform is based on a commercial squirrel-cage IM of 7.5 kW (M2AA 132M4, ABB) which is connected mechanically to a synchronous AC servo motor of 10.6 kW (190U2, Unimotor) in its shaft, to implement the load torque (controlled by torque). Table 1, shows the parameters of the IM installed in the experiment platform. Both machines are connected to a DC bus of 540 V by using their respective three-phase Voltage Source Inverters (VSI) with a switching frequency of 10 kHz (SVPWM modulator frequency). This way, the control and monitoring tasks are done from a Personal Computer, which contains the MatLab/Simulink software and dSControl application to control the DS1103 controller board real-time interface of dSpace. The mechanical speed of the machine is calculated by using a FPGA module and the measurements of an incremental encoder of 4096 impulses per revolution.

Figure 2. Experiment platform for IM with load torque.

Table 1. Parameters of the M2AA 132M4 ABB Induction Motor 7.5 (rpm) and 1445 (rpm).

Symbol	Rated Value
B_v	0.0105 [Kg m/(rad/s)]
J	0.0503 [Kg m^2]
L_m	0.1125 [H]
L_s	0.1138 [H]
L_r	0.1152 [H]
σ	0.0346
R_r	0.400 [Ω]
R_s	0.729 [Ω]
p	4 poles
$\omega_m(n)$	151.32 [rad/s] (1445[rpm])
ϕ_r	0.9030 [Wb]
I_{sd}	8.026 [A]
I_{sq}	20 [A]
I_s	15.3 [A]
V	380 [V]
P_N	7500 [W]
μ	87%

Three different design cases such as D1, D2 and PI have been applied to the same induction motor to validate the proposed enhanced regulator (D2 design). To run the experiments, the IM is using the values of the nominal parameters for the IM, which have been listed in Table 1, and the IM has been tested at three different speeds. Furthermore, parameters values of the ISMC speed regulator are: $K_\omega = 1600$ and $\beta_\omega = 80$. On the other hand, the uncertainty test takes a moment of inertia which is 60% lower than the nominal value (J = 0.0201 kg·m^2), while the rest of the machine parameters will be nominal values, then these parameters are $K_\omega = 1700$ and $\beta_\omega = 20$. Regarding the adjustment of the two PI current regulators, both have been tuned by using a bandwidth of 3000 rad/s and a phase margin of 90 ($Kp_{is_{d,q}} = 11.81, Ki_{is_{d,q}} = 2187$) [27]. The mechanical speed PI regulator which is employed to compare with the enhanced ISMC speed regulator (D2 design), have been tuned taking a bandwidth of 300 rad/s and a margin phase of 82 ($Kp_{\omega_m} = 5.64, Ki_{\omega_m} = 238$). This way as faster as possible dynamics have been obtained experimentally by using PI regulators.

Figure 3(3-1) shows the performance of the machine by using a simulation test when the motor is working with a square speed reference of 1000 rpm with a period of 2 s, and the load torque is applied to the system in two steps: 10 Nm at the starting point and 30 nm at $t = 1.5$ s. In the first graph, (a), the reference and real rotor speed can be observed, and the second graph (b), shows the rotor speed error. As it can be seen, in the presence of a sudden change in the load torque the mechanical rotor speed tracks the desired speed properly with low error in steady-state (less than 1 rpm). The third graph, (c), shows the electromagnetic torque and load torque, where it can be appreciated that electromagnetic torque is smooth and consequently does not present any chattering. In (d) graph it can be seen that the stator torque current tracks the reference correctly. Moreover, this current is proportional to the electromagnetic torque which is also smooth and limited to its rated value (20 A). Graph (e) shows how the rotor flux tracks its reference adequately. Furthermore, the good response of the rotor flux current can be presented as graph (f). Finally, as the torque current is limited, the stator currents are also limited to a similar value that has been shown in the (g), keeping protected the stator windings against over-currents.

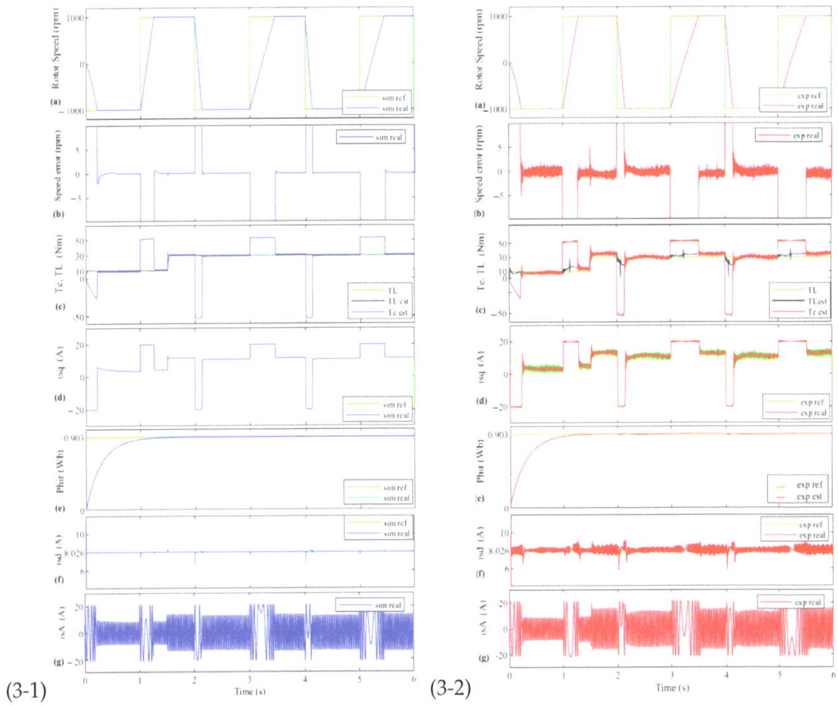

Figure 3. IM performance by using 1000 rpm reference speed and two load torque step changes: (**3-1**) simulation and (**3-2**) experimental (D2 design): (**a**) Rotor speed; (**b**) Speed error; (**c**) T_e, T_L; (**d**) Torque current; (**e**) Rotor flux; (**f**) Rotor flux current and (**g**) Stator current.

Figure 3(3-2) shows the performance of the machine by using the corresponding experimental test to the simulation test shown in Figure 3(3-1). In the Figure 3(3-2), the graphs (a) and (b) show that the speed tracking and the accuracy are very similar to the simulation case, getting fast dynamics and speed error in steady-state is less than 2 rpm. Regarding the electromagnetic torque (c), it can be compared with its simulation case, concluding that both are very similar, smooth, and efficient. Moreover, the estimated load torque is very similar to the load torque, which that demonstrates the estimator works properly. Graph (d) shows the reference and the real stator torque currents, which can be observed that the current tracking is very satisfactory and very similar to the simulation case (the form is proportional to the electromagnetic torque and limited to 20 A). In Figure 3(3-2), e and f graphs show how the rotor flux and rotor flux current adequately track their references. It can be observed that the two stator currents, i_{sd} and i_{sq} are decoupled. Finally, due to the limited torque current, the stator currents are also limited to similar values (g). By comparing simulation and experimental case in Figure 3, it can be seen that the simulation and the platform experiment have the same behavior, which is proof of the experimental validation of the presented ISMC speed controller (D2 design) and good system modeling.

Figure 4 shows the experimental tests for performance comparison between the proposed ISMC speed controller (D2 design) and the conventional ISMC (D1 design) method by using the same speed and load conditions in the test of Figure 3. This illustration provides the sliding surface convergence and tracking speed performance for both controllers. It is evident in the zoom mode graphs that using the $arctan()$ function of the speed error, instead of the $sign()$ function, in the surface design, provides faster error convergence to zero. Moreover, the PI speed regulator performance is also added to show that despite its performance being good its response is slower than the other two ISMC (D1 and D2 designs).

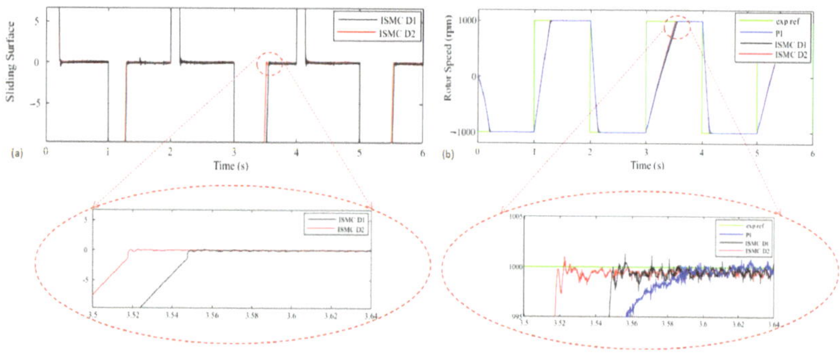

Figure 4. Experimental tests for performance comparison between the proposed ISMC (D2 design) regulator and the conventional ISMC by using 1000 rpm reference speed (D1 design) and also with PI regulator.

The performance of the motor when it is working at rated speed is shown in Figure 5 by employing a square speed reference of 1445 rpm and two load torque steps, starting at 0 s (10 nm) and $t = 1.5$ s (plus 20 nm). Graph Figure 5a shows how the actual speed tracks its reference accordingly, and graph Figure 5b shows that the speed error is limited to 3–4 rpm (0.27%), which means getting high accuracy in the presence of load disturbance. It can be observed in graph Figure 5c that the electromagnetic torque is smooth and effective, and also the load torque is estimated suitably. Figure 5d shows good tracking of the torque current. In Figure 5e, the precise rotor flux current tracking can be seen.

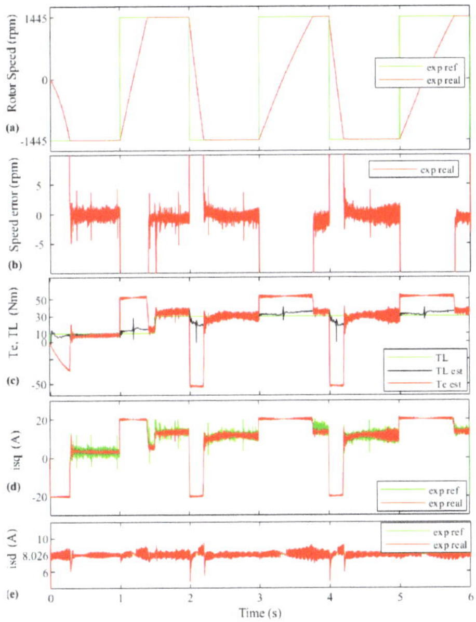

Figure 5. Experimental results by using 1445 rpm reference speed and two load torque step changes (D2 design): (**a**) Rotor speed; (**b**) Speed error; (**c**) T_e, T_L; (**d**) Torque current and (**e**) Rotor flux current.

Figure 6 shows the experimental tests for performance comparison between the two ISMC and the PI speed regulators by using the same speed and load conditions as in the experiment of Figure 5. Worth noting that, in the tuning process of the PI controller, experimentally maximum bandwidth has been assigned to the regulator to have the fastest possible response in the rotor speed convergence. As it can be seen, the ISMC is converging to a reference signal faster than the PI controller even when the motor is working at a high speed, such as in the same way as is shown in the previous graph of Figure 5. Furthermore, at $t = 1.5$ s (when the second step of load torque is applied to the motor), the effect of the load torque is being rejected more efficiently employing the proposed ISMC (D2 design) compared with the conventional ISMC (D1 design) and the PI regulators: recovering period is faster, and the response presents less oscillation.

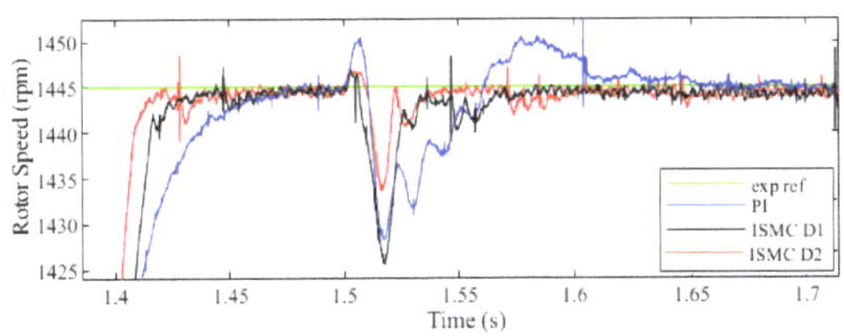

Figure 6. Experimental tests for performance comparison between the D1, D2 and the PI regulators by using 1445 rpm reference speed.

In Figure 7, the performance of the motor when it is working at low speed (100 rpm) with load disturbance ($T_L = 10$ Nm at $t = 0$ s and $T_L = 30$ Nm at $t = 3.5$ s) can be seen. It can be observed that the speed tracking Figure 7a is very satisfactory. Furthermore, the accuracy in Figure 7b is excellent, getting an error of less than 2 rpm (0.16%). The electromagnetic torque necessary to get these good results is smooth, and consequently, it does not present the chattering phenomenon (Figure 7c).

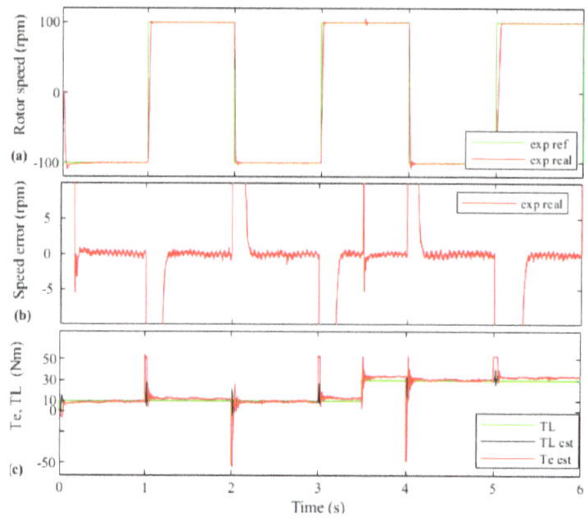

Figure 7. Experimental results by using 100 rpm reference speed and two load torque steps changes (D2 design): (**a**) Rotor speed; (**b**) Speed error and (**c**) T_e, T_L.

Figure 8 shows the experimental tests for performance comparison between the enhanced ISMC (D2 design) and the PI speed regulators while using the same speed and load conditions as in Figure 8 test. The proposed ISMC is converging to a reference signal faster than the PI controller. Furthermore, at $t = 3.5$ s (after applying the second step of load torque) to the motor, the behavior of the enhanced ISMC is more robust due to the fast and more effective rejection of load torque.

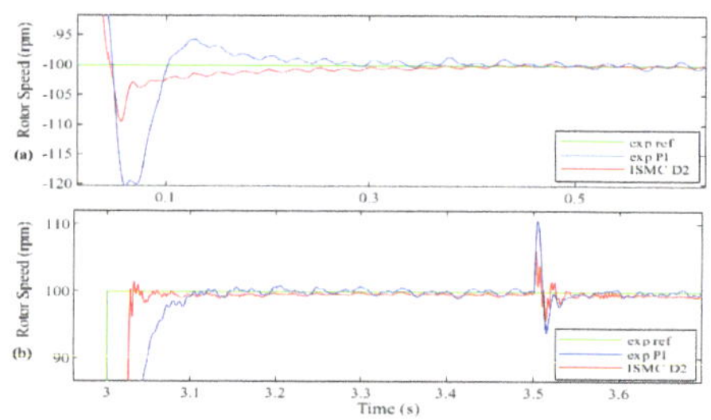

Figure 8. Experimental tests for performance comparison between the proposed ISMC (D2 design) and the PI regulators by using 100 rpm reference speed.

Figure 9(9-1) shows the simulation performance of the machine by using enhanced ISMC (D2 design), which takes a considerably minor J parameter (60% lower), as mentioned before. The IM has been tested by employing 1200 rpm reference speed. In Figure 9(9-2), the result of experimental robustness performance corresponding to the provided simulation is demonstrated. Graph (a) shows the motor speed response. However, the speed tracking is good. Furthermore, the speed error (graph (b)) can be considered very satisfactory, which is around 2 rpm (0.16%). Finally, electromagnetic torque, stator torque current, rotor flux current, and three-phase of stator currents have been shown in graphs (c), (d), (e) and (f), respectively, and they can be considered very proper. Therefore, the robustness of the speed controller has been tested by changing an important parameter of the motor specification in the speed controller (J).

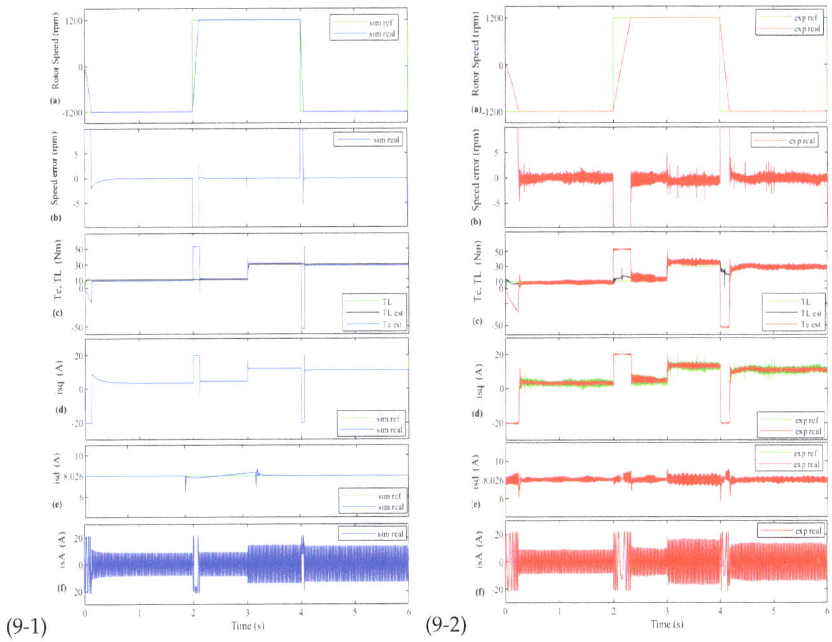

Figure 9. IM performance with 60% of uncertainties in J at 1200 rpm reference speed and two load torque step changes, (**9-1**) Simulation and (**9-2**) Experimental (D2 design): (**a**) Rotor speed; (**b**) Speed error; (**c**) T_e, T_L; (**d**) Torque current; (**e**) Rotor flux current and (**f**) Stator current.

4. Conclusions

In this paper, ISMC is applied to the IM vector control system to regulate the mechanical rotor speed and reject the load disturbances and parametric variations. The proposed controller incorporates an integral part in the sliding surface to eliminate static machine errors and enhance regulator accuracy. Indeed, the stability analysis of the controller has been done based on the Lyapunov function approach. The MatLab/Simulink's simulation and real tests utilizing a commercial IM have confirmed its experimental validation. In addition, the controller has good performance in practice because the speed tracking objective is achieved. The obtained accuracy for the speed regulator system can be considered excellent, getting a small speed error in stationary, which is between 0.16% and 0.27% for low, medium, and high speeds. Furthermore, it has been demonstrated that the presented enhanced ISMC speed regulator is more robust, efficient, and faster than the conventional ISMC and PI speed regulators. Finally, the regulators' capability of rectifying system chattering effectively under important system mechanical uncertainties (60%) and load torque disturbance demonstrates good robustness of the controlled system.

Author Contributions: Conceptualization, F.S., P.A., J.A.C. and O.B.; methodology, F.S. and P.A.; software, F.S., P.A. and J.A.C.; validation, F.S., P.A. and J.A.C.; formal analysis, O.B.; investigation, F.S. and P.A.; resources, O.B.; data curation, P.A.; writing—original draft preparation, F.S.; writing—review and editing, F.S., P.A., J.A.C. and O.B.; visualization, J.A.C. and O.B.; supervision, P.A. All authors have read and agreed to the published version of the manuscript.

Funding: The University of the Basque Country (UPV/EHU) [grant number PIF 18/127] has funded the research in this paper.

Acknowledgments: The authors wish to express their gratitude to the Basque Government, through the project EKOHEGAZ (ELKARTEK KK-2021/00092), and to the UPV/EHU, through the project GIU20/063, for supporting this work.

Conflicts of Interest: The authors declare no conflict of interest.

Nomenclature

Symbols of Induction Motor

B_v	Viscous friction coefficient
J	Moment of inertia
L_m	Magnetizing inductance
L_s	Stator inductance
L_r	Rotor inductance
R_r	Rotor resistance
R_s	Stator resistance
p	Number of poles
σ	Coefficient of magnetic dispersion
T_e	Electromagnetic torque
T_L	Load or disturbance torque
ω_m	Mechanical rotor speed
ω_s	Synchronous speed
ψ_r	Rotor flux
I	Stator rated current

References

1. Nguyen, P.; Quang, J.D. *Vector Control of Three-Phase AC Machines*; Springer: Heidelberg, Germany, 2008.
2. Talla, J.; Leu, V.Q.; Šmídl, V.; Peroutka, Z. Adaptive Speed Control of Induction Motor Drive With Inaccurate Model. *IEEE Trans. Ind. Electron.* **2018**, *65*, 8532–8542. [CrossRef]
3. Li, L.; Wang, M.; Yang, R.; Fu, Y.; Zhu, D. Adaptive Damping Variable Sliding Mode Control for an Electrohydrostatic Actuator. *Actuators Multidiscip. Digit. Publ. Inst.* **2021**, *10*, 83. [CrossRef]
4. El Kharki, A.; Boulghasoul, Z.; Et-Taaj, L.K.; oussi, Z.; Elbacha, A. Adaptive Speed Control of Induction Motor Drive With Inaccurate Model. In Proceedings of the 2019 4th World Conference on Complex Systems (WCCS), Ouarzazate, Morocco, 22–25 April 2019; pp. 1–8.
5. Happyanto, D.C.; Fauzi, R.; Hair, J. Backstepping development as controller in fast response three phase induction motor based on indirect field oriented control. In Proceedings of the 2016 International Electronics Symposium (IES), Denpasar, Indonesia, 29–30 September 2016; Volume 65, pp. 25–30.
6. Ortega, C.; Arias, A.; Espina, J. Predictive Direct Torque Control of Matrix Converter Fed Permanent Magnet Synchronous Machines. *Asian J. Control* **2014**, *16*, 70–79. [CrossRef]
7. Wang, M.; Wang, Y.; Yang, R.; Fu, Y.; Zhu, D. A Sliding Mode Control Strategy for an ElectroHydrostatic Actuator with Damping Variable Sliding Surface. *Actuators Multidiscip. Digit. Publ. Inst.* **2021**, *10*, 3. [CrossRef]
8. Cheng, X.; Liu, H.; Lu, W. Chattering-suppressed sliding mode control for flexible-joint robot manipulators. *Actuators Multidiscip. Digit. Publ. Inst.* **2021**, *10*, 288. [CrossRef]
9. Lu, Y.; Tan, C.; Ge, W.; Li, B.; Lu, J. Improved Sliding Mode-Active Disturbance Rejection Control of Electromagnetic Linear Actuator for Direct-Drive System. *Actuators Multidiscip. Digit. Publ. Inst.* **2021**, *10*, 138. [CrossRef]
10. Du, C.; Yang, C.; Li, F.; Gui, W. A Novel Asynchronous Control for Artificial Delayed Markovian Jump Systems via Output Feedback Sliding Mode Approach. *IEEE Trans. Syst. Man Cybern. Syst.* **2019**, *49*, 364–374. [CrossRef]
11. Brandtstädter H.; Tokio M.; Buss, V.I.U. Sliding Mode Control of Electromechanical Systems. Ph.D. Thesis, Technische Universität München, Munich, Germany.
12. Muñoz-Vázquez, A.J.; Parra-Vega, V.; Sánchez-Orta, A. Continuous fractional sliding mode-like control for exact rejection of non-differentiable Hölder disturbances. *J. Math. Control Inf.* **2015**, *34*, dnv064. [CrossRef]
13. Ahifar, A.; Ranjbar, N.A.; Rahmani, Z. Finite-time terminal synergetic control of a class of nonlinear systems with unmatched uncertainties. *J. Math. Control Inf.* **2020**, *37*, 765–776. [CrossRef]
14. Kikuuwe, R.; Prieto, P.J.; López-Rentería, J.A. Chattering of proxy-based sliding mode control in the presence of parasitic dynamics. *J. Math. Control Inf.* **2021**, *38*, 177–191. [CrossRef]
15. Morawiec, M.; Lewicki, A. Speed Observer Structure of Induction Machine Based on Sliding Super-Twisting and Backstepping Techniques. *IEEE Trans. Ind. Inform.* **2021**, *17*, 1122–1131. [CrossRef]
16. Chen, S.; Chiang, H.; Liu, T.; Chang, C. Precision Motion Control of Permanent Magnet Linear Synchronous Motors Using Adaptive Fuzzy Fractional-Order Sliding-Mode Control. *IEEE/ASME Trans. Mechatron.* **2019**, *24*, 741–752. [CrossRef]

17. Morawiec, M.; Lewicki, A.; Wilczyński, F. Speed observer of induction machine based on backstepping and sliding mode for low-speed operation. *Asian J. Control* **2021**, *23*, 636–647. [CrossRef]
18. Ilten, E.; Demirtas, M. Fractional order super-twisting sliding mode observer for sensorless control of induction motor, Compel. *Int. J. Comput. Math. Electr. Electron. Eng.* **2019**, *38*, 878–892. [CrossRef]
19. Veselic, B.; Perunicic-Drazenovic, B.; Milosavljevic, C. High-Performance Position Control of Induction Motor Using Discrete-Time Sliding-Mode Control. *IEEE Trans. Ind. Electron.* **2008**, *55*, 3809–3817. [CrossRef]
20. Comanescu, M. Minimum Time Speed Control of the Induction Motor Drive Using Discrete Time Sliding Mode. In Proceedings of the 2020 International Symposium on Power Electronics, Electrical Drives, Automation and Motion (SPEEDAM), Sorrento, Italy, 20–22 June 2020; pp. 213–218.
21. Barambones, O.; Garrido, A.J.; Maseda, F.J. Integral sliding-mode controller for induction motor based on field-oriented control theory. *IET Digit. Library* **2007**, *1*, 786–794. [CrossRef]
22. Comanescu, M. An Induction-Motor Speed Estimator Based on Integral Sliding-Mode Current Control. *IEEE Trans. Ind. Electron.* **2009**, *56*, 3414–3423. [CrossRef]
23. Oliveira, C.; Aguiar, M.; Monteiro, J.; Pereira, W.; Paula, G.; Almeida, T. Vector Control of Induction Motor Using an Integral Sliding Mode Controller with Anti-windup. *J. Control Autom. Electr. Syst.* **2016**, *27*, 169–178. [CrossRef]
24. Gou, L.; Wang, C.; Zhou, M.; You, X. Integral Sliding Mode Control for Starting Speed Sensorless Controlled Induction Motor in the Rotating Condition. *J. IEEE Trans. Power Electron.* **2016**, *35*, 4105–4116. [CrossRef]
25. Slotine, J.J.E.; Li, W. *Applied Nonlinear Control*; Prentice Hall: Englewood Cliffs, NJ, USA, 1991.
26. Khalil, H.K.; Quang, J.D. *Nonlinear Control*; Pearson Higher: London, UK, 2014.
27. Mohan, N. *Advanced Electric Drives: Analysis, Control, and Modeling Using MATLAB/Simulink*; John Wiley & Sons: Hoboken, NJ, USA, 2014.

actuators

Article

A Deep Trajectory Controller for a Mechanical Linear Stage Using Digital Twin Concept

Kantawatchr Chaiprabha and Ratchatin Chancharoen *

Department of Mechanical Engineering, Faculty of Engineering, Chulalongkorn University, Bangkok 10330, Thailand
* Correspondence: ratchatin.c@chula.ac.th; Tel.: +6682-218-6643

Abstract: An industrial linear stage is a device that is commonly used in robotics. To be precise, an industrial linear stage is an electro-mechanical system that includes a motor, electronics, flexible coupling, gear, ball screw, and precision linear bearing. A tight fit can provide better precision but also generates a difficult-to-model friction that is highly nonlinear and asymmetrical. Herein, this paper proposes an advanced trajectory controller based on a digital twin framework incorporated with artificial intelligence (AI), which can effectively control a precision linear stage. This framework offers several advantages: detection of abnormalities, estimation of performance, and selective control over any situation. The digital twin is developed via Matlab's Simscape and runs concurrently having a real-time controller.

Keywords: motion control; trajectory following controller; digital twins; bond graph; anomaly detection; adaptive controller

Citation: Chaiprabha, K.; Chancharoen, R. A Deep Trajectory Controller for a Mechanical Linear Stage Using Digital Twin Concept. *Actuators* **2023**, *12*, 91. https://doi.org/10.3390/act12020091

Academic Editors: Oscar Barambones, Jose Antonio Cortajarena and Patxi Alkorta

Received: 9 January 2023
Revised: 13 February 2023
Accepted: 16 February 2023
Published: 17 February 2023

Copyright: © 2023 by the authors. Licensee MDPI, Basel, Switzerland. This article is an open access article distributed under the terms and conditions of the Creative Commons Attribution (CC BY) license (https://creativecommons.org/licenses/by/4.0/).

1. Introduction

Motion control is a fundamental part in robotics that deals with the design and control of the movement of robots, machines, and other mechanical systems [1]. The development of linear motion is highly intriguing and is favored in many industrial applications [2]. Among motion control devices, an industrial linear stage is widely used in machines to produce precise linear motion in various applications, including manufacturing [3–7], testing and inspection [7–9], and material handling [10,11]. In a typical design, a linear stage is driven by various motors: a stepper motor [8], or brushed motor [3,7,11], ball screw [5] and precision linear bearing [12]. A precision linear bearing has a rigid structure. All components are precision made such that the position of a carriage can be sensed and controlled with an encoder at the rear end of the motor. This design offers high load capacity [6], accuracy [4,6,13], repeatability, and speed [14].

In recent years, computing power has dramatically increased having a smaller footprint. A machine is connected to cloud services [15,16]. Both actuation and sensing have also advanced [17,18], leading to a digital twin framework [19] that uses a virtual model for a variety of purposes such as design [20–22], simulation [23–25], analysis [26,27], and monitoring [28–30]. Similar to model predictive control [31,32], which is one approach for this framework [31,33,34], a digital twin simulates a physical plant in simulation time where internal states and outcomes can be projected [20,32,35]. For a complex system, this is very practical. Moreover, the comparison between digital and physical twin can be used to monitor the health of the physical twin or anomaly detection [36,37]. While many studies have explored the use of digital twin technology for trajectory planning and optimization [38–45], the majority of them have focused on machines and cells with multi-degrees-of-freedom motion [38–45]. Thus, a joint level digital twin is yet to be explored. In this work, the benefits of a digital twin for a trajectory controller at joint level have been studied for generating linear motion.

In this project, a new class of trajectory controller is proposed based on a digital twin system. Advances in technology mentioned earlier make it possible to construct a digital model of a physical linear stage that can be run in real-time. This leads to the next era of motion control in the way that an anomaly can be sensed [46]. Artificial intelligence can be incorporated with digital twin to practically demonstrate its power. In addition, a shuffle and adaptive controller can be selected to suit a situation so that we may tradeoff the positioning precision with something else, e.g., energy consumption, lifespan, smoothness, and safety [47]. This work focuses on: (A) a proposed trajectory controller based on a digital twin concept, and its successful implementation, (B) detection of anomalies, and (C) controller adaptability to suit the situation.

2. Theoretical Background

Figure 1, an investigation plant is an electro-mechanical linear stage, which is driven by a 15 W permanent magnet DC (PMDC) motor. The linear stage is installed having a 2 mm pitch lead screw and precision linear bearing. The purpose of a lead screw is to convert the rotational motion of a motor to linear translation [12,13,48]. The PMDC motor is powered by a VNH5019 H-bridge motor driver. The motor is coupled with a 500 PPR encoder to measure the shaft displacement. Thus, the carriage's displacement can be determined via the kinematic relationship of the screw mechanism.

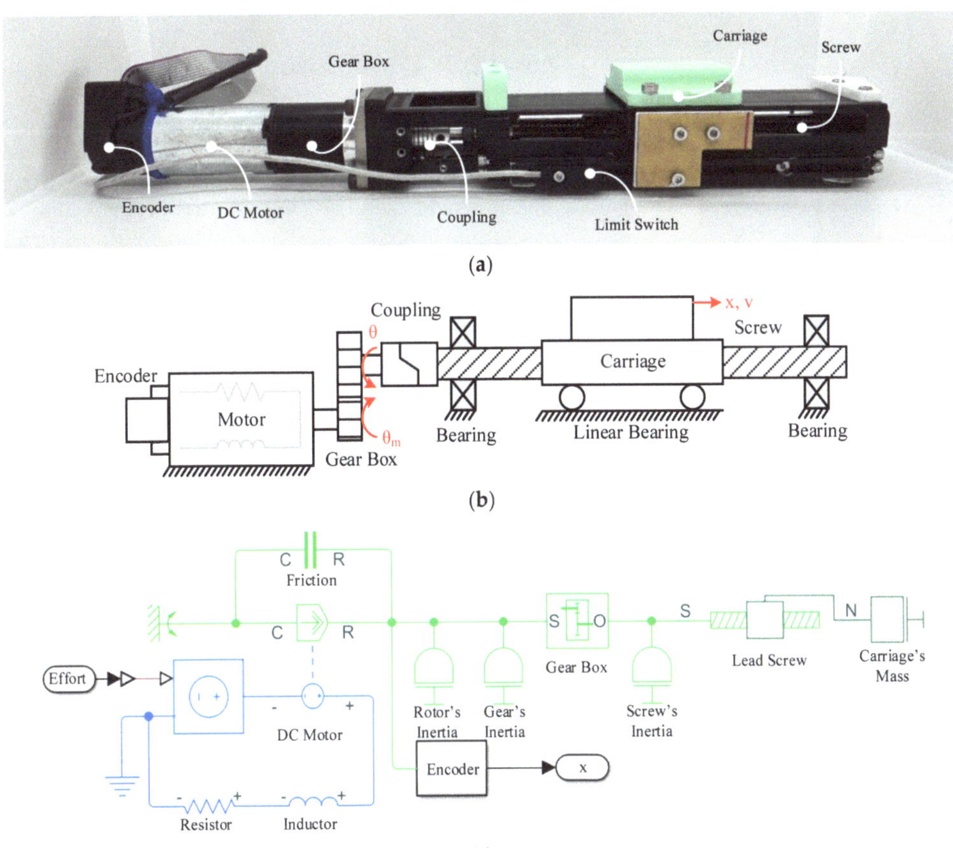

Figure 1. Investigation of the linear stage: (**a**) Electro−mechanical linear stage (Photograph), (**b**) Schematic diagram of the linear stage, (**c**) Simscape model.

In this conventional linear stage design, both the gear and screw mechanisms are used to provide a boost to the PMDC motor and the overfitted linear bearing is used to ensure precise linear motion. In this design, the PMDC motor can drive the carriage at near-zero back-drivability. Disturbance at the carriage side is suppressed and the closed loop control of the motor is eased. Moreover, the gear, screw, and bearing comes with a significant amount of viscous friction that drains out the kinetic energy into heat during the motion [49]. Such friction is good for the robustness and stability of the closed loop control system in most applications.

In the investigation plant, the MAXON 236662 motor is a graphite brushed cylindrical motor having an alnico magnet. The gear is MAXON 166160 planetary gear with 21:1 mechanical advantage. The drive circuit is VHN5019: 24 V that accepts a pulse-width modulation (PWM) command up to 20 kHz. In the linear stage, Misumi LX26 is equipped with a lead screw having 2 mm pitch and a diameter of 8 mm. Such a linear stage can efficiently and effectively drive the linear motion whereas the PWM commands the drive circuit. The drive circuit is highly correlated to the dynamical motion of the carriage on the linear stage.

The dynamical simulation of the electro-mechanical system has considerably advanced over recent years. For instance, advances in solving algorithms have taken place [50,51]. Both computing power and massive data handling have been significantly enhanced. These developments support a realistic and accurate complex multi-domain simulation. Parallel computing makes it possible to perform complex dynamical simulation that may be used in real-time applications.

In this work, Matlab's Simscape is employed as in block diagrams [52] and bond graphs [49,53]. The linear dynamical system can be written in S-domain where its input-output relation is written algebraically. A framework of block diagrams is an effective way to handle dynamical devices in a complex system [54–57] in a graphical environment [58]. A single ordinary differential equation (ODE) is set for an entire simulation but can easily be modified. Matlab's Simulink is a block diagram framework with hundreds of predefined blocks and effective working environments. The bond graph modelling framework is complementarily used with block diagrams [55,59]. The difference is that bond graphs use both across and through variables to connect blocks that effectively represent each physical device and transmission signals can flow in both directions in the model. In this way, we can model a complex system in an abstraction layer. The mathematics of each device is packed within the block and are thus easy to manage [55–57,60].

With advances in simulation, the digital twin of the electro-mechanical linear stage can be modeled and real-time simulated. In this framework, an electrical domain is shown in blue. Mechanical rotational and translational domains are represented by light green and dark green, respectively. The connection between the components is via an energy flow pipe whereby energy can flow bi-directionally. Thus, this framework can easily be applied and adapted to several scenarios within a single simulation model. Moreover, this framework creates an abstraction in which detailed parameters of each component provide a better focus on how physical components are connected and cooperate.

3. Design and Construction of a Dynamical Digital Twin

3.1. Mathematics

In the electro-mechanical linear stage, the variables (a function of time) are listed in Table 1.

Table 1. State variables in the electro-mechanical linear stage.

Variable	Description	Unit
$V(t)$	Voltage applied across motor	V
$i_a(t)$	Current flow in motor	A
$T(t)$	Motor's torque	Nm
$T_f(t)$	Friction torque	Nm
$P(t)$	Load force	N
$\theta(t)$	Screw angle	rad
$x(t)$	Carriage displacement	m

In this system, a controllable voltage source applied voltage $V(t)$ across a DC brushed motor. Applying Kirchhoff's voltage law to the DC brushed motor model with gear yields:

$$V(t) = L\frac{\partial}{\partial t}i_a(t) + Ri_a(t) + K_b K_g \frac{d}{dt}\theta(t). \tag{1}$$

where the motor's current $i_a(t)$ flows through the motor's coil. The current $i_a(t)$ generates the torque output via the Lorentz force where the torque $T(t)$ of the motor is proportional to the current $i_a(t)$, accordingly:

$$T(t) = K_a\, K_g\, i_a(t). \tag{2}$$

Considering the kinematic coupling between the motor's shaft and lead screw's carriage, the motion is governed by:

$$x(t) = r_s\, \tan(\lambda)\, \theta(t) \tag{3}$$

where the rotational motion converts to translation motion by the screw mechanism [61].

The dynamics of the screw mechanism can be realized by applying Newton's second law on the Wedge model. The equation is shown in rotational form, as in Equation (4):

$$\left(J + m\, {r_s}^2\, \xi\, \tan(\lambda)\right)\frac{d^2}{dt^2}\theta(t) = T(t) - T_f(t) - r_s\, \xi\, P(t) \tag{4}$$

where the ξ is the efficiency of ball screw [61].

3.2. *Identification of Parameters*

Table 2, theparameters in the mentioned dynamical equations are identified either from data sheets of the hardware or from the experimental system. Herein, the overall friction's behavior is complex and thus determined by experimentation (Figure 2a) [62–64].

Table 2. Parameters in the linear stage.

Parameter	Description	Quantity
L	Motor's inductance	0.556 mH
R	Motor's resistance	0.399 Ohm
K_b	Motor's velocity constant	0.03525 V/(rad/s)
K_a	Motor's torque constant	0.03525 Nm/A
K_g	Gear ratio	21
J	Moment of inertia	46.1 g/cm^2
λ	Screw angle	0.0794 rad
m	Carriage's mass	1.37 kg
r_s	Shaft radius	4 mm
ξ	Screw's efficiency	0.79
μ_s	Coulomb friction coefficient	0.02

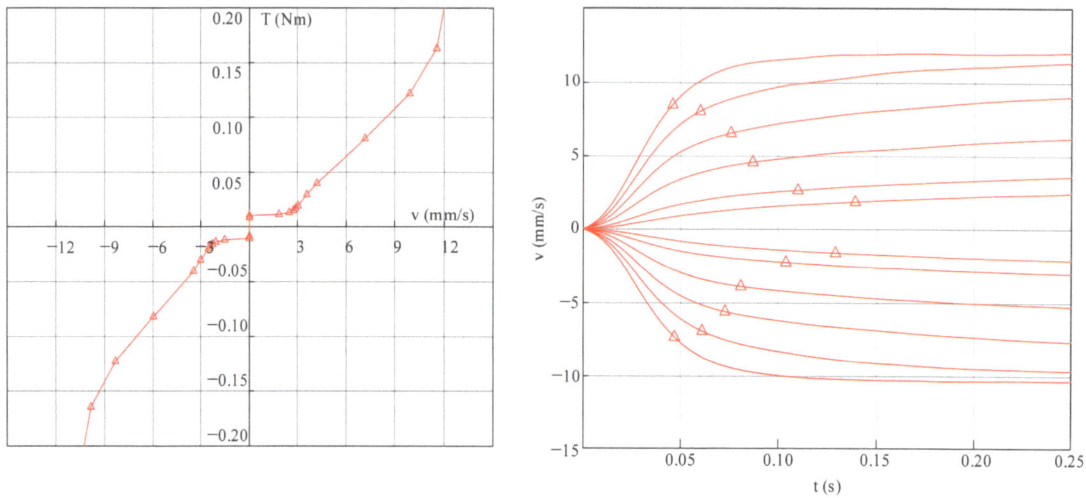

(a) (b)

Figure 2. (**a**) Friction model of the system. (**b**) Velocity of carriage at 24.0, 19.2, 14.4, 9.6, 4.8, 2.4, −2.4, −4.8, −9.6, −14.4, −19.2, −24 V applied to the PMDC motor, from top to bottom. The triangles showing 70% of the terminal velocity indicate the speed response of the system.

The process of identification regards the friction model was designed to apply constant voltage to the motor. Voltage was set at a low voltage to determine the breakaway friction behavior. The position of the linear stage was acquired through the 500 PPR motor's encoder with 1 kHz sampling frequency. After that, the time series of positions was fitted via smoothing-spline, and then the velocity was estimated.

Combining Equations (1) and (2), motor torque can be determined as in Equation (5):

$$T(t) = K_a K_g \frac{(V(t) - K_b K_g \frac{d}{dt}\theta(t))}{R}. \tag{5}$$

where $i_a(t)$ assumes to be constant. Therefore, the inductance term can be negligible.

In Figure 2a, the graph shows the relationship between velocity and motor torque, as calculated in Equation (5). Static friction torque can be determined using the data when the system remains stationary; later, low voltage is applied continuously. At a velocity between −2.5 and 2.5 mm/s, the torque due to friction remains the same. However, beyond this interval, viscous friction was dominant as the lubricant layer between the bearing surfaces formed. This viscous interval was observed linearly until maximum voltage in both directions was reached.

In Figure 2b, open-loop transient behavior of the linear stage is shown. In the time between 0 and 0.02 s, the graph illustrates the increase in acceleration as it corresponds to the acceleration of the carriage due to the open-loop voltage reference, as seen in the trend of growth in velocity. As the system gains more velocity, the viscous friction builds up and resists the effort as a result of the decrease in acceleration, which can be clearly seen by the triangle marks. The maximum velocity that this system can reach is around 12.2 mm/s and −10.3 mm/s. Thus, the existence of asymmetric friction is confirmed.

The model of the physical electro-mechanical linear stage is used to investigate predicted behavior when there are parameter changes and/or disturbances. Various controller designs can be simulated to predict the resulting outcome before a suitable one is chosen to run the physical plant.

4. A Typical Trajectory Controller

The linear stage has to be controlled such that its carriage moves to the desired position. The controller is closed-loop; the position of the carriage is monitored in real-time and used as feedback to locate the desired position. The difference, called an error, is fed into the controller to generate a command to actuate the linear stage in such a way that the carriage arrives at the desired position. A stiff PD controller is used in this investigation. However, since the friction in the investigation plant is considerably high, it plays as a natural derivative gain.

4.1. Regulatory Control

The investigation plant is experimentally driven with an electromechanical linear stage having no load. The stage is commanded to go to 1 mm position from zero position. The high gain proportional controller (K = 0.1) computes the effort beyond the hardware limit (24 V). In this period, the controller is likely to be open-loop with maximum capacity and its behavior is demonstrated in Figure 2b. It is noted that the settling time of the open loop velocity control is between 0.05–0.15 s.

In Figure 3, the maximum velocity of the linear stage in this combination is 11 mm/s. The velocity ramped up from stationary to its maximum capacity in 0.15 s (Figure 3b). Once the carriage is close to the target position where the computed control effort is less than the maximum capacity (24 V), the controlled system demonstrated the second-order dynamical system (Figure 3b from 0.08 s). In this case, an overshot of 16.25%, a settling time of 0.24 s, and a near zero steady state error are achieved. In most applications, the industrial linear stage proves to be good.

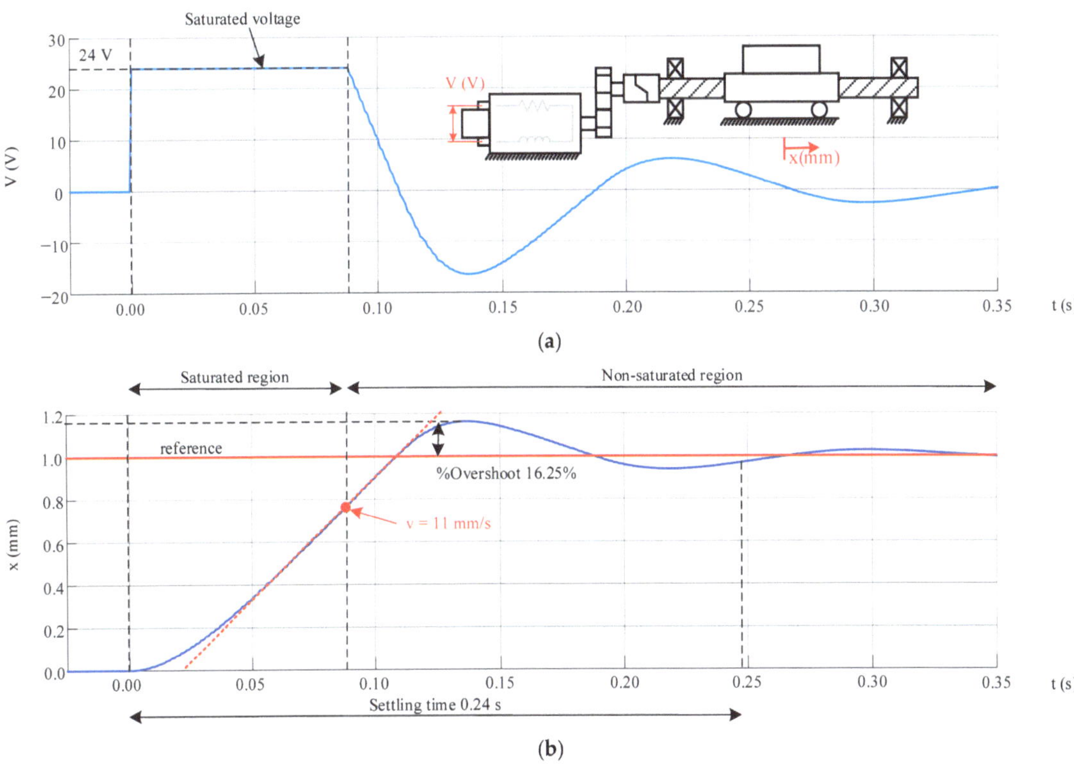

Figure 3. Closed−loop step responses: (**a**) Voltage input from the 24 V power supply, and (**b**) Carriage position corresponding to the input from the controller.

4.2. Trajectory Control

Trajectory control is more challenging since we are faced with uncertainties, disturbances, and constraints on the input. In this work, the linear stage was commanded to follow a sine sweep trajectory. The reference trajectory is sinusoidal and sweeps from 0.5 Hz to 2 Hz in 25 s. During low frequency (Figure 4 Zone I), the actual trajectory perfectly tracks the reference. However, fluctuation is noted, especially when the carriage changes its direction of motion. When frequency is high (Figure 4 Zone II), fluctuation lessens while tracking error rises. If the frequency is high (Figure 4 Zone III), fluctuation disappears but the tracking error increases along with the actuation effort.

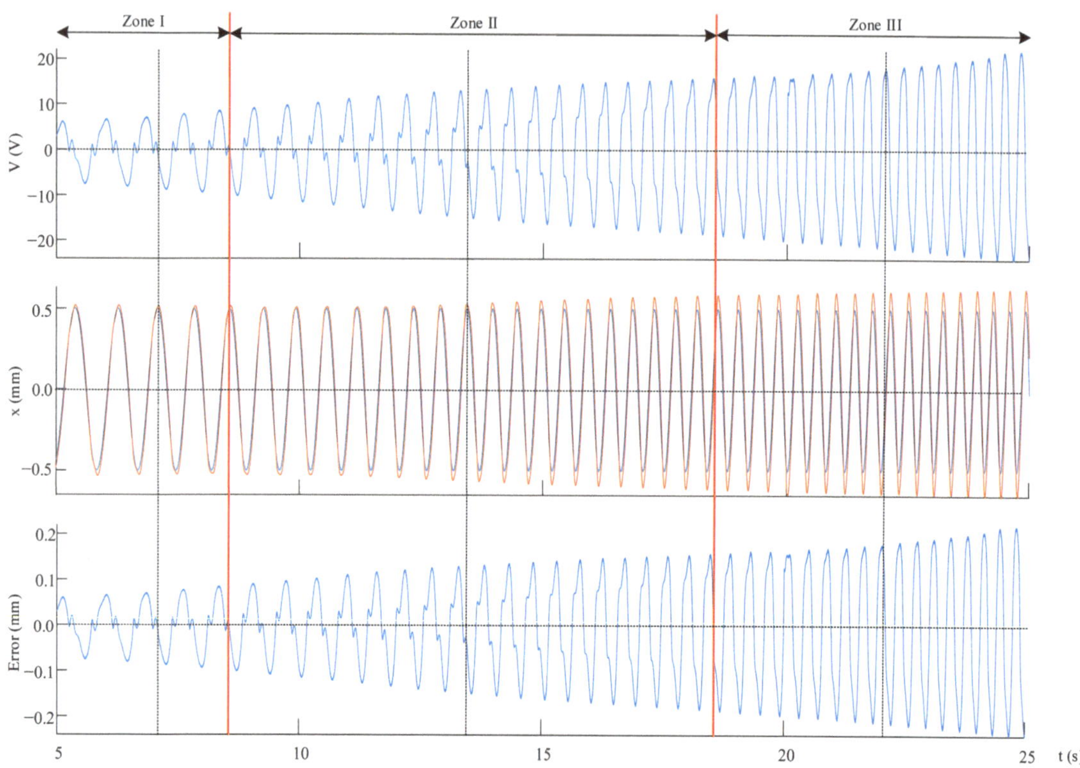

Figure 4. Closed−loop trajectory: experimental results.

4.3. *Effect of the Controller Gain on the Soft/Stiff Behavior*

Our trajectory controller is the stiff PD controller. The stiff controller attempts to follow a trajectory while facing uncertainties, disturbances, and nonlinear behavior. Fluctuation can occur during the trajectory. The reference trajectory is sinusoidal at 1 Hz. In Figure 5a, the effect of the soft/stiff controller's behavior is demonstrated. There is a tradeoff between the tracking error and the positioning fluctuations due to the stiffness of the controller. The resulting behavior is further analyzed. The closed-loop control system is approximated as a linear system and its input–output data pairs are used to find the dominant zero(s)/root(s) in Figure 5b. This locus plot clearly explains the behavior related to the stiff controller [52]. If the controller is stiff, the roots stay on the left side (on the real axis). When stiffness is lessened, the roots walk into the right (slower). However, if it is too soft, two roots split and go into the imaginary zone. In a real application, precision positioning or tracking is not needed all the time during the mission. For instance, if we command a linear stage to go to its home position, at its fastest speed, to reset the zero position, open loop velocity control where the command effort is at its full capacity is suitable. If there is uncertainty or high disturbance, a soft controller may be more suitable. A soft controller is also good in terms of energy consumption and service life.

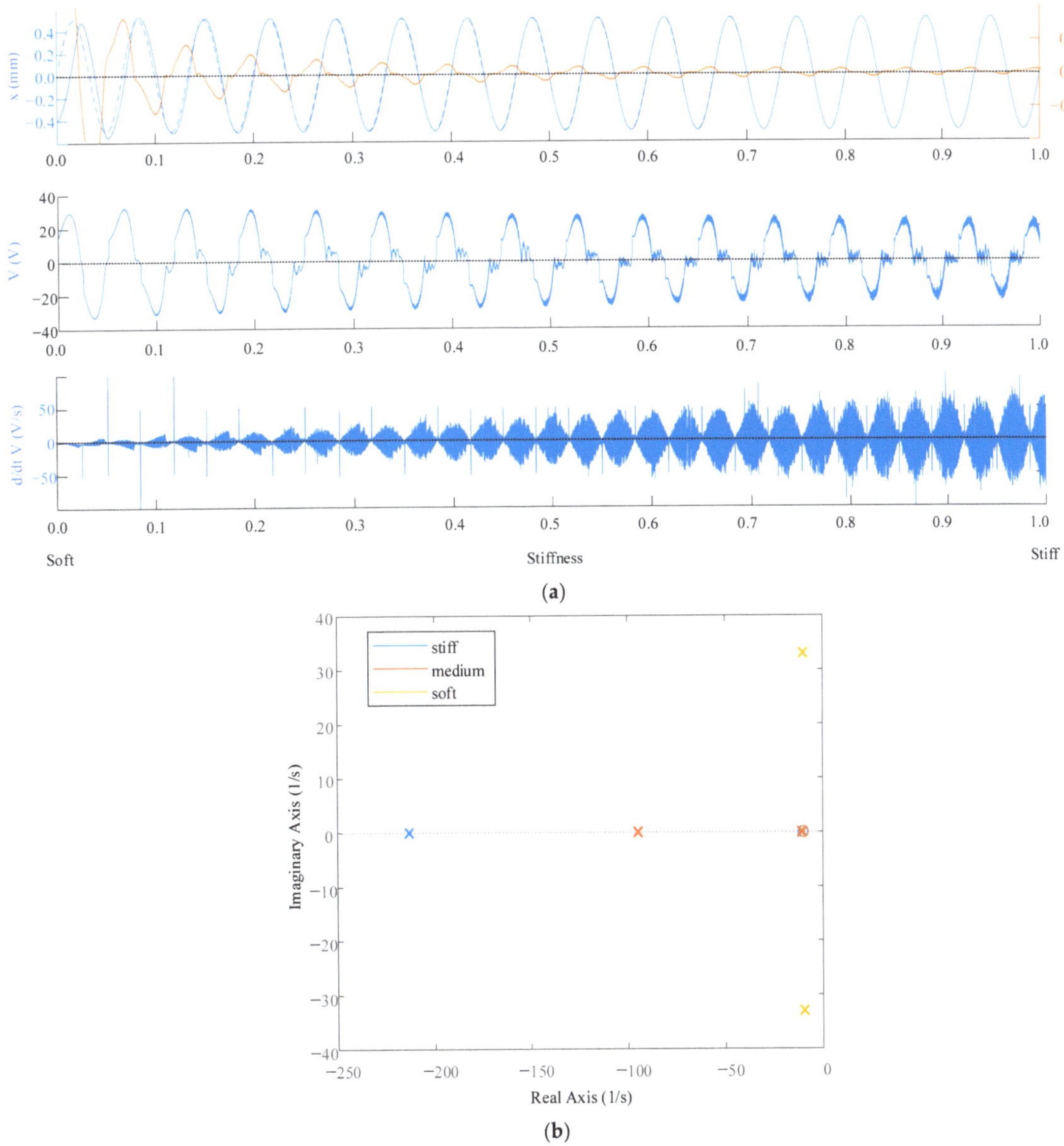

Figure 5. Effect of soft/stiff controller on the behavior of the closed−loop: experimental results.

The electro-mechanical linear stage can be controlled by an adaptive controller to suit a different type of motion. In our physical plant, the controller can be either an open loop with a programmable voltage or a closed loop with a programmable stiffness.

5. The Proposed Trajectory Controller Design with Digital Twin

Figure 6, the proposed trajectory controller is designed to have a digital twin framework. The digital twin is constructed with Matlab's Simscape. The mathematical model of motor, gear, ball screw, and overall friction are modeled in the digital twin. The reference trajectory is fed via both the physical controller and digital controller. The physical controller drives the physical twin, and the trajectory is measured by an encoder. Meanwhile, the digital controller drives the digital twin along. In this way, the real trajectory can be compared and analyzed with a predicted trajectory for anomaly detection. The digital twin provides an opportunity to try out different control strategies in the system to test its performance. Furthermore, the performance of a new controller can be analyzed. If there is a change in the physical plant or there is a disturbance, a comparison is made between the actual trajectory and predicted trajectory whereby the aggregated data is fed through a classifier to adjust the digital twin or switch to a suitable controller.

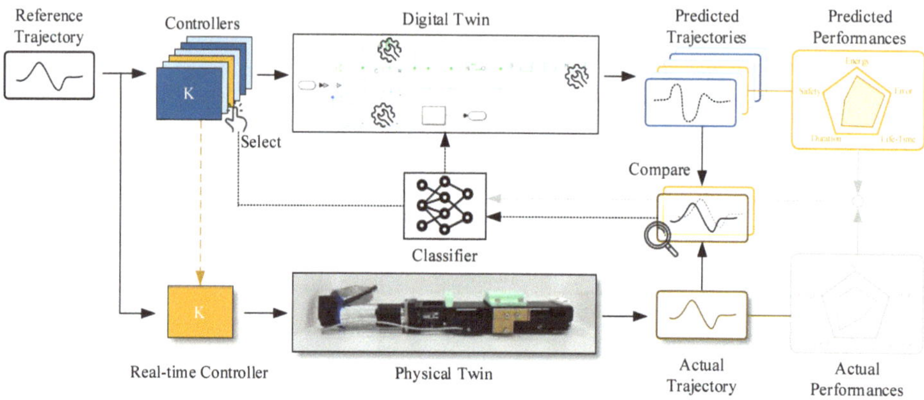

Figure 6. The proposed trajectory controller based on a digital twin framework.

In this work, the actual trajectory is compared with the predicted one in runtime. The joint controller is updated at 1 kHz while the comparison is carried out at 50 Hz. The difference is fed into the classifier. If the difference goes beyond the threshold, the classifier switches the controller to a suitable one and may update the applied parameters.

6. The Resulting Behavior of the Proposed Controller

The existing industrial linear stage is excellent for most applications but can be enhanced by having a digital twin framework. In this paper, anomaly detection and adaptive control are demonstrated.

6.1. Anomaly Detection

If there is unusual or unexpected behavior during operation, the resulting trajectory can differ from the expected one. Anomaly detection is normally used to detect potential problems or issues that can impact the performance or reliability of the linear stage. In our demonstration, the stiff controller follows a 1 Hz sinusoidal. At some point, we obstructed the ongoing motion. The tracking error image is obviously seen at one cycle time (Figure 7). When the obstruction was removed, the trajectory went back to normal. The projected behavior and/or the empirical historical data can be used as a normal reference. The empirical historical data is used in the experiment, as demonstrated in Figure 7.

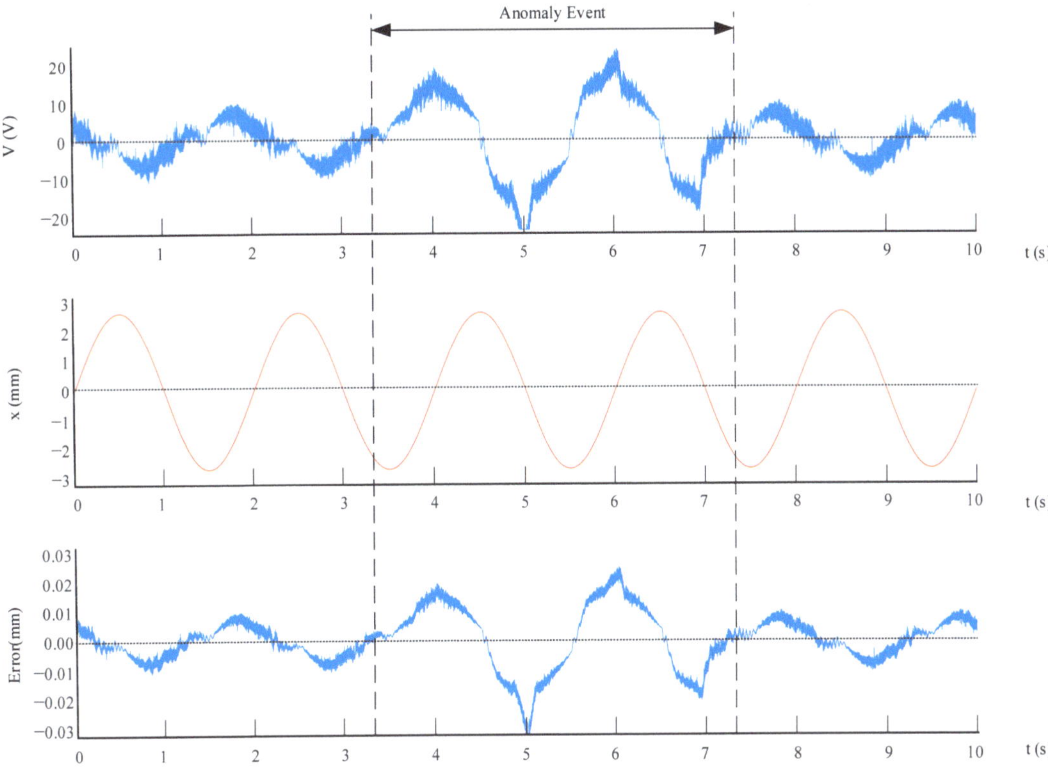

Figure 7. Anomaly detection.

6.2. Controller Adaptability

The next experiment is to demonstrate the adaptability of the proposed controller. Previously, detection of anomalies was carried out using an error threshold method, which compared errors found with empirical error data. The stiff controller followed a 0.5 Hz sinusoidal (Figure 8). At 3.3 s, we intentionally placed an obstruction in front of the carriage. Subsequently, the anomaly was successfully detected via the threshold technique, and the classifier switched the controller to the soft controller.

In this experiment, both anomaly detection and use of controllers: soft and stiff are demonstrated. During normal operation, the controller is stiff: such behavior is shown in Figure 8 before 3.3 s: observable maximum tracking error is 0.051 mm. In Table 3, afterhitting the obstacle at 3.3 s, the proposed anomaly detection detects this event with 0.095 s delay. The controller is switched to the soft controller where the observable maximum error is 0.535 mm. Subsequently, the effort continuously rose until it reached a maximum voltage of 24 V. When the controller switched from anomaly detection, the effort dropped and resulted in a peak voltage of 16.77 V.

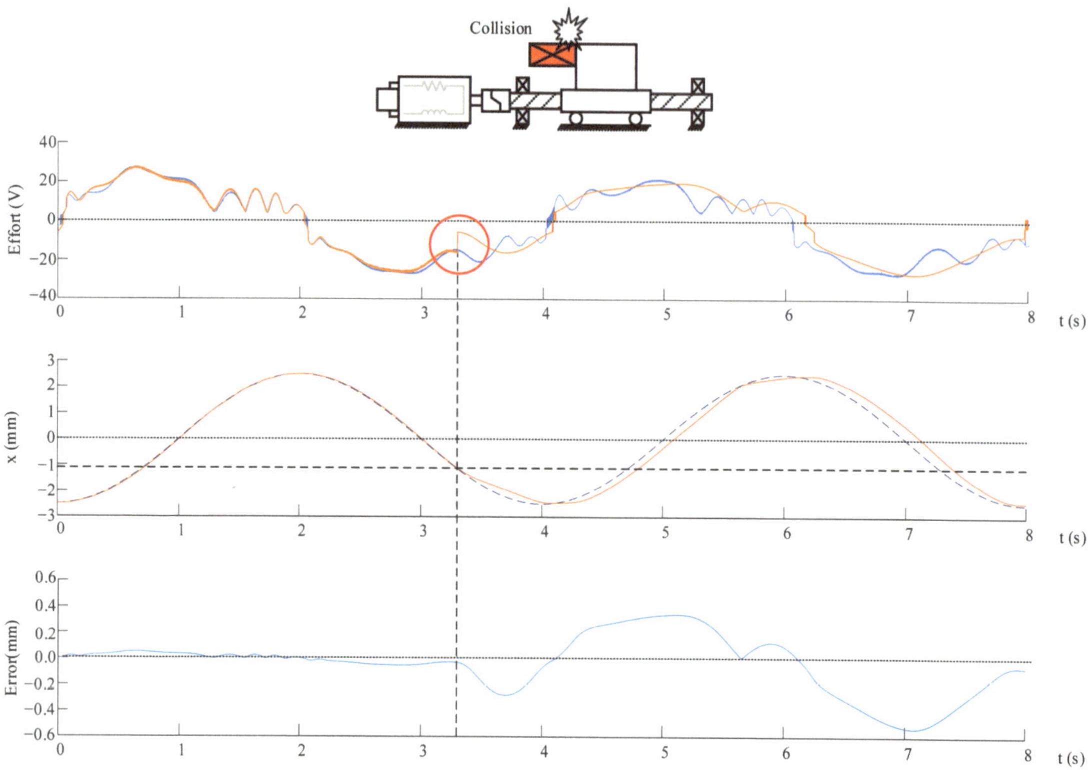

Figure 8. Anomaly detection and controller's adaptability.

Table 3. Observable performance of the proposed controller.

Controller-Type	Maximum Error (mm)	Hitting Peak Voltage (V)
Stiff	0.051	24
Soft	0.535	16.77

7. Conclusions

A trajectory controller having a digital twin framework for the electromechanical linear stage was successfully designed. Its capabilities are demonstrated. The controller can be programmed as an open loop control with a programmable voltage or a closed loop control with a programmable stiffness. The digital twin for the linear stage is successfully implemented via Matlab's Simscape along with a Beaglebone board. The digital twin runs along with the physical linear stage in real-time to equip the existing high precision controller with new capabilities. In this work, anomaly detection and adaptive soft-stiff controller are demonstrated. The controller was able to detect the anomaly, and consequently switched to soft controller; the voltage decreased to 69.86%, but the increment in maximum tracking error was 0.484 mm. In further work, it is proposed to use digital twin technology as a pathway for a novel system that we are researching, which incorporates an AI-embedded drilling and air supply for a pneumatic robot.

Author Contributions: Conceptualization, R.C.; methodology, K.C. and R.C.; software, K.C.; validation, K.C. and R.C.; formal analysis, K.C. and R.C.; investigation, R.C.; resources, R.C.; data curation, K.C.; writing—original draft preparation, K.C. and R.C.; writing—review and editing, K.C. and R.C.; visualization, K.C.; supervision, R.C.; project administration, R.C.; funding acquisition, R.C. All authors have read and agreed to the published version of the manuscript.

Funding: This project is funded by National Research Council of Thailand (NRCT).

Data Availability Statement: Not applicable.

Conflicts of Interest: The authors declare no conflict of interest. Furthermore, the funders had no role in the design of the study; in the collection, analysis, or interpretation of data; in the writing of the manuscript, or in the decision to publish the results.

References

1. Xu, Z.; Song, T.; Guo, S.; Peng, J.; Zeng, L.; Zhu, M. Robotics technologies aided for 3D printing in construction: A review. *Int. J. Adv. Manuf. Technol.* **2022**, *118*, 3559–3574. [CrossRef]
2. Humaidi, A.J.; Hameed, A.H. PMLSM position control based on continuous projection adaptive sliding mode controller. *Syst. Sci. Control Eng.* **2018**, *6*, 242–252. [CrossRef]
3. Phanomchoeng, G.; Chancharoen, R. Hybrid motor system for high precision position control of a heavy load plant. *Eng. J.* **2019**, *23*, 161–173. [CrossRef]
4. Chen, C.L.; Jang, M.J.; Lin, K.C. Modeling and high-precision control of a ball-screw-driven stage. *Precis. Eng.* **2004**, *28*, 483–495. [CrossRef]
5. Altintas, Y.; Verl, A.; Brecher, C.; Uriarte, L.; Pritschow, G. Machine tool feed drives. *CIRP Ann.* **2011**, *60*, 779–796. [CrossRef]
6. Cheng, Q.; Qi, B.; Liu, Z.; Zhang, C.; Xue, D. An accuracy degradation analysis of ball screw mechanism considering time-varying motion and loading working conditions. *Mech. Mach. Theory* **2019**, *134*, 1–23. [CrossRef]
7. García-Martínez, J.R.; Cruz-Miguel, E.E.; Carrillo-Serrano, R.V.; Mendoza-Mondragón, F.; Toledano-Ayala, M.; Rodríguez-Reséndiz, J. A PID-Type Fuzzy Logic Controller-Based Approach for Motion Control Applications. *Sensors* **2020**, *20*, 5323. [CrossRef] [PubMed]
8. Plengkham, S.; Kukiattikoon, C.; Phanomchoeng, G.; Boonyasuppayalorn, S. Development of an Automated Viral Plaque Assay Program and ELISpot Counting Program on 96-Well Plate Platform. In Proceedings of the International Conference on Electrical, Computer, and Energy Technologies, Prague, Czech Republic, 20–22 July 2022.
9. Phanomchoeng, G.; Kukiattikoon, C.; Plengkham, S.; Boonyasuppayakorn, S.; Salakij, S.; Poomrittigul, S.; Wuttisittikulkij, L. Machine-learning-based automated quantification machine for virus plaque assay counting. *PeerJ Comput. Sci.* **2022**, *8*, 878. [CrossRef]
10. Huang, T.; Kang, Y.; Du, S.; Zhang, Q.; Luo, Z.; Tang, Q.; Yang, K. A survey of modeling and control in ball screw feed-drive system. *Int. J. Adv. Manuf. Technol.* **2022**, *121*, 2923–2946. [CrossRef]
11. Phanomchoeng, G.; Saadi, M.; Sasithong, P.; Tangmongkhonsuk, J.; Wijayasekara, S.K.; Wuttisittikulkij, L. Hardware software co-design of a farming robot. *Eng. J.* **2020**, *24*, 199–208. [CrossRef]
12. Butler, Q.; Ziada, Y.; Stephenson, D.; Gadsden, S.A. Condition Monitoring of Machine Tool Feed Drives: A Review. *J. Manuf. Sci. Eng.* **2022**, *144*, 103454. [CrossRef]
13. Feng, G.-H.; Pan, Y.-L. Investigation of ball screw preload variation based on dynamic modeling of a preload adjustable feed-drive system and spectrum analysis of ball-nuts sensed vibration signals. *Int. J. Mach. Tools Manuf.* **2012**, *52*, 85–96. [CrossRef]
14. Pravika, M.; Jacob, J.; Paul Joseph, K. Design of linear electromechanical actuator for automatic ambulatory Duodopa pump. *Eng. Sci. Technol. Int. J.* **2022**, *31*, 101056. [CrossRef]
15. Oztemel, E.; Gursev, S. Literature review of Industry 4.0 and related technologies. *J. Intell. Manuf.* **2020**, *31*, 127–182. [CrossRef]
16. Kang, H.S.; Lee, J.Y.; Choi, S.; Kim, H.; Park, J.H.; Son, J.Y.; Kim, B.H.; Noh, S.D. Smart manufacturing: Past research, present findings, and future directions. *Int. J. Precis. Eng. Manuf. -Green Technol.* **2016**, *3*, 111–128. [CrossRef]
17. Yoshida, S.; Xie, H.; Miyata, K. Omnidirectional Haptic Stimulation System via Pneumatic Actuators for Presence Presentation. *Sensors* **2023**, *23*, 584. [CrossRef]
18. Sithiwichankit, C.; Chanchareon, R. Adaptive Pincer Grasping of Soft Pneumatic Grippers Based on Object Stiffness for Modellable and Controllable Grasping Quality. *Robotics* **2022**, *11*, 132. [CrossRef]
19. Chancharoen, R.; Chaiprabha, K.; Wuttisittikulkij, L.; Asdornwised, W.; Saadi, M.; Phanomchoeng, G. Digital Twin for a Collaborative Painting Robot. *Sensors* **2023**, *23*, 17. [CrossRef]
20. Tao, F.; Zhang, H.; Liu, A.; Nee, A.Y.C. Digital Twin in Industry: State-of-the-Art. *IEEE Trans. Ind. Inform.* **2019**, *15*, 2405–2415. [CrossRef]
21. Schleich, B.; Anwer, N.; Mathieu, L.; Wartzack, S. Shaping the digital twin for design and production engineering. *CIRP Ann. Manuf. Technol.* **2017**, *66*, 141–144. [CrossRef]
22. Liu, Q.; Leng, J.; Yan, D.; Zhang, D.; Wei, L.; Yu, A.; Zhao, R.; Zhang, H.; Chen, X. Digital twin-based designing of the configuration, motion, control, and optimization model of a flow-type smart manufacturing system. *J. Manuf. Syst.* **2021**, *58*, 52–64. [CrossRef]
23. Caputo, F.; Greco, A.; Fera, M.; Macchiaroli, R. Digital twins to enhance the integration of ergonomics in the workplace design. *Int. J. Ind. Ergon.* **2019**, *71*, 20–31. [CrossRef]
24. Zhao, P.; Liu, J.; Jing, X.; Tang, M.; Sheng, S.; Zhou, H.; Liu, X. The Modeling and Using Strategy for the Digital Twin in Process Planning. *IEEE Access* **2020**, *8*, 41229–41245. [CrossRef]
25. Lu, Y.; Min, Q.; Liu, Z.; Wang, Y. An IoT-enabled simulation approach for process planning and analysis: A case from engine re-manufacturing industry. *Int. J. Comput. Integr. Manuf.* **2019**, *32*, 413–429. [CrossRef]

26. Deebak, B.D.; Al-Turjman, F. Digital-twin assisted: Fault diagnosis using deep transfer learning for machining tool condition. *Int. J. Intell. Syst.* **2022**, *37*, 10289–10316. [CrossRef]
27. Hosamo, H.H.; Nielsen, H.K.; Kraniotis, D.; Svennevig, P.R.; Svidt, K. Digital Twin framework for automated fault source detection and prediction for comfort performance evaluation of existing non-residential Norwegian buildings. *Energy Build.* **2023**, *281*, 112732. [CrossRef]
28. Li, C.; MahaDeVan, S.; Ling, Y.; Choze, S.; Wang, L. Dynamic Bayesian network for aircraft wing health monitoring digital twin. *AIAA J.* **2017**, *55*, 930–941. [CrossRef]
29. Rodriguez-Guerra, J.; Calleja, C.; Macarulla, A.M.; Pujana, A.; Elorza, I.; Ramos, M. On Fault-Tolerant Control Systems: A Novel Reconfigurable and Adaptive Solution for Industrial Machines. *IEEE Access* **2020**, *8*, 39322–39335. [CrossRef]
30. Asavasirikulkij, C.; Mathong, C.; Chancharoen, R.; Asdornwised, W. A Study of Digital Twin and Its Communication Protocol in Factory Automation Cell. In Proceedings of the International Conference on Emerging Technologies for Communications, Online, 1–3 December 2021.
31. Thelen, A.; Zhang, X.; Fink, O.; Lu, Y.; Ghosh, S.; Youn, B.D.; Todd, M.D.; Mahadevan, S.; Hu, C.; Hu, Z. A comprehensive review of digital twin—Part 1: Modeling and twinning enabling technologies. *Struct. Multidiscip. Optim.* **2022**, *65*, 354. [CrossRef]
32. Jones, D.; Snider, C.; Nassehi, A.; Yon, J.; Hicks, B. Characterising the Digital Twin: A systematic literature review. *CIRP J. Manuf. Sci. Technol.* **2020**, *29*, 36–52. [CrossRef]
33. McClellan, A.; Lorenzetti, J.; Pavone, M.; Farhat, C. A physics-based digital twin for model predictive control of autonomous unmanned aerial vehicle landing. *Philos. Trans. R. Soc. A Math. Phys. Eng. Sci.* **2022**, *380*, 20210204. [CrossRef] [PubMed]
34. Cavone, G.; Bozza, A.; Carli, R.; Dotoli, M. MPC-Based Process Control of Deep Drawing: An Industry 4.0 Case Study in Automotive. *IEEE Trans. Autom. Sci. Eng.* **2022**, *19*, 1586–1598. [CrossRef]
35. Zhou, J.; Zhang, S.; Gu, M. Revisiting digital twins: Origins, fundamentals, and practices. *Front. Eng. Manag.* **2022**, *9*, 668–676. [CrossRef]
36. Lu, Q.; Xie, X.; Parlikad, A.K.; Schooling, J.M. Digital twin-enabled anomaly detection for built asset monitoring in operation and maintenance. *Autom. Constr.* **2020**, *118*, 103277. [CrossRef]
37. Huang, H.; Yang, L.; Wang, Y.; Xu, X.; Lu, Y. Digital Twin-driven online anomaly detection for an automation system based on edge intelligence. *J. Manuf. Syst.* **2021**, *59*, 138–150. [CrossRef]
38. Xu, J.; Guo, T. Application and research on digital twin in electronic cam servo motion control system. *Int. J. Adv. Manuf. Technol.* **2021**, *112*, 1145–1158. [CrossRef]
39. Gao, L.; Zhao, L.J.; Yang, G.S.; Ma, C.J. A digital twin-driven trajectory tracking control method of a lower-limb exoskeleton. *Control Eng. Pract.* **2022**, *127*, 105271. [CrossRef]
40. Haber, R.; Strzelczak, S.; Miljkovic, Z.; Castano, F.; Fumagalli, L.; Petrovic, M. Digital twin-based Optimization on the basis of Grey Wolf Method. A Case Study on Motion Control Systems. In Proceedings of the 2020 IEEE Conference on Industrial Cyberphysical Systems, Tampere, Finland, 10–12 June 2020; pp. 469–474.
41. Wenna, W.; Weili, D.; Changchun, H.; Heng, Z.; Feng, H.; Yao, Y. A digital twin for 3D path planning of large-span curved-arm gantry robot. *Robot. Comput. Integr. Manuf.* **2022**, *76*, 102330. [CrossRef]
42. Kaczmarek, W.; Lotys, B.; Borys, S.; Laskowski, D.; Lubkowski, P. Controlling an industrial robot using a graphic tablet in offline and online mode. *Sensors* **2021**, *21*, 2439. [CrossRef] [PubMed]
43. Shi, H.; Li, J.; Guo, L.; Mei, X. Control Performance Evaluation of Serial Urology Manipulator by Virtual Prototyping. *Chin. J. Mech. Eng.* **2021**, *34*, 25. [CrossRef]
44. González Ojeda, I.D.J.; Bengoa, P.; Ibarguren, A.; Antolín-Urbaneja, J.C.; Angelakis, P.; Gkournelos, C.; Lotsaris, K.; Makris, S.; Martínez De Lahidalga, S. Robot Coordination: Aeronautic Use Cases Handling Large Parts. *Designs* **2022**, *6*, 116. [CrossRef]
45. Martinez, M.A.G.; Sanchez, I.Y.; Chavez, F.M. Development of automated virtual CNC router for application in a remote mechatronics laboratory. In Proceedings of the International Conference on Electrical, Computer, Communications and Mechatronics Engineering, Mauritius, Mauritius, 7–8 October 2021.
46. Chandola, V.; Banerjee, A.; Kumar, V. Anomaly detection: A survey. *ACM Comput. Surv.* **2009**, *41*, 15. [CrossRef]
47. Dai, W.; Vyatkin, V.; Christensen, J.H.; Dubinin, V.N. Bridging Service-Oriented Architecture and IEC 61499 for Flexibility and Interoperability. *IEEE Trans. Ind. Inform.* **2015**, *11*, 771–781. [CrossRef]
48. Li, P.; Jia, X.; Feng, J.; Davari, H.; Qiao, G.; Hwang, Y.; Lee, J. Prognosability study of ball screw degradation using systematic methodology. *Mech. Syst. Signal Process.* **2018**, *109*, 45–57. [CrossRef]
49. Karnopp, D.C.; Margolis, D.L.; Rosenberg, R.C. *System Dynamics: Modeling, Simulation, and Control of Mechatronic Systems*, 5th ed.; Wiley: Hoboken, NJ, USA, 2012.
50. Farrés, A.; Laskar, J.; Blanes, S.; Casas, F.; Makazaga, J.; Murua, A. High precision symplectic integrators for the Solar System. *Celest. Mech. Dyn. Astron.* **2013**, *116*, 141–174. [CrossRef]
51. Borutzky, W. *Bond Graph Methodology: Development and Analysis of Multidisciplinary Dynamic System Models*; Springer: Berlin/Heidelberg, Germany, 2010; pp. 1–662.
52. Dorf, R.C.; Bishop, R.H. *Modern Control Systems*, 10th ed.; Pearson Prentice Hall: London, UK, 2005.
53. Thoma, J.U.; Perelson, A.S. Introduction to Bond Graphs and Their Applications. *IEEE Trans. Syst. Man Cybern.* **1976**, *SMC-6*, 797–798. [CrossRef]
54. Das, S. Modeling and Simulation of Mechatronic Systems using Simscape. *Synth. Lect. Mech. Eng.* **2020**, *5*, 1–171. [CrossRef]

55. Wei, Y.; Hu, T.; Yue, P.; Luo, W.; Ma, S. Study on the construction theory of digital twin mechanism model for mechatronics equipment. *Int. J. Adv. Manuf. Technol.* **2022**. [CrossRef]
56. Chaber, P.; Domański, P.D.; Dąbrowski, D.; Ławryńczuk, M.; Nebeluk, R.; Plamowski, S.; Zarzycki, K. Digital Twins in the Practice of High-Energy Physics Experiments: A Gas System for the Multipurpose Detector. *Sensors* **2022**, *22*, 678. [CrossRef]
57. Das, S. *Modeling for Hybrid and Electric Vehicles Using Simscape*; Springer: Berlin/Heidelberg, Germany, 2021.
58. Zeb, K.; Ali, S.M.; Khan, B.; Mehmood, C.A.; Tareen, N.; Din, W.; Farid, U.; Haider, A. A survey on waste heat recovery: Electric power generation and potential prospects within Pakistan. *Renew. Sustain. Energy Rev.* **2017**, *75*, 1142–1155. [CrossRef]
59. Cipek, M.; Pavković, D.; Petrić, J. A control-oriented simulation model of a power-split hybrid electric vehicle. *Appl. Energy* **2013**, *101*, 121–133. [CrossRef]
60. Emara, M.B.; Youssef, A.W.; Mashaly, M.; Kiefer, J.; Shihata, L.A.; Azab, E. Digital Twinning for Closed-Loop Control of a Three-Wheeled Omnidirectional Mobile Robot. *Procedia CIRP* **2022**, *107*, 1245–1250. [CrossRef]
61. Vahid-Araghi, O.; Golnaraghi, F. *Friction-Induced Vibration in Lead Screw Drives*; Springer: Berlin/Heidelberg, Germany, 2011; pp. 1–214.
62. Liu, X.; Li, Y.; Cheng, Y.; Cai, Y. Sparse identification for ball-screw drives considering position-dependent dynamics and nonlinear friction. *Robot. Comput. Integr. Manuf.* **2023**, *81*, 102486. [CrossRef]
63. Xie, Z.; Wang, Y.; Ni, J.; Yin, J.; Yu, D.; Liu, J. Friction torque analysis of planetary roller screw based on the creepage theory. *Tribol. Int.* **2023**, *178*, 108059. [CrossRef]
64. Zhang, Y.; Cui, P.; Yang, J.; Wang, Z.; Pu, W. Dynamic grease lubrication and friction behavior of ball screw mechanism in high-frequency reciprocating motion. *Tribol. Int.* **2023**, *178*, 108068. [CrossRef]

Disclaimer/Publisher's Note: The statements, opinions and data contained in all publications are solely those of the individual author(s) and contributor(s) and not of MDPI and/or the editor(s). MDPI and/or the editor(s) disclaim responsibility for any injury to people or property resulting from any ideas, methods, instructions or products referred to in the content.

actuators

MDPI

Article

Robust Liquid Level Control of Quadruple Tank System: A Nonlinear Model-Free Approach

Zahraa Sabah Hashim [1], Halah I. Khani [1], Ahmad Taher Azar [2,3,*], Zafar Iqbal Khan [2], Drai Ahmed Smait [4], Abdulkareem Abdulwahab [5], Ali Mahdi Zalzala [6], Anwar Ja'afar Mohamad Jawad [7], Saim Ahmed [2], Ibraheem Kasim Ibraheem [8], Aws Abdulsalam Najm [9], Suliman Mohamed Fati [2], Mohamed Tounsi [2] and Ahmed Redha Mahlous [2]

1 Medical Instrumentation Techniques Engineering Department, College of Engineering and Information Technology, AlShaab University, Baghdad 10001, Iraq
2 College of Computer and Information Sciences, Prince Sultan University, Riyadh 11586, Saudi Arabia
3 Faculty of Computers and Artificial Intelligence, Benha University, Benha 13518, Egypt
4 The University of Mashreq, Baghdad 10001, Iraq
5 Air Conditioning and Refrigeration Techniques Engineering Department, Al-Mustaqbal University College, Babylon 10001, Iraq
6 Department of Electronics and Communication, College of Engineering, Uruk University, Baghdad 10001, Iraq
7 Department of Computer Techniques Engineering, Al-Rafidain University College, Baghdad 10001, Iraq
8 Department of Computer Techniques engineering, Dijlah University College, Baghdad 10001, Iraq; ibraheemki@coeng.uobaghdad.edu.iq
9 Department of Electrical Engineering, College of Engineering, University of Baghdad, Baghdad 10001, Iraq
* Correspondence: aazar@psu.edu.sa or ahmad.azar@fci.bu.edu.eg or ahmad_t_azar@ieee.org

Abstract: In this paper, two new versions of modified active disturbance rejection control (MADRC) are proposed to stabilize a nonlinear quadruple tank system and control the water levels of the lower two tanks in the presence of exogenous disturbances, parameter uncertainties, and parallel varying input set-points. The first proposed scheme is configured from the combination of a modified tracking differentiator (TD), modified super twisting sliding mode (STC-SM), and modified nonlinear extended state observer (NLESO), while the second proposed scheme is obtained by aggregating another modified TD, a modified nonlinear state error feedback (MNLSEF), and a *fal*-function-based ESO. The MADRC schemes with a nonlinear quadruple tank system are investigated by running simulations in the MATLAB/SIMULINK environment and several comparison experiments are conducted to validate the effectiveness of the proposed control schemes. Furthermore, the genetic algorithm (GA) is used as a tuning algorithm to parametrize the proposed MADRC schemes with the integral time absolute error (ITAE), integral square of the control signal (ISU), and integral absolute of the control signal (IAU) as an output performance index (OPI). Finally, the simulation results show the robustness of the proposed schemes with a noticeable reduction in the OPI.

Keywords: four-tank system; modified active disturbance rejection control (MADRC); water level control

Citation: Hashim, Z.S.; Khani, H.I.; Azar, A.T.; Khan, Z.I.; Smait, D.A.; Abdulwahab, A.; Zalzala, A.M.; Jawad, A.J.M.; Ahmed, S.; Ibraheem, I.K.; et al. Robust Liquid Level Control of Quadruple Tank System: A Nonlinear Model-Free Approach. *Actuators* **2023**, *12*, 119. https://doi.org/10.3390/act12030119

Academic Editors: Oscar Barambones, Jose Antonio Cortajarena and Patxi Alkorta

Received: 4 February 2023
Revised: 4 March 2023
Accepted: 9 March 2023
Published: 11 March 2023

Copyright: © 2023 by the authors. Licensee MDPI, Basel, Switzerland. This article is an open access article distributed under the terms and conditions of the Creative Commons Attribution (CC BY) license (https://creativecommons.org/licenses/by/4.0/).

1. Introduction

Industrial processes are physical systems that comprise a group of operations to complete a specific requirement. Interaction during the industrial process is essential to most industrial processes; this interaction may cause multiple variables which increase the complexity of nonlinear systems. One of these industrial processes is the chemical industry, which has become significant to other industries in the last century such as the transport and pharmaceutical industries, in turn contributing to the development of the economy [1]. The four-tank system is an example of the industrial chemical process proposed by [1] at the end of 1995. It is considered one of the multivariable systems with strong nonlinearity and is used as a laboratory process for understanding the control concept for the multivariable control system [1,2].

At present, several control strategies have been proposed to study and control the performance of the four-tank system, from conventional simple methods to more accurate and complex methods. The author of [3] presented four controllers: the linear quadratic regulator (LQR), linear quadratic Gaussian regulator (LQGR), H_2 controller, and H_∞ controller. Simulation results showed the effectiveness of the LQR in providing a good percentage with regard to settling time. However, it showed a noticeable overshoot in the output response. Recently, some researchers considered the problem of disturbance and delay in the four-tank system and studied the performance of the system under these conditions. Authors in [4] developed a decentralized nonlinear robust model predictive control (MPC). A comparison between the decentralized MPC, a centralized MPC, and a cascaded PI controller was conducted. Moreover, the author of [5] proposed a second-order sliding mode controller, which is a twisting algorithm (TA) based controller. A comparison between the TA and a conventional sliding mode controller (SMC) was undertaken, and the simulation results showed that the TA performed better than the SMC in chattering reduction and disturbance rejection. In addition, the author of [6] designed and implemented two controllers: the adaptive pole placement controller (APPC) and robust adaptive sliding mode controller (ASMC). A comparison between the APPC, ASMC, and the conventional PID was conducted under different conditions such as with reference tracking, exogenous disturbance applied to the four-tank system, and parameter uncertainties. The results showed that the ASMC has better transient and disturbance attenuation and a faster response than both APPC and PID. Furthermore, the author of [7] presented a developed version of the four-tank system with SMC-based feedback linearization to stabilize the system within a specific range. The time-delayed four-tank system has been controlled using various techniques, such as SMC, H_∞ observer-based robust control, fuzzy control, and neural control [8–12], while some authors have used the disturbance rejection technique as in [13], wherein the author presented the design of a nonlinear disturbance observer-based port-controlled Hamiltonian (PCH) with a basic feedback controller. Simulation results showed that the proposed method had a better response and was more robust to the disturbance than the terminal sliding mode control. A nonlinear disturbance observer has been introduced to estimate disturbance along with a novel input/output feedback linearization controller. In [14], a comparison between this proposed method, PID, and a disturbance observer-based sliding mode controller (DOBSMC) was conducted and showed that the proposed method improves the robustness of the system against disturbance and has superior performance compared with both PID and DOBSMC. The authors in [15] studied a sliding mode observer (SMO) to estimate the valve ratios and higher-order sliding mode controller (HOSM) to ensure accurate performance and to attenuate chattering. The simulation results of the sliding mode observer based on the higher-order sliding mode controller showed that the proposed method was stable and accurately estimated the unknown parameter. Moreover, a comparison between the super twisting controller (STA) and the conventional sliding mode controller has been conducted and showed the robustness and the smoothing feature of the STA. In addition, the authors in [16] proposed a new linear active disturbance rejection control (LADRC) with a nonlinear function. A comparison between the PID, LADRC, and ADRC was conducted. The simulation results showed that the proposed LADRC provides good steady-state performance, fast-tracking, and eliminates disturbance more accurately than PID and conventional LADRC. Finally, in [17], the authors proposed two disturbance rejection control laws which were designed and tuned using Embedded Model Control to solve two problems, the first being the regulation of the water levels of the lower tanks, and the second being the regulation of the water levels of the four tanks. Although all the above studies proposed excellent and accurate controllers for the four-tank system, there are still some drawbacks in their work. Some of the above studies used a linearized model of the four-tank system [1,2]. As a result, these controllers were incapable of following the nonlinear dynamics of the system, especially in practical implementation. Moreover, the controllers could not handle the nonlinearity or cancel the effect of the applied disturbance in a sufficient way. Thus, the main goal of this research is to design robust control laws with the nonlinear four-tank system and in consideration of the problems of disturbance, uncertainty, and reference tracking.

Motivated by the aforementioned studies, two schemes of modified active disturbance rejection control (MADRC) are proposed with the nonlinear model of the four-tank system. The modification part of the MADRC is presented as follows:

i. The proposed tracking differentiator is used in the control unit to provide the error signal and its derivative.
ii. The proposed super twisting sliding mode controller (STC-SM), nonlinear proportional derivative (NLPD), and modified nonlinear state error feedback (MNLSEF) are used as nonlinear state error feedback (NLSEF) instead of the conventional NLSEF proposed by [18].
iii. The modified nonlinear extended state observer (MNLESO) and *fal* function ESO are used instead of the linear extended state observer (LESO) [19].

The advantages of our proposal are that it may overcome problems presented previously such as nonlinearity, strong interacting, disturbance, uncertainty of the parameters and that it is able to track any applied reference.

To the best of our knowledge, a super twisting controller with improved active disturbance rejection control has not yet been proposed in the literature. This model solves the problem of the four-tank system in a significant and accurate way, which is our incentive for continuing this research endeavor.

The rest of this paper is organized as follows: Section 2 presents the problem statement. The modeling of the four-tank system is presented in Section 3, and the design of the modified ADRC is introduced in Section 4. The convergence of the proposed control schemes is then demonstrated in Section 5. Section 6 demonstrates the simulation results and provides a discussion. Finally, the conclusions of this paper are given in Section 7.

2. Problem Statement

Suppose the nonlinear model of the four-tank system under the presence of the disturbance can be written as

$$\begin{cases} \dot{x} = f(x_1, \cdots, x_4) + g(x)u + d(t) \\ y = x \end{cases} \tag{1}$$

where $\vec{x} \in \mathbb{R}^4$, $\vec{x} = \begin{bmatrix} x_1 \\ x_2 \\ x_3 \\ x_4 \end{bmatrix}$, $\vec{x} = [x_1\ x_2\ x_3\ x_4]^T$ is the water level in the first, second, third, and fourth tank, respectively, $\vec{y} \in \mathbb{R}^2, y = [x_1, x_2]$ are the measured output, and $\vec{u} \in \mathbb{R}^2, \vec{u} = [u_1, u_2]$ are the control input required to be designed to stabilize and track the desired setpoint value of the water level of the two lower tanks and to eliminate the effect of the unknown exogenous disturbance $d(t)$ and system parameter uncertainty applied to the four-tank system.

3. Mathematical Modeling of Nonlinear Four-Tank System

As can be seen from Figure 1, the four-tank system consists of two water pumps, two two-way valves, a source tank, and four tanks. The water pump$_1$ (M_1) draws water from the source tank and distributes it to tank$_1$ and tank$_4$ via valve$_1$. Similarly, the water pump$_2$ (M_2) draws water from the source tank and distributes it to tank$_2$ and tank$_3$ via valve$_2$. It is important to note the amount of water delivered to the tanks depends on the valve constant (i.e., γ_1 and γ_2). The nonlinear mathematical model of the four-tank system is given in [1]:

$$\begin{cases} \dot{h}_1 = -\frac{a_1}{A_1}\sqrt{2gh_1} + \frac{a_3}{A_1}\sqrt{2gh_3} + \frac{\gamma_1 k_{FT_1}}{A_1}(u_1 + d_1) \\ \dot{h}_2 = -\frac{a_2}{A_2}\sqrt{2gh_2} + \frac{a_4}{A_2}\sqrt{2gh_4} + \frac{\gamma_2 k_{FT_2}}{A_2}(u_2 + d_2) \\ \dot{h}_3 = -\frac{a_3}{A_3}\sqrt{2gh_3} + \frac{(1-\gamma_2)k_{FT_2}}{A_3}u_2 \\ \dot{h}_4 = -\frac{a_4}{A_4}\sqrt{2gh_4} + \frac{(1-\gamma_1)k_{FT_1}}{A_4}u_1 \\ y_1 = k_c h_1 \\ y_2 = k_c h_2 \end{cases} \quad (2)$$

where A_i is the cross-sectional area of tank$_i$; a_i is the cross-sectional area of the outlet hole; h_i is the water level in tank$_i$; $i = \{1, \cdots, 4\}$; u_1 and u_2 are the voltages applied to M_1 and M_2, respectively; g is the acceleration of gravity; K_c is the calibrated constant; k_{FT_1} and k_{FT_2} are pump proportionality constants; and d_1 and d_2 are the causes of the exogenous disturbances by the flow rate.

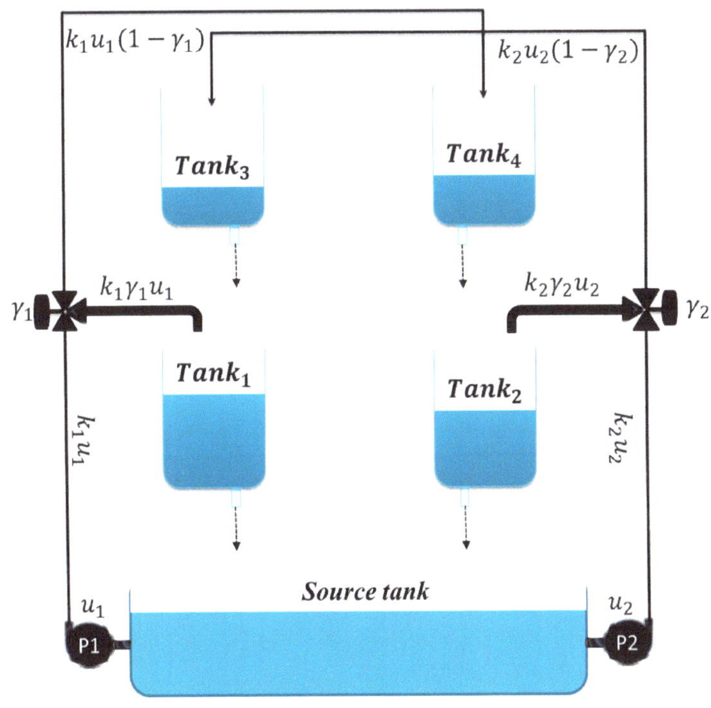

Figure 1. Four-tank system schematic diagram.

4. The proposed Modified Active Disturbance Rejection Control (MADRC)

In general, active disturbance rejection control (ADRC) is considered a disturbance elimination technique and one of the most reliable, robust, and accurate techniques in the field of disturbance/uncertainty attenuation [20,21]. At present, the ADRC that was first proposed by [18] has been widely utilized in different fields, with many researchers improving the conventional ADRC and using it in various applications [22,23]. The design of an ADRC depends on its relative degree. In this section, the design of the proposed ADRC, which consists of two options, a nonlinear controller and a tracking differentiator; and the proposed nonlinear ESO, is introduced and examined. The two schemes of the modified ADRC are introduced as follows:

4.1. The First MADRC Scheme

The first scheme of the modified active disturbance rejection control (MADRC) consists of the proposed super twisting sliding mode controller (STC-SM) and a nonlinear proportional derivative controller (NLPD) as a nonlinear state error feedback (NLSEF); a tracking differentiator; and a modified nonlinear extended state observer (MNLESO). The three main parts of the first MADRC scheme can be expressed in sequence as follows:

(a) The proposed tracking differentiator (TD)

The tracking differentiator is the part of the ADRC used to generate the reference signal and the reference signal derivative, which must offer a tuned and efficient response. The proposed TD is used to provide the smooth error signal and its derivative. The proposed TD for relative degree one ($\rho = 1, \rho \leq n$) can be expressed as follows:

$$\left\{ \begin{array}{ll} \left\{ \begin{array}{l} \dot{\tilde{e}}_{i1}(t) = \tilde{e}_{i2}(t) \\ \dot{\tilde{e}}_{i2}(t) = -a_1 R^2 \left(\dfrac{\tilde{e}_{i1}(t) - \tilde{e}_i(t)}{1 + |\tilde{e}_{i1}(t) - \tilde{e}_i(t)|} \right) - a_2 R \tilde{e}_{i2}(t) \end{array} \right. & (a) \\ \left\{ \begin{array}{l} \dot{\tilde{e}}_{i1}(t) = \tilde{e}_{i2}(t) \\ \dot{\tilde{e}}_{i2}(t) = -a_1 R^2 \left(\dfrac{2}{1 + e^{-(\tilde{e}_{i1}(t) - \tilde{e}_i(t))}} - 1 \right) - a_2 R \tilde{e}_{i2}(t) \end{array} \right. & (b) \end{array} \right. \tag{3}$$

where ρ and n are the relative degree of the system and the system order, respectively. $\tilde{e}_{i1}$ is the tracking error, $\tilde{e}_{i2}$ is its derivative, and $\tilde{e}_i$ is the input to the tracking differentiator. R, a_1 and a_2 are tuning parameters. It is important to note that in the proposed TD, the nonlinear function presented in [24] is utilized instead of the *sign* function that is used in the classical model [18].

(b) The proposed nonlinear controllers

Two different nonlinear controllers are proposed instead of the conventional nonlinear state error feedback (NLSEF) whose mathematical expressions are illustrated in Table 1.

Table 1. Proposed nonlinear controller mathematical expressions.

Nonlinear Controller	Mathematical Expression
STC-SM	$\begin{cases} \varsigma_i = \kappa \tilde{e}_{i1} + \dot{\tilde{e}}_{i1} \\ u_{0i_{STC-SM}} = \kappa_i \lvert\varsigma_i\rvert^{\wp_i} sign(\varsigma_i) + \xi_i \text{tank}\left(\frac{\varsigma_i}{\delta}\right) \end{cases} \quad (4)$
NLPD	$\begin{cases} u_{0i_{NLPD}} = u_{i1} + u_{i2} \\ u_{i1} = \frac{k_{i1}}{1 + \exp(\tilde{e}_{i1}{}^2)} \lvert\tilde{e}_{i1}\rvert^{\alpha_{i1}} sign(\tilde{e}_{i1}) \\ u_{i2} = \frac{k_{i2}}{1 + \exp\left(\dot{\tilde{e}}_{i1}{}^2\right)} \left\lvert\dot{\tilde{e}}_{i1}\right\rvert^{\alpha_{i2}} sign\left(\dot{\tilde{e}}_{i1}\right) \end{cases} \quad (5)$

where $i \in \{1,2\}$ is the number of subsystems of the four-tank system, ς_i is the sliding surface, $(\kappa_i, \wp_i, \xi_i, \delta)$ are the proposed super twisting sliding mode controller (STC-SM) tuning parameters, $\tilde{e}_{i1}$ and $\dot{\tilde{e}}_{i1}$ are the tracking error and its derivative, and $(k_{i1}, k_{i2}, \alpha_{i1}, \alpha_{i2})$ are the proposed NLPD parameters.

(c) The proposed Modified Nonlinear Extended State Observer (MNLESO)

The MNLESO is an improved version of the NLESO proposed by [25]. Two MNLESO schemes are proposed in this paper and can be expressed as presented in Table 2.

Table 2. MNLESO mathematical expressions.

MNLESO Schemes	Mathematical Expression
1st **scheme**	$\begin{cases} \begin{cases} \dot{z}_{i1}(t) = z_{i2}(t) + \beta_{i1}\hat{e}_{i1}(t) \\ \dot{z}_{i2}(t) = \beta_{i2}\hat{e}_{i2}(t) \end{cases} \\ \begin{cases} \hat{e}_{i1}(t) = sign(e_i)\lvert e_i\rvert^{\dashv_i} + e_i \\ \hat{e}_{i2}(t) = sign(e_i)\lvert e_i\rvert^{2\dashv_i - 1} + e_i \end{cases} \end{cases}$ (6)
2nd **scheme**	$\begin{cases} \begin{cases} \dot{z}_{i1}(t) = z_{i2}(t) + \beta_{i1}\hat{e}_{i1}(t) \\ \dot{z}_{i2}(t) = \beta_{i2}\hat{e}_{i2}(t) \end{cases} \\ \begin{cases} \hat{e}_{i1}(t) = sign(e_i)\lvert e_i\rvert^{\dashv_i} + \mathcal{A}_i e_i \\ \hat{e}_{i2}(t) = sign(e_i)\lvert e_i\rvert^{\frac{\dashv_i}{2}} + \mathcal{A}_i e_i \end{cases} \end{cases}$ (7)

where $z_{i1}(t)$ is the estimated state; $z_{i2}(t)$ is the estimated total disturbance; β_{i1}, β_{i2} are the observer gain selected such that the characteristic polynomial $s^2 + \beta_{i1}s + \beta_{i2}$ is Hurwitz [21]; $e_i = h_i - z_{i1}$ is the estimated error; h_i is the output water level; and $\dashv_i$ and $\mathcal{A}_i$ are tuning parameters.

4.2. The Second MADRC Scheme

This subsection presents the second MADRC scheme, which consists of the modified NLSEF with the proposed TD and fal function ESO. The three main parts of the second scheme are as follows:

(a) The proposed Tracking Differentiator TD

The mathematical representation of the proposed TD for the second MADRC scheme is given as

$$\begin{cases} \dot{\tilde{e}}_{i1}(t) = \tilde{e}_{i2}(t) \\ \dot{\tilde{e}}_{i2}(t) = -a_1 R^2 \left(\frac{(\tilde{e}_{i1}(t) - \tilde{e}_i(t)) + 2(\tilde{e}_{i1}(t) - \tilde{e}_i(t))^3}{1 + \lvert (\tilde{e}_{i1}(t) - \tilde{e}_i(t)) + 2(\tilde{e}_{i1}(t) - \tilde{e}_i(t))^3 \rvert} \right) - a_2 R\tilde{e}_{i2}(t) \end{cases} \tag{8}$$

where $\tilde{e}_{i1}(t)$ and $\tilde{e}_{i2}(t)$ are the output tracking error and its derivative, respectively; $\tilde{e}_i(t) = r_i - z_{i1}$ is the input error, r_i is the reference signal; and a_1, a_1, and R are the proposed tracking differentiator tuning parameters.

(b) The modified NLSEF (MNLSEF)

The MNLSEF is the improved version of the classical NLSEF. The MNLSEF is proposed for unit relative degree systems such as the four-tank system and can be presented as follows:

$$\begin{cases} fal(\tilde{e}_{i1}, \alpha_{i1}, \delta_{i1}) = \begin{cases} \tilde{e}_{i1}/\left(\delta_{i1}{}^{1-\alpha_{i1}}\right) & , x \le \delta_{i1} \\ \lvert \tilde{e}_{i1}\rvert^{\alpha_{i1}} sign(\tilde{e}_{i1}) & , x > \delta_{i1} \end{cases} \\ fal(\tilde{e}_{i2}, \alpha_{i2}, \delta_{i2}) = \begin{cases} \tilde{e}_{i2}/\left(\delta_{i2}{}^{1-\alpha_{i2}}\right) & , x \le \delta_{i2} \\ \lvert \tilde{e}_{i2}\rvert^{\alpha_i} sign(\tilde{e}_{i2}) & , x > \delta_{i2} \end{cases} \end{cases} \tag{9}$$

$$\begin{cases} u_{0_{NLSEF_i}} = fal(\tilde{e}_{i1}, \alpha_{i1}, \delta_{i1}) + fal(\tilde{e}_{i2}, \alpha_{i2}, \delta_{i2}) \\ u_{NLSEF_i} = u_{0_{NLSEF_i}} - \frac{z_{i2}}{b_{0i}} \end{cases} \tag{10}$$

where $i = \in \{1, 2\}$ denotes the number of the subsystem of the four-tank system; $\tilde{e}_{i1} = r_i - z_{i1}$ is the reference error and $\tilde{e}_{i2}$ is its derivative; $\delta_{i(1,2)}$ and $\alpha_{i(1,2)}$ are the controller parameters; $u_{0_{NLSEF_i}}$ and u_{NLSEF_i} are the control output of the NLSEF control and input of the four-tank system, respectively; and $fal(\cdot)$ is a continuous nonlinear function.

(c) The proposed fal-function ESO

In this proposed NLESO, the fal function is used as a nonlinear function. The mathematical representation of the proposed fal function ESO is given in Table 3.

Table 3. The fal function ESO mathematical expression.

fal ESO Schemes	Mathematical Expression	
Symmetrical *fal* (S-*fal ADRC*)	$\begin{cases} \begin{cases} \dot{z}_{i1}(t) = z_{i2}(t) + \beta_{i1}\hat{e}_{i1}(t) \\ \dot{z}_{i2}(t) = \beta_{i2}\hat{e}_{i1}(t) \end{cases} \\ \hat{e}_{i1}(t) = fal(e_i, \alpha_{i_{ESO}}, \delta_{i_{ESO}}) \end{cases}$	(11)
Different *fal* (D-*fal ADRC*)	$\begin{cases} \begin{cases} \dot{z}_{i1}(t) = z_{i2}(t) + \beta_{i1}\hat{e}_{i1}(t) \\ \dot{z}_{i2}(t) = \beta_{i2}\hat{e}_{i2}(t) \end{cases} \\ \begin{cases} \hat{e}_{i1}(t) = fal(e_{i1}, \alpha_{i1_{ESO}}, \delta_{i1_{ESO}}) \\ \hat{e}_{i2}(t) = fal(e_{i2}, \alpha_{i2_{ESO}}, \delta_{i2_{ESO}}) \end{cases} \end{cases}$	(12)

where i denotes the number of the subsystem. In the case of the four-tank system, $i \in \{1,2\}$, $e_i(t) = y(t) - z_{1i}(t)$ and $fal(\cdot)$ is a continuous nonlinear function. It is important to note that when the estimation error entered into the *fal* function is the same for all the states, then it may be called the symmetrical *fal* function. However, if a separate *fal* function is used for the estimation error of each state of the system, then it is called the different *fal* function. The final schematic diagram of the MADRC with the four-tank system is shown in Figure 2.

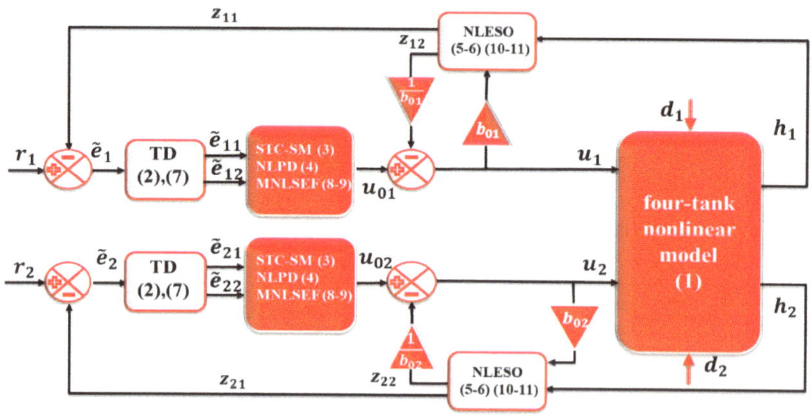

Figure 2. The complete MADRC four-tank system schematic diagram.

5. The Convergence of the Proposed STC-SM

To check the convergence of the proposed STC-SM in finite time, some assumptions and theorem are introduced as follows:

Assumption 1. *According to [26], to ensure the stability of the system, the Lyapunov function derivative must be negative definite or negative semi-definite. Thus, the sliding surface must be* $\varsigma \geq 0$. *Moreover, for simplicity, let* $tanh(\mathbb{x}) = 1$ *for all* $\mathbb{x} \geq 1$.

Theorem 1. *The nonlinear system is asymptotically stable if* $\kappa > 0$ *and* $\xi > 0$ *[22]. The proposed super twisting sliding mode controller (STC-SM) converges in finite time if* $ln|X|$ *is defined for the positive value of* X*, and since* $\wp > 0$ *and* $\varsigma > 0$*, then* X *is always positive [27]. Then, the finite time* $t_{finite} = \frac{ln|X|}{((\wp)\kappa\varsigma^{\wp}(0))}$.

Proof. To check the stability of the proposed STC-SM, assume a second-order system is given as

$$\begin{cases} \dot{x}_1(t) = x_2 \\ \dot{x}_2(t) = f(x_1(t), x_2(t)) + gu + d(t) \\ y(t) = x_1(t) \end{cases} \tag{13}$$

and the reference error is given as

$$e(t) = y_{ref}(t) - y(t) \tag{14}$$

Differentiating $\varsigma = \kappa e + \dot{e}$ yields

$$\dot{\varsigma} = \kappa \dot{e}(t) + \ddot{e}(t) \tag{15}$$

where

$$\begin{cases} \dot{e}(t) = \dot{y}_{ref}(t) - \dot{y}(t);\ \dot{y}(t) = x_2 \\ \ddot{e}(t) = \ddot{y}_{ref}(t) - \ddot{y}(t);\ \ddot{y}(t) = f(x_1(t), x_2(t)) + gu + d(t) \end{cases} \tag{16}$$

Substituting Equation (16) in Equation (15) yields

$$\dot{\varsigma} = \kappa\left(\dot{y}_{ref}(t) - x_2\right) + \left(\ddot{y}_{ref}(t) - f(x_1(t), x_2(t)) - gu - d(t)\right) \tag{17}$$

Rearranging Equation (17) yields

$$\dot{\varsigma} = \Gamma(\dot{y}(t), \ddot{y}(t), t) - gu \tag{18}$$

where $\Gamma(\dot{y}(t), \ddot{y}(t), t) = \kappa(\dot{y}_{ref}(t) - x_2) + \ddot{y}_{ref}(t) - f(x_1(t), x_2(t)) - d(t)$ represents the overall disturbance. Substituting Equation (4) in Equation (18) and assuming $g = 1$ and $\Gamma(\dot{y}(t), \ddot{y}(t), t) = 0$ for simplicity,

$$\dot{\varsigma} = -\kappa|\varsigma|^{\rho} sign(\varsigma) - \xi \tanh\left(\frac{\varsigma}{\delta}\right) \tag{19}$$

Using the Lyapunov stability approach [26], let $V_{STC-SM} = \frac{1}{2}\varsigma^T\varsigma$ and $\dot{V}_{STC-SM} = \varsigma\dot{\varsigma}$

$$\dot{V}_{SM-STC} = -\varsigma\left[\kappa|\varsigma|^{\rho} sign(\varsigma) + \xi \tanh\left(\frac{\varsigma}{\delta}\right)\right]$$

where

$$|\varsigma| = f(x) = \begin{cases} -\varsigma, \ \varsigma < 0 \\ \varsigma, \ \varsigma > 0 \end{cases},$$

$$sign(\varsigma) = \begin{cases} +1, \varsigma > 0 \\ 0, \ \varsigma = 0 \\ -1, \varsigma < 0 \end{cases},$$

$$\tanh\left(\frac{\varsigma}{\delta}\right) = \tanh(\mathbb{x}) = \begin{cases} \dfrac{e^2 - 1}{e^2 + 1}, \ \mathbb{x} = 1 \\ 0, \ \mathbb{x} = 0 \\ \dfrac{1 - e^2}{-1 - e^2}, \ \mathbb{x} = -1 \end{cases}$$

Then, according to Assumption 1, $\dot{V}_{STC-SM}$ will be in the following form:

$$\begin{aligned} \dot{V}_{STC-SM} &= -\kappa \parallel \varsigma \parallel^2 \ \varsigma^{\rho-1} - \xi\varsigma \\ \dot{V}_{STC-SM} &< -\kappa \parallel \varsigma \parallel^2 \varsigma^{\rho-1} - \xi\varsigma \end{aligned}$$

The system is asymptotically stable if $\kappa > 0$ and $\xi > 0$. □

To check the convergence of the proposed STC-SM in finite time, integrating both sides of Equation (19) with respect to time yields

$$\int \frac{d\varsigma}{dt} = -\int (\kappa\, \varsigma(t)^{\rho} + \xi) \tag{20}$$

Rearranging Equation (20) yields

$$\int \frac{-d\varsigma}{(\kappa\, \varsigma^{\rho}(t) + \xi)} = \int dt \Rightarrow \int -d\varsigma(\kappa\, \varsigma^{\rho}(t) + \xi)^{-1} = t + C_1 \tag{21}$$

Simplifying Equation (21) yields

$$\frac{-1}{((\mathcal{p})\kappa\varsigma^{\mathcal{p}-1}(t))}\ln|X| + \frac{1}{((\mathcal{p})\kappa\varsigma^{\mathcal{p}-1}(0))}\ln|X| + \mathcal{C}_2 = t + \mathcal{C}_1 \quad (22)$$

where $\mathcal{C}_1$ and $\mathcal{C}_2$ are the integration constant and X is a variable equal to $(\kappa\,\varsigma^{\mathcal{p}}(t) + \xi)$. For simplicity, let $\mathcal{C}_1 = \mathcal{C}_2 = 0$ and at $t = t_{finite}$, $\varsigma\left(t_{finite}\right) = 0$. According to Theorem 1, the finite time equation can be expressed as

$$t_{finite} = \frac{\ln|X|}{((\mathcal{p})\kappa\varsigma^{\mathcal{p}}(0))} \quad (23)$$

6. Simulation Results and Discussion

This section presents the simulation results and discussion. The four-tank model with the modified ADRC was tested and implemented using MATLAB/SIMULINK. Moreover, all the obtained results of the proposed method were compared with different methods. In this study, the genetic algorithm was utilized as an optimization technique to tune the parameters. The genetic algorithm (GA) is an optimization algorithm that was first proposed by Holland in [28]. There are three steps to generating the next generation from the current generation, the implementation of which is used to select the best generation with the best genes. Mutation is the step that generates a new offspring with a random mutation in its genes. Finally, crossover is the step that generates offspring by exchanging the genes of the parents randomly until crossover is available. In addition, the four-tank parameters used for simulation are listed in Table 4. Furthermore, a summary of all the proposed methods used in this work along with the other methods is given in Table 5. The optimization processes were achieved by means of function (GA) within the MATLAB simulation. Figure 3 shows the MADRC with GA.

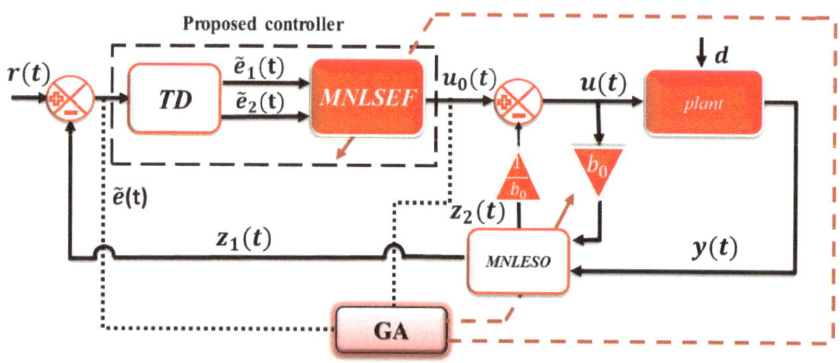

Figure 3. MADRC with GA.

Table 4. Sampled parameters of the four-tank system [1].

Parameters	Description	Value	Unit	Reference
$h_{1_{des}}$	The water level of tank_1	16	cm	Estimated
$h_{2_{des}}$	The water level of tank_2	13	cm	Estimated
h_3	The water level of tank_3	9.5	cm	Estimated
h_4	The water level of tank_4	6	cm	Estimated
a_1	The cross-section area of the outlet hole of tank_1	0.071	cm^2	[1]

Table 4. *Cont.*

Parameters	Description	Value	Unit	Reference
$\mathbf{a_2}$	The cross-section area of the outlet hole of $tank_2$	0.056	cm^2	[1]
$\mathbf{a_3}$	The cross-section area of the outlet hole of $tank_3$	0.071	cm^2	[1]
$\mathbf{a_4}$	The cross-section area of the outlet hole of $tank_4$	0.056	cm^2	[1]
A_1	The cross-section area of $tank_1$	28	cm^2	[1]
A_2	The cross-section area of $tank_2$	32	cm^2	[1]
A_3	The cross-section area of $tank_3$	28	cm^2	[1]
A_4	The cross-section area of $tank_4$	32	cm^2	[1]
γ_1	The ratio of the flow in the $valve_1$	0.7	unitless	[1]
γ_2	The ratio of the flow in the $valve_2$	0.6	unitless	[1]
k_{FT_1}	Pump proportionality constant	3.33	$\frac{cm^3}{volt.s}$	[1]
k_{FT_2}	Pump proportionality constant	3.35	$\frac{cm^3}{volt.s}$	[1]
g	Gravity constant	981	volt/cm	[1]
K_c	The calibrated constant	1	cm/s^2	Estimated
h_{max}	The maximum height	25	cm	Estimated

Table 5. Summary of all the proposed methods used in this work.

Scheme	TD	SEF	ESO
Linear active disturbance rejection control LADRC	-	LPID that can be given as $u_{0_{PID}} = k_p\tilde{e}_i + k_i\int_0^T \tilde{e}_i\, dt + k_d\frac{d\tilde{e}_i}{dt}$ (24)	Linear extended state observer (LESO) [19]
ADRC	-	NLSEF [18] $\begin{cases} u_{0i_{NLSEF}} = fal(\tilde{e}_i, \alpha_{i1}, \delta_{i1}) \\ fal(\tilde{e}_i, \alpha_{i1}, \delta_{i1}) = \begin{cases} \frac{\tilde{e}_i}{\delta_{i1}^{1-\alpha_{i1}}}, & \lvert x\rvert < \delta_{i1} \\ \lvert\tilde{e}_i\rvert^{\alpha_{i1}} sign(\tilde{e}_i), & \lvert x\rvert \geq \delta_{i1} \end{cases} \end{cases}$ (26)	$\begin{cases} \dot{z}_{i1}(t) = z_{i2}(t) + \beta_{i1}(e_i) \\ \dot{z}_{i2}(t) = \beta_{i2}(e_i) \end{cases}$ (25)
Improved active disturbance rejection control (IADRC)	-	Improved nonlinear state error feedback (INLSEF) [29] $\begin{cases} u_{i1} = k_{i11} + \frac{k_{i12}}{1+\exp(\mu_{i1}\tilde{e}_i^2)}\lvert\tilde{e}_i\rvert^{\alpha_{i1}} sign(\tilde{e}_i) \\ u_{INLPID} = u_{i1} \end{cases}$ (27) where $\tilde{e}_i$ is the error and $(k_{i11}, k_{i12}, \mu_{i1}, \alpha_{i1})$ are the controller parameters	Sliding mode extended state observer (SMESO) [30] $\begin{cases} \dot{z}_{i1}(t) = z_{i2}(t) + \beta_{i1}(k(e_i(t))e_i(t)) \\ \dot{z}_{i2}(t) = \beta_{i1}(k(e_i(t))e_i(t) \\ k(e_i(t)) = k_{\alpha_i}\lvert e_i\rvert^{\alpha_i - 1} + k_\beta\lvert e_i\rvert^{\beta_i} \end{cases}$ (28) where $k(e_i(t))$ is a nonlinear function
S *fal*-ADRC	-	NLSEF Equation (26)	Symmetrical *fal* ESO Equation (11)
D *fal*-ADRC	-	NLSEF Equation (26)	Different *fal* ESO Equation (12)
1st MADRC scheme (NLP-ADRC) and (STC-ADRC)	Equation (3a)	Proposed NLPD Equation (5)	MNLESO 1st scheme Equation (7)
	Equation (3b)	Proposed STC-SM Equation (6)	MNLESO 2nd scheme Equation (8)
2nd MADRC scheme (D *fal*-ADRC-TD)	Equation (8)	Proposed MNLSEF Equations (9) and (10)	Different *fal* ESO Equation (12)

The multi-objective performance index (OPI) was used in this work to tune the parameters of the modified ADRC (MADRC) and the methods in Table 4 in order to find the optimal value. It is expressed as follows:

$$\begin{cases} OPI_1 = w_{11} \times \dfrac{ITAE_1}{\mathcal{N}_{11}} + w_{12} \times \dfrac{IAU_1}{\mathcal{N}_{12}} + w_{13} \times \dfrac{ISU_2}{\mathcal{N}_{13}} \\ OPI_2 = w_{21} \times \dfrac{ITAE_2}{\mathcal{N}_{21}} + w_{22} \times \dfrac{IAU_2}{\mathcal{N}_{22}} + w_{32} \times \dfrac{ISU_2}{\mathcal{N}_{23}} \\ \qquad OPI = W_1 \times OPI_1 + W_2 \times OPI_2 \end{cases} \tag{29}$$

The value of W_1 and W_2 were set to 0.5, and the weighted factor of each subsystem was set to $w_{11} = w_{21} = 0.4$, $w_{12} = w_{22} = 0.2$, $w_{13} = w_{23} = 0.4$. The nominal values of the individual objectives that contain the (OPI) were set to $\mathcal{N}_{11} = 1.814362$, $\mathcal{N}_{12} = 4389.20$, $\mathcal{N}_{13} = 305.59$, $\mathcal{N}_{21} = 1.77746$, $\mathcal{N}_{22} = 4332.233$, and $\mathcal{N}_{23} = 285.2937$. After the tuning process was conducted using GA [31,32], the parameter values of the modified ADRC and all the methods mentioned previously in Table 5 were obtained and are given in Tables 6–13.

Table 6. LADRC parameters.

ADRC Parts	Parameter	Value	Parameter	Value
LPID	k_{p_1}	18.6300	k_{p_2}	26.6550
	k_{i_1}	0.0002	k_{i_2}	0.0024
	k_{d_1}	2.5300	k_{d_2}	3.0500
LESO	ω_{01}	43.130000	ω_{02}	15.910000
	b_{01}	0.124875	b_{02}	0.094219

Table 7. ADRC parameters.

ADRC Parts	Parameter	Value	Parameter	Value
NLSEF	α_1	0.7763	α_2	0.4167
	δ_1	0.0140	δ_2	1.8958
LESO	ω_{01}	149.345000	ω_{02}	173.005000
	b_{01}	1.706625	b_{02}	1.287656

Table 8. IADRC parameters.

ADRC Parts	Parameter	Value	Parameter	Value
INLP (INSEF)	k_{111}	6.2650	k_{212}	7.0400
	k_{121}	1.4124	k_{222}	0.0142
	μ_{11}	8.5790	μ_{22}	5.6130
	α_{11}	0.6812	α_{22}	0.6625
SMESO	k_{α_1}	0.3675	k_{α_2}	0.8579
	α_1	0.9733	α_2	0.6265
	k_{β_1}	0.6713	k_{β_2}	0.6812
	β_1	0.2221	β_2	0.7062
	ω_{01}	133.200000	ω_{02}	163.840000
	b_{01}	0.666000	b_{02}	0.502500

Table 9. NLPD-ADRC parameters (1st MADRC scheme).

ADRC Parts	Parameter	Value	Parameter	Value
NLPD	k_{11}	12.676500	k_{12}	24.414000
	α_{11}	0.351000	α_{12}	0.453100
	k_{21}	22.057500	k_{22}	21.553500
	α_{21}	0.931500	α_{22}	0.69130
TD	R	55.380000	a_2	7.842000
	a_1	0.142000	–	–
1st MNLESO scheme	ω_{01}	341.190000	ω_{02}	538.050000
	a_1	0.696800	a_2	0.587100
	b_{01}	2.414250	b_{02}	1.821562

Table 10. STC-ADRC parameters (1st MADRC scheme).

ADRC Parts	Parameter	Value	Parameter	Value
STC-SM	κ_1	0.552000	κ_2	0.587800
	ξ_1	3.405000	ξ_2	3.889500
	p_1	0.704480	p_2	0.695040
	δ	7.631000	–	–
TD	R	188.580000	a_2	3.896000
	a_1	3.052000	–	–
NLESO	ω_{01}	103.000000	ω_{02}	80.300000
	a_1	0.905498	a_2	0.873169
	$\mathcal{A}_1$	0.524300	$\mathcal{A}_2$	0.102500
	b_{01}	3.230100	b_{02}	2.688375

Table 11. Sfal -ADRC parameters.

ADRC Parts	Parameter	Value	Parameter	Value
NLSEF	α_1	0.962100	α_2	0.542400
	δ_1	0.532800	δ_2	0.693800
S-*fal*	$\alpha_{1_{ESO}}$	0.097200	$\alpha_{2_{ESO}}$	0.547200
	$\delta_{1_{ESO}}$	0.765600	$\delta_{2_{ESO}}$	0.369000
LESO	ω_{01}	261.300000	ω_{02}	224.220000
	b_{01}	1.914750	b_{02}	1.444687

Table 12. Dfal -ADRC parameters.

ADRC Parts	Parameter	Value	Parameter	Value
NLSEF	α_1	0.962900	α_2	0.968200
	δ_1	0.910600	δ_2	0.139200
D-*fal*	$\alpha_{11_{ESO}}$	0.151200	$\alpha_{21_{ESO}}$	0.443600
	$\delta_{11_{ESO}}$	0.903800	$\delta_{21_{ESO}}$	0.217900
	$\alpha_{12_{ESO}}$	0.066200	$\alpha_{22_{ESO}}$	0.045300
	$\delta_{12_{ESO}}$	0.485300	$\delta_{22_{ESO}}$	0.024700
LESO	ω_{01}	230.430000	ω_{02}	266.220000
	b_{01}	2.289375	b_{02}	2.167031

Table 13. $Dfal$ -ADRC-TD parameters (2nd MADRC scheme).

ADRC Parts	Parameter	Value	Parameter	Value
MNLSEF	α_{11}	0.802800	α_{21}	0.646400
	δ_{11}	0.277500	δ_{21}	0.090400
	α_{12}	0.170300	α_{22}	0.698500
	δ_{12}	0.457400	δ_{22}	0.153900
TD	R	6.050000	a_2	13.404000
	a_1	0.887000	—	—
D-*fal*	$\alpha_{11_{ESO}}$	0.843900	$\alpha_{21_{ESO}}$	0.724800
	$\delta_{11_{ESO}}$	0.521700	$\delta_{21_{ESO}}$	0.634300
	$\alpha_{12_{ESO}}$	0.231600	$\alpha_{21_{ESO}}$	0.141500
	$\delta_{12_{ESO}}$	0.023300	$\delta_{21_{ESO}}$	0.086500
LESO	ω_{01}	183.570000	ω_{02}	292.230000
	b_{01}	4.620375	b_{02}	3.548906

To check the effectiveness of the proposed method, three tests were applied to the four-tank model as follows:

A. Case study 1. Exogenous disturbance

In this test, a step function was applied as a desired reference for both subsystems. Moreover, to investigate the effectiveness and robustness of the designed ADRC against the applied disturbance, a step function was applied as an exogenous disturbance. The water levels of both subsystems while applying an exogenous disturbance after 40 s of starting the simulation for the first subsystem and after 60 s of starting the simulation for the second subsystem are shown in Figure 4. The output response of the first subsystem is given in Figure 4a. The results show that in applying disturbance for the first subsystem at $t = 40$ s, LADRC, ADRC, and IADRC exhibited an output response with an undershoot which reached nearly 0.1265%, 0.375%, 0.1875%, and 0.125% of the steady-state value, respectively, and lasted about 1.2 s for LADRC, 2.1 s for ADRC, and 0.5 s for IADRC until the output response reached its steady state. For the second subsystem as shown in Figure 4b, at $t = 60$s the output response exhibited an undershoot which reached nearly 0.307%, 0.315%, and 0.153% of its steady-state value for LADRC, ADRC and IADRC, respectively, and lasted about 1.9 s for LADRC, 1.92 s for ADRC, and 0.5 s for IADRC until it reached its steady state. However, the proposed methods rejected the disturbance very quickly. It is observed that the output response of the proposed methods (i.e., NLPD-ADRC, STC-ADRC) is faster, smoother, and without overshooting when compared with the other methods. It took less than approximately 2 s to reach the steady state (desired value), while a longer settling time was clearly observed in the output responses of the other methods.

The output response when using the second MADRC scheme is shown in Figure 5. As can be seen from Figure 5a, the output response reached a steady state with a smooth and fast response. However, it took approximately 1 s and 1.5 s for $Sfal$-ADRC and $Dfal$-ADRC to attenuate the disturbance and return to the steady state, respectively. Furthermore, under the effect of the disturbance, the output showed an undershoot of 0.3125% and 0.25% of the steady state value for $Sfal$-ADRC and $Dfal$-ADRC, respectively. By contrast, the output response of the proposed method (i.e., $Dfal$-ADRC-TD) is smooth and more accurate, representing an improvement in terms of disturbance attenuation compared with $Dfal$-ADRC and $Sfal$-ADRC, which proves the effectiveness of the designed controller and observer. Further, as can be seen from Figure 5b, the output response of the proposed method (i.e., $Dfal$-ADRC-TD) under the disturbance effect shows an improvement and robustness in canceling the disturbance effect in a short time, while the other methods, LADRC, ADRC, and $Sfal$-ADRC, take about 1.9 s, 1.92 s, and 2 s to remove the disturbance effect and return to the steady state, respectively. Moreover, the output responses of the methods under the effect of the disturbance showed an undershoot of 0.307%, 0.315%,

0.305%, 0.153%, and 0.315% of the steady-state value for LADRC, ADRC, IADRC, and Sfal-ADRC, respectively.

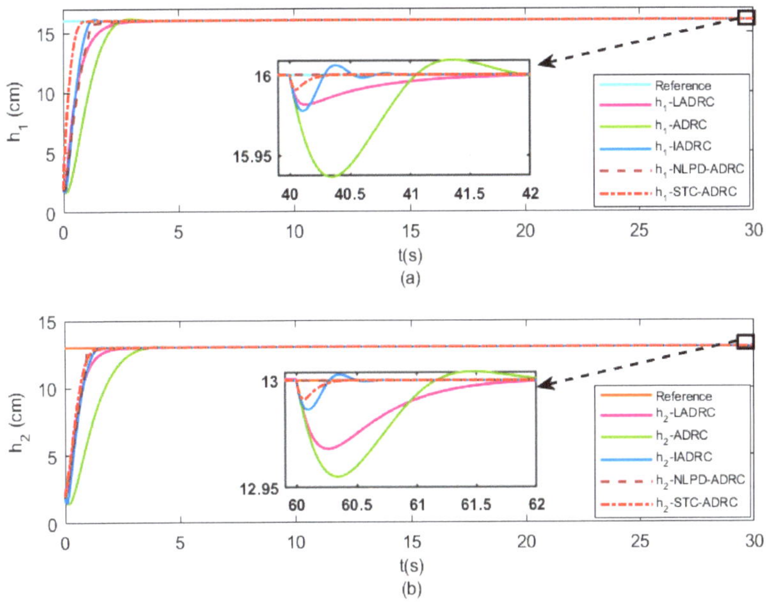

Figure 4. The water level when using the 1st MADRC scheme. (**a**) 1st subsystem. (**b**) 2nd subsystem.

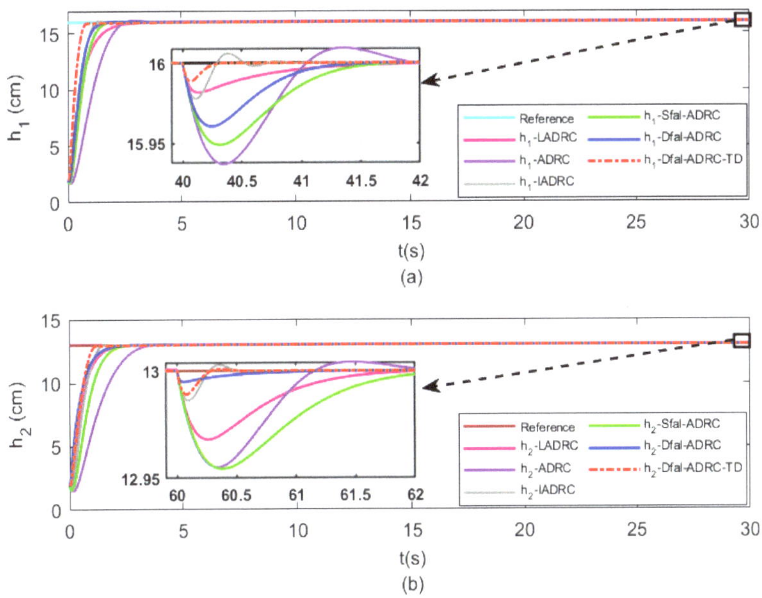

Figure 5. The water level when using the 2nd MADRC scheme. (**a**) 1st subsystem. (**b**) 2nd subsystem.

Figure 6a,b show the control signal of the first subsystem and the second subsystem respectively. It is observed that the first MADRC scheme (i.e., STC-ADRC) is chattering-free with a smooth response; moreover, the NLPD-ADRC shows a reduction in chattering, while the other methods, such as ADRC, show chattering in the control signal. This proves the effectiveness of the proposed method.

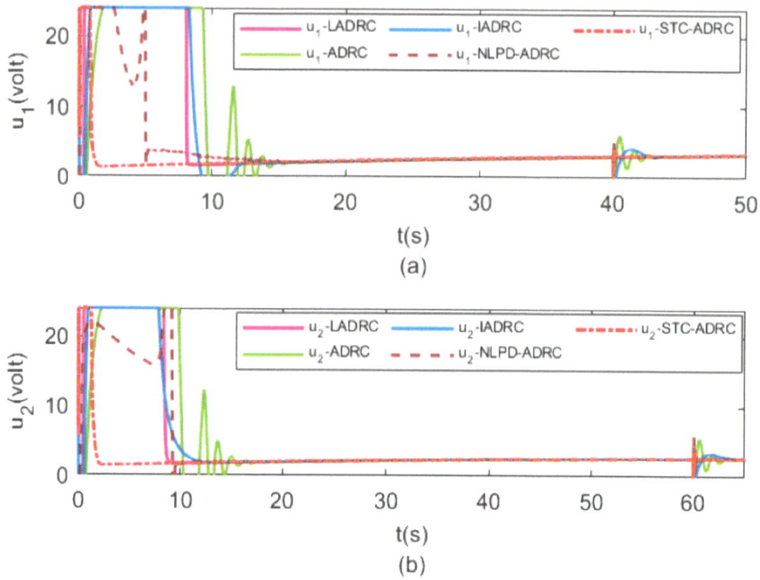

Figure 6. The control signal when using the 1st MADRC scheme. (**a**) 1st subsystem. (**b**) 2nd subsystem.

Figure 7a,b show the control signal of the first subsystem and the second subsystem, respectively. It is observed that the second MADRC scheme (i.e., $Dfal$-ADRC-TD) is chattering-free with a smooth response, while the other methods, such as ADRC, show chattering in the control signal. This proves the effectiveness of the proposed method.

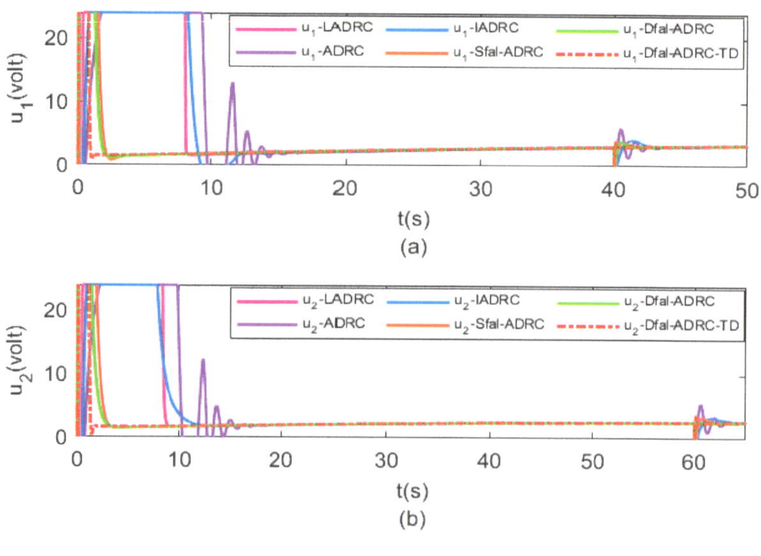

Figure 7. The control signal when using the 2nd MADRC scheme. (**a**) 1st subsystem. (**b**) 2nd subsystem.

B. Case study 2. Parameter uncertainty

In this test, the system's parameter uncertainty is taken into consideration to observe its effect on the nonlinear system. One of the parameters that affects the performance of the four-tank model is uncertainty in the design of the outlet hole of the first tank. Figure 8 show the output response of the first tank under the presence of the parameter uncertainties ($\Delta_{a_1} = +9\%$) after 10 s from starting the simulation. It appears that the proposed methods (i.e., NLPD-ADRC, STC-ADRC) can cope with parameter variation easily, while the other methods give undershoots of 0.1875%, 0.5%, 0.31255% and 0.5% of the steady state, respectively, and it takes LADRC, ADRC, and IADRC approximately 1.95 s, 2.96 s, 1.07 s, and 1.04 s to weaken and reduce the uncertainty effect, respectively. By contrast, the proposed methods attenuate the effect of the parameter variation in less than 1 s.

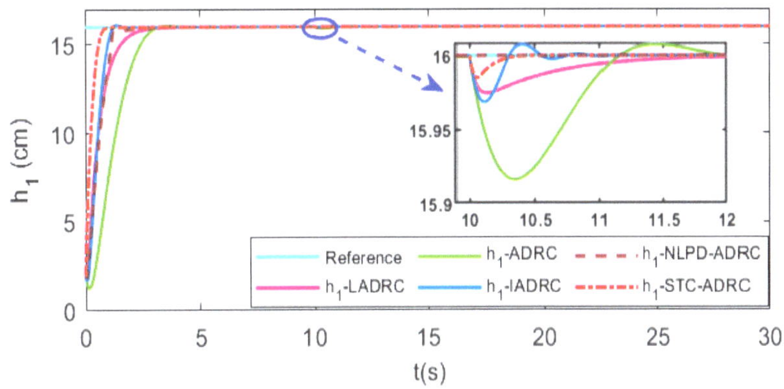

Figure 8. The water level in tank $_1$ with uncertainty in the outlet hole a_1. Using the 2nd MADRC scheme.

C. Case study 3. Reference tracking

In this test, a step function was applied for both subsystems with different amplitude and time to check the robustness and validation of the designed MNLESO and controllers in tracking references applied at different times.

Figure 9a shows the output response of the first subsystem when using NLPD-ADRC and STC-ADRC, while Figure 9b shows the output response of the second subsystem. It is clear for both subsystems that the proposed methods (i.e., NLPD-ADRC and STC-ADRC) exhibit a smooth, fast-tracking, chattering-free, and accurate response without any visible over or undershoot (STC-ADRC), which reflects the effectiveness of the designed ESO and controller in coping with the time-varying reference.

Figure 10a shows the output response of the first subsystem when using Sfal-ADRC, Dfal-ADRC, and Dfal-ADRC-TD, while Figure 10b shows the output response of the second subsystem. It is clear for both subsystems that the proposed method (Dfal-ADRC-TD) exhibits a smooth, fast-tracking, and accurate response without any peaking, which proves the effectiveness of the designed ESO and controller under the time-varying references.

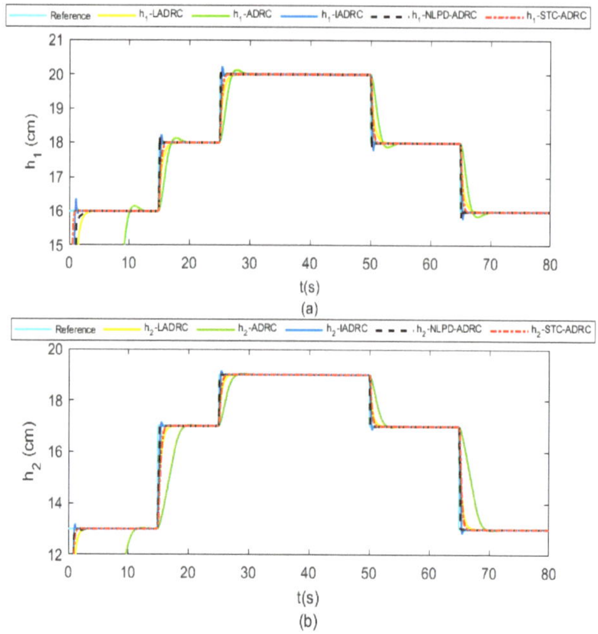

Figure 9. The water level under different time-varying references using the 1st MADRC scheme. (**a**) 1st subsystem. (**b**) 2nd subsystem.

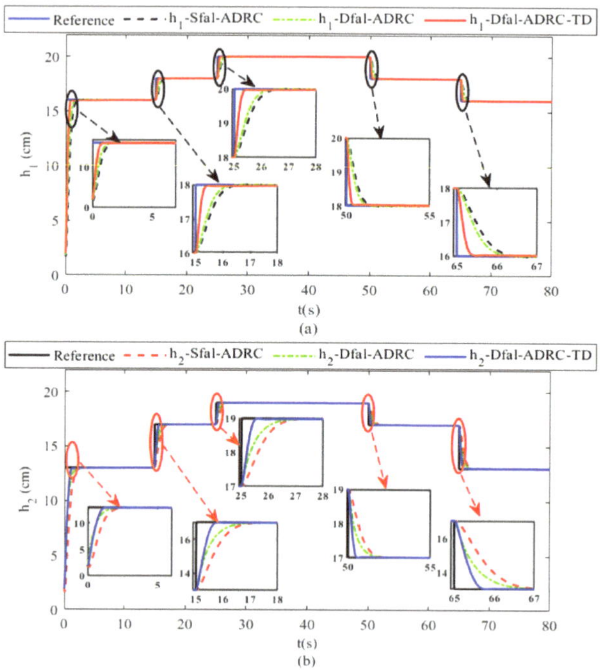

Figure 10. The water level under different time-varying references using the 2nd MADRC scheme. (**a**) 1st subsystem. (**b**) 2nd subsystem.

Table 14 below proves that the STC-ADRC method provides the best result in terms of OPI reduction and steady-state error, while the other methods, such as NLPD-TD, D *fal*-ADRC, and D *fal*-ADRC-TD exhibit noticeable reduction in the OPI.

Table 14. Performance indices.

PI	$ITAE_1$	$ITAE_2$	IAU_1	IAU_2	ISU_1	ISU_2	OPI
LADRC	5.0488	7.5142	13115.098	16124.001	16194.336	20840.561	23.2190
ADRC	10.7318	13.4633	976.4132	1021.5292	54.0156	46.700165	2.48116
IADRC	2.6097	2.6842	2518.480536	2695.5036	678.3466	658.7566	1.578022
$Sfal-$ ADRC	3.8810	5.5178	850.1411	908.2580	75.9051	64.6534	0.993720
$Dfal-$ ADRC	2.5716	2.4849	682.9920	528.7538	59.2170	29.5555	0.675591
$Dfal-$ ADRC $-$ TD	2.2750	2.5949	350.4789	328.2163	21.8594	12.0197	0.546157
NLPD $-$ ADRC	3.7407	2.2867	635.1552	696.2649	30.7188	37.6227	0.893849
STC $-$ ADRC	0.714658	1.8739	289.7189	426.9655	32.3728	22.6835	0.213131

Remark 1. *The intelligent PID (iPID) is a PID controller where the nonlinearity, unknown parts of the plant, and time-varying parameters are considered but do not appear in the modeling (see [33]). The ADRC is a robust control which estimates the total disturbance and rejects it in an online manner (see, for example, Equation (7) and [18,20,21]).*

7. Conclusions

Two MADRC schemes were proposed in this paper with the nonlinear model of the four-tank system and under different conditions such as exogenous disturbance, parameter uncertainty, and reference tracking. The first MADRC scheme was designed using the STC-SM and NLPD with an NLESO and the proposed TD. The simulation results showed the robustness of the first MADRC scheme (i.e., STC-ADRC and NLPD-ADRC) in terms of disturbance and uncertainty reduction and the desired output tracking. Another modified ADRC scheme (i.e., second MADRC scheme) was also proposed with NLSEF-TD as a new controller and a new NLSEO using the fal as a nonlinear function. In conclusion, the simulation results of the proposed method (i.e., $Dfal$-ADRC-TD) exhibited better results in terms of disturbance cancelation and output response performance with minimum OPI compared with both $Sfal$-ADRC and $Dfal$-ADRC. Finally, with respect to future work, we intend to extend the current work by including delays in the four-tank system and a condition wherein the four-tank system operates in the non-minimum phase mode.

Author Contributions: Conceptualization, Z.S.H., A.T.A. and I.K.I.; Methodology, Z.S.H., H.I.K., A.T.A., Z.I.K., D.A.S., A.A., A.M.Z., A.J.M.J., S.A., I.K.I., A.A.N., S.M.F. and M.T.; Software, Z.S.H., H.I.K., D.A.S., A.A., A.M.Z., A.J.M.J., S.A., A.A.N., S.M.F., A.R.M. and M.T.; Validation, H.I.K., Z.I.K., D.A.S., A.A., A.J.M.J., A.A.N., A.R.M. and S.M.F.; Formal analysis, Z.S.H., H.I.K., A.T.A., Z.I.K., D.A.S., A.A., A.M.Z., A.J.M.J., S.A., I.K.I., A.A.N., S.M.F., A.R.M. and M.T.; Investigation, Z.I.K., A.M.Z., S.A., I.K.I., A.R.M. and M.T.; Resources, Z.S.H., H.I.K., Z.I.K., D.A.S., A.A., A.M.Z., A.J.M.J., S.A., A.A.N., S.M.F., A.R.M. and M.T.; Data curation, Z.I.K., D.A.S., A.A., A.M.Z., A.J.M.J., S.A., A.A.N. and S.M.F.; Writing—original draft, Z.S.H., H.I.K., A.T.A. and I.K.I.; Writing—review & editing, Z.S.H., H.I.K., A.T.A., Z.I.K., D.A.S., A.A., A.M.Z., A.J.M.J., S.A., I.K.I., A.A.N., S.M.F., A.R.M. and M.T.; Visualization, A.T.A., A.R.M. and M.T.; Supervision, A.T.A. and I.K.I. All authors have read and agreed to the published version of the manuscript.

Funding: This research was funded by Prince Sultan University, Riyadh, Saudi Arabia.

Institutional Review Board Statement: Not applicable.

Informed Consent Statement: Not applicable.

Data Availability Statement: Not applicable.

Acknowledgments: The authors would like to thank Prince Sultan University, Riyadh, Saudi Arabia for funding the article processing charges (APCs) of this publication. Special acknowledgment is given to the Automated Systems & Soft Computing Lab (ASSCL), Prince Sultan University, Riyadh, Saudi Arabia.

Conflicts of Interest: The authors declare no conflict of interest.

References

1. Johansson, K.H. The quadruple-tank process: A multivariable laboratory process with an adjustable zero. *IEEE Trans. Control. Syst. Technol.* **2000**, *8*, 456–465. [CrossRef]
2. Johansson, K.H.; Nunes, J.L.R. A multivariable laboratory process with an adjustable zero. In Proceedings of the 1998 American Control Conference. ACC (IEEE Cat. No. 98CH36207), Philadelphia, PA, USA, 26 June 1998; Volume 4, pp. 2045–2049. [CrossRef]
3. Altabey, W.A. Model Optimal Control of the Four Tank System. *Int. J. Syst. Sci. Appl. Math.* **2016**, *1*, 30–41.
4. Sorcia-Vázquez, F.; García-Beltrán, C.; Valencia-Palomo, G.; Brizuela-Mendoza, J.; Rumbo-Morales, J. Decentralized robust tube-based model predictive control: Application to a four-tank-system. *Rev. Mex. Ing. Química* **2020**, *19*, 1135–1151. [CrossRef]
5. Chaudhari, V.; Tamhane, B.; Kurode, S. Robust Liquid Level Control of Quadruple Tank System-Second Order Sliding Mode. *IFAC Pap. OnLine* **2020**, *53*, 7–12. [CrossRef]
6. Osman, A.; Kara, T.; Arıcı, M. Robust Adaptive Control of a Quadruple Tank Process with Sliding Mode and Pole Placement Control Strategies. *IETE J. Res.* **2021**. [CrossRef]
7. Biswas, P.P.; Srivastava, R.; Ray, S.; Samanta, A.N. Sliding mode control of quadruple tank process. *Mechatronics* **2009**, *19*, 548–561. [CrossRef]
8. Shah, D.H.; Patel, D.M. Design of sliding mode control for quadruple-tank MIMO process with time delay compensation. *J. Process Control* **2019**, *76*, 46–61. [CrossRef]
9. Thamallah, A.; Sakly, A.; M'Sahli, F. A new constrained PSO for fuzzy predictive control of Quadruple-Tank process. *Measurement* **2018**, *136*, 93–104. [CrossRef]
10. Eltantawie, M.A. Decentralized neuro-fuzzy controllers of nonlinear quadruple tank system. *SN Appl. Sci.* **2018**, *1*, 39. [CrossRef]
11. Son, N.N. Level Control of Quadruple Tank System Based on Adaptive Inverse Evolutionary Neural Controller. *Int. J. Control Autom. Syst.* **2020**, *18*, 2386–2397. [CrossRef]
12. Naami, G.; Ouahi, M.; Rabhi, A.; Tadeo, F.; Tuan, V.L.B. Design of robust control for uncertain fuzzy quadruple-tank systems with time-varying delays. *Granul. Comput.* **2022**, *7*, 951–964. [CrossRef]
13. Meng, X.; Yu, H.; Xu, T.; Wu, H. Sliding mode disturbance observer-based the port-controlled Hamiltonian control for a four-tank liquid level system subject to external disturbances. In Proceedings of the Chinese Control and Decision Conference (CCDC), Hefei, China, 22–24 August 2020; pp. 1720–1725. [CrossRef]
14. Meng, X.; Yu, H.; Zhang, J.; Xu, T.; Wu, H.; Yan, K. Disturbance Observer-Based Feedback Linearization Control for a Quadruple-Tank Liquid Level System. *ISA Trans.* **2022**, *122*, 146–162. [CrossRef] [PubMed]
15. Gurjar, B.; Chaudhari, V.; Kurode, S. Parameter estimation based robust liquid level control of quadruple tank system -Second order sliding mode approach. *J. Process Control* **2021**, *104*, 1–10. [CrossRef]
16. Meng, X.; Yu, H.; Zhang, J.; Xu, T.; Wu, H. Liquid Level Control of Four-Tank System Based on Active Disturbance Rejection Technology. *Measurement* **2021**, *175*, 109146. [CrossRef]
17. Huang, C.; Canuto, M.E.; Novara, C. The four-tank control problem: Comparison of two disturbance rejection control solutions. *ISA Trans.* **2017**, *71*, 252–271. [CrossRef] [PubMed]
18. Han, J. From PID to Active Disturbance Rejection Control. *IEEE Trans. Ind. Electron.* **2009**, *56*, 900–906. [CrossRef]
19. Yoo, D.; Yau, S.S.T.; Gao, Z. On convergence of the linear extended state observer. In Proceedings of the 2006 IEEE Conference on Computer Aided Control System Design, 2006 IEEE International Conference on Control Applications, 2006 IEEE International Symposium on Intelligent Control, Munich, Germany, 4–6 October 2006; pp. 1645–1650. [CrossRef]
20. Sira-Ramírez, H.; Luviano-Juárez, A.; Ramírez-Neria, M.; Zurita-Bustamante, E.W. *Active Disturbance Rejection Control of Dynamic Systems: A Flatness-Based Approach*; Butterworth-Heinemann: Oxford, UK, 2017.
21. Guo, B.Z.; Zhao, Z. *Active Disturbance Rejection Control for Nonlinear Systems: An Introduction*; John Wiley & Sons: New York, NY, USA, 2017.
22. Najm, A.A.; Ibraheem, I.K. Altitude and attitude stabilization of UAV quadrotor system using improved active disturbance rejection control. *Arab. J. Sci. Eng.* **2020**, *45*, 1985–1999. [CrossRef]
23. Hou, G.; Ke, Y.; Huang, C. A flexible constant power generation scheme for photovoltaic system by error-based active disturbance rejection control and perturb & observe. *Energy* **2021**, *237*, 121646. [CrossRef]
24. Abdul-Adheem, W.R.; Ibraheem, I.K.; Humaidi, A.J.; Alkhayyat, A.; Maher, R.A.; Abdulkareem, A.I.; Azar, A.T. Design and Analysis of a Novel Generalized Continuous Tracking Differentiator. *Ain Shams Eng. J.* **2021**, *13*, 101656. [CrossRef]
25. Guo, B.Z.; Zhao, Z.L. On the convergence of an extended state observer for nonlinear systems with uncertainty. *Syst. Control Lett.* **2011**, *60*, 420–430. [CrossRef]
26. Khalil, K.H. *Nonlinear Control*; Global Edition; Pearson Education: London, UK, 2015.
27. Anton, H.; Bivens, I.C.; Davis, S. *Calculus*, 11th ed.; John Wiley & Sons: New York, NY, USA, 2020.

28. Holland, J. *Adaptation in Natural and Artificial Systems*; University of Michigan Press: Ann Arbor, MI, USA, 1975.
29. Abdul-Adheem, W.R.; Ibraheem, I.K. From PID to Nonlinear State Error Feedback Controller. *Int. J. Adv. Comput. Sci. Appl.* **2017**, *8*, 312–322. [CrossRef]
30. Abdul-Adheem, W.R.; Ibraheem, I.K. Improved Sliding Mode Nonlinear Extended State Observer based Active Disturbance Rejection Control for Uncertain Systems with Unknown Total Disturbance. *Int. J. Adv. Comput. Sci. Appl.* **2016**, *7*, 80–93. [CrossRef]
31. Najm, A.A.; Ibraheem, I.K.; Azar, A.T.; Humaidi, A. Genetic optimization-based consensus control of multi-agent 6-DoF UAV system. *Sensors* **2020**, *20*, 3576. [CrossRef] [PubMed]
32. Aziz, A.S.A.; Hassanien, A.E.; Azar, A.T.; Hanafy, S.E. Genetic Algorithm with Different Feature Selection Techniques for Anomaly Detectors Generation. In Proceedings of the 2013 Federated Conference on Computer Science and Information Systems (FedCSIS), Kraków, Poland, 8–11 September 2013.
33. Fliess, M.; Join, C. Model-free control. *Int. J. Control* **2013**, *86*, 2228–2252. [CrossRef]

Disclaimer/Publisher's Note: The statements, opinions and data contained in all publications are solely those of the individual author(s) and contributor(s) and not of MDPI and/or the editor(s). MDPI and/or the editor(s) disclaim responsibility for any injury to people or property resulting from any ideas, methods, instructions or products referred to in the content.

actuators

Article

Tracking Control of Robot Manipulator with Friction Compensation Using Time-Delay Control and an Adaptive Fuzzy Logic System

Yao Sun, Xichang Liang * and Yi Wan

School of Mechanical Engineering, and Key Laboratory of High Efficiency and Clean Mechanical Manufacture of Ministry of Education, National Demonstration Center for Experimental Mechanical Engineering Education, Shandong University, Jinan 250061, China; yao@mail.sdu.edu.cn (Y.S.)
* Correspondence: liangxichang@sdu.edu.cn

Abstract: This paper aims to highlight the critical role of robot manipulators in industrial applications and elucidate the challenges associated with achieving high-precision control. In particular, the detrimental effects of nonlinear friction on manipulators are discussed. To overcome this challenge, a novel friction compensation controller (FCC) that combines time-delay estimation (TDE) and an adaptive fuzzy logic system (AFLS) is proposed in this paper. The friction compensation controller is designed to take advantage of the time-delay estimation algorithm's strengths in eliminating and estimating unknown dynamic functions of the system using information from the previous sampling period. Simultaneously, the adaptive fuzzy logic system compensates for the hard nonlinearities in the system and suppresses the errors generated by time-delay estimation, thus improving the accuracy of the robotic arm's tracking. The numerical experimental results demonstrate that the proposed friction compensation controller can significantly enhance the tracking accuracy of the robotic arm, with the addition of the adaptive fuzzy logic system improving time delay estimation's performance by an average of 90.59%. Moreover, the proposed controller is more straightforward to implement than existing methods and performs exceptionally well in practical applications.

Keywords: robot manipulator; high-precision control; time-delay estimation (TDE); adaptive fuzzy logic system (AFLS); friction compensation

Citation: Sun, Y.; Liang, X.; Wan, Y. Tracking Control of Robot Manipulator with Friction Compensation Using Time-Delay Control and an Adaptive Fuzzy Logic System. *Actuators* **2023**, *12*, 184. https://doi.org/10.3390/act12050184

Academic Editors: Oscar Barambones, Jose Antonio Cortajarena and Patxi Alkorta

Received: 23 March 2023
Revised: 12 April 2023
Accepted: 21 April 2023
Published: 25 April 2023

Copyright: © 2023 by the authors. Licensee MDPI, Basel, Switzerland. This article is an open access article distributed under the terms and conditions of the Creative Commons Attribution (CC BY) license (https://creativecommons.org/licenses/by/4.0/).

1. Introduction

Manipulators play a crucial role in various industrial processes, such as handling [1–3], assembly [4–6], and assistance [7–9]. To achieve these operations, high-precision tracking control of the robot manipulator is essential. However, the complex dynamics of the mechanical arm, which arise from highly nonlinear, time-varying parameters, dynamic coupling, and uncertainty, pose significant challenges to achieving high-precision control [10–12]. Model-based controllers, such as sliding-mode control, can improve the performance of mechanical arms, but a precise calculation of nonlinear dynamic models is intricate, thereby limiting the potential of model-based controllers in practical applications.

The time-delay estimation (TDE) technique [13–17] is a model-free control approach that was first proposed in the 1980s. TDE utilizes information from the previous time period to eliminate and estimate unknown dynamic functions of the system [18,19]. Within a sampling period, TDE assumes that the system's dynamic changes are not significant. TDE technology has been used to develop a simple, robust, and efficient time-delay control (TDC) method, which does not require prior knowledge or offline identification. As a result, it has found widespread use in the control of robot manipulators and chaotic systems [20,21].

TDC typically consists of two components: TDE elements and expected error dynamic injection elements [22]. From the perspective of TDE, the dynamic nonlinear factors of robot manipulators can be classified into two types: soft nonlinearity [23] and hard nonlinearity [24]. Soft nonlinearity can be completely eliminated using TDE, but hard nonlinearity, such as static and Coulomb friction, can adversely affect the tracking accuracy of TDC. Given that the sampling period cannot be infinitely small, dynamic characteristics may change rapidly even within a single sampling period. When using TDE to estimate such friction, TDE errors can lead to increased tracking errors, thereby hindering high-precision tracking control of robot manipulators [25].

In recent research, a third element has been introduced into TDC to compensate for hard nonlinearities and suppress TDE errors. Jin [26] proposed an ideal velocity feedback (IVF) term for suppressing TDE errors and demonstrated its effectiveness compared to adaptive friction compensation (AFC). To further improve the tracking accuracy of robot arms, Jin proposed a high-precision position tracking control method that uses terminal sliding mode (TSM) as the third element to suppress TDE errors and provide a faster convergence rate. Although the validation results show that the performance of TDE-TSM is better than that of TDC and TDE-IVF, TDE-TSM has two main drawbacks. One is the jitter problem caused by TSM, which is highly undesirable due to the sign function present in the TSM element. The other problem is the long computation time of TSM, which requires the calculation of fractional power functions and can take several tens of milliseconds for some worst-case controller hardware. To address these issues and achieve high-precision tracking control, Bae et al. [27] proposed a controller that uses a fuzzy logic system (FLS) as the third element, marking the first time that TDE was combined with intelligent technology. However, the FLS is simple and standard, requiring careful parameter tuning and experience.

In this paper, we propose a friction compensation controller (FCC) aimed at improving the tracking accuracy of robot arms and making the controller easier to use in practical applications. The controller adds an adaptive fuzzy logic system (AFLS) as the third element to handle strong nonlinearities and TDE errors, while an adaptive rule is designed to update the parameters of the fuzzy logic system online. The controller consists of three elements: the TDE element for canceling soft nonlinearities, the injection element for dynamically calculating target error, and the AFLS element for suppressing TDE errors. By using TDE, this controller is easier to implement in practical applications. The design of AFLS ensures the high-precision tracking of the robot arm.

The main contributions of this paper can be summarized as follows. Firstly, a novel friction compensation controller (FCC) algorithm is proposed to mitigate the adverse effects of nonlinear friction on manipulators. The FCC algorithm combines time delay estimation (TDE) and an adaptive fuzzy logic system (AFLS). TDE uses information from the previous sampling period to eliminate and estimate unknown dynamic functions of the system, while AFLS compensates for the strong nonlinearities in the system and suppresses errors generated by TDE. Secondly, the proposed FCC is designed to significantly improve the tracking accuracy of robot arms. Numerical experiments show that the performance of TDE can be improved by an average of 90.59% with the addition of AFLS. Lastly, the proposed controller is easier to implement than existing methods and exhibits exceptional performance in practical applications. Hence, the algorithm proposed in this paper is a practical choice to address the challenges of achieving high-precision control in industrial applications.

The structure of this paper is as follows: in Section 2, we provide a review of traditional iterative learning control (TDC) and highlight its associated issues. In Section 3, we propose a novel control algorithm based on TDC and adaptive fuzzy logic systems (AFLS) and mathematically prove its convergence. Section 4 presents a performance comparison between the proposed controller and multiple TDE-based controllers. Finally, we summarize the experimental results and draw conclusions in the Section 5 of this paper.

2. Review and Problem of TDC

2.1. Review of TDC

The dynamic behaviour of the robot manipulator can be accurately described by means of the following equation based on Assumptions 1–3.

$$M(\theta)\theta + C(\theta,\theta)\theta + G(\theta) + F(\theta,\theta) + \tau_d = \tau \tag{1}$$

where $\theta, \dot{\theta}, \ddot{\theta} \in R^n$ are the position, velocity, and acceleration of the joints; $\tau \in R^n$ denotes the torque; and $M(\theta) \in R^{n\times n}$ represents the inertia matrix; $C(\theta, \dot{\theta}) \in R^n$ stands for the Coriolis and centrifugal matrix; $G(\theta) \in R^n$ is the gravitational vector; $F \in R^n$ is the friction term, and $\tau_d \in R^n$ denotes the disturbance torques.

After defining a constant diagonal matrix $\overline{M}$, the equation mentioned above can be rewritten as:

$$u = \overline{M}\ddot{\theta} + \overline{H} \tag{2}$$

$$\overline{H} = (M(\theta) - \overline{M})\ddot{\theta} + H(\theta, \dot{\theta}, \ddot{\theta}) \tag{3}$$

where $H(\theta, \dot{\theta}, \ddot{\theta})$ is the sum of unknown nonlinear dynamics of the manipulator, and

$$H(\theta, \dot{\theta}, \ddot{\theta}) = C(\theta, \dot{\theta})\dot{\theta} + G(\theta) + F(\theta, \dot{\theta}) + \tau_d \tag{4}$$

The tracking error is defined as follows.

$$e = \theta_d - \theta \tag{5}$$

where θ_d represents the desired position. The velocity and acceleration error are defined as $\dot{e} = \dot{\theta}_d - \dot{\theta}$, $\ddot{e} = \ddot{\theta}_d - \ddot{\theta}$, respectively.

The control objective of the TDC is to attain the following error dynamic:

$$\ddot{e} + K_D\dot{e} + K_P e = 0 \tag{6}$$

where K_D and K_P are the constant diagonal gain. Then, the TDC applied in tracking control of the robot manipulator is designed as follows.

$$u = \overline{M}\gamma_0 + \hat{H}(\theta, \dot{\theta}, \ddot{\theta}) \tag{7}$$

$$\gamma_0 = \ddot{\theta}_d + K_D\dot{e} + K_P e \tag{8}$$

where $\hat{H}(\theta, \dot{\theta}, \ddot{\theta}$ is the estimation value of $\overline{H}$ in Equation (3). The TDE technique is utilized to estimate the unknown function $\overline{H}$, which can be expressed as follows:

$$\hat{H}(\theta, \dot{\theta}, \ddot{\theta}) = \overline{H}_{t-L} = u_{t-L} - \overline{M}\ddot{\theta}_{t-L} \tag{9}$$

where $\bullet_{t-L}$ denotes the value of $\bullet$ at time $t - L$.

During the controller design process, the following assumptions were made.

Assumption 1. *The system has a fast enough processing speed to handle a small sampling time, the sampling time can be set to a small value without causing significant delays or errors in the data acquisition process.*

Assumption 2. *The system is subject to bounded external disturbances, meaning that the external disturbances acting on the system are limited and will not cause the system to become unstable.*

Assumption 3. *The desired trajectory $\theta_d \in \mathbb{R}^n$ for each joint is assumed to be both smooth and continuous. This assumption implies that the first and second-time derivatives, $\dot{\theta}_d$ and $\ddot{\theta}_d$, exist for all time intervals and are continuous and bounded. The smoothness and continuity of θ_d is* *essential for ensuring the continuity and stability of the motion profile of the joint. The continuity of $\dot{\boldsymbol{\theta}}_d$ and $\ddot{\boldsymbol{\theta}}_d$ ensures that the velocity and acceleration of the joint remain bounded and the motion is predictable.*

Then, according to Assumption 1, the following equation can be satisfied.

$$\overline{\boldsymbol{H}} \approx \overline{\boldsymbol{H}}_{t-L} \tag{10}$$

The final form of TDC can be obtained by combining Equations (7)–(9) as follows.

$$\boldsymbol{u} = \overline{\boldsymbol{M}}(\ddot{\boldsymbol{\theta}}_d + \boldsymbol{K}_D\dot{\boldsymbol{e}} + \boldsymbol{K}_P\boldsymbol{e}) + \boldsymbol{u}_{t-L} - \overline{\boldsymbol{M}}_{t-L} \tag{11}$$

According to the study in [28], the stability condition of the controller is:

$$\left|\phi_i\left(\boldsymbol{M}^{-1}\overline{\boldsymbol{M}} - \boldsymbol{I}_n\right)\right| < 1 \tag{12}$$

where $\boldsymbol{I}_n$ is the $n \times n$ identity matrix and $\phi_i\left(\boldsymbol{M}^{-1}\overline{\boldsymbol{M}} - \boldsymbol{I}_n\right)$ is the $i-$th eigenvalue of $\boldsymbol{M}^{-1}\overline{\boldsymbol{M}} - \boldsymbol{I}_n$. The above function can be satisfied by the choice of $\overline{\boldsymbol{M}}$.

2.2. TDE Error Due to Friction

This paper discusses the use of time-delay estimation (TDE), which is based on the assumption that the nonlinearity in the system dynamics does not vary significantly. TDE can achieve perfect time-delay estimation performance as the sampling time $\boldsymbol{L}$ approaches zero. However, in practical digital implementations, the minimum value of $\boldsymbol{L}$ is limited. Therefore, the estimation performance of TDE depends on the finite $\boldsymbol{L}$, which can be represented by the following relationship.

$$\overline{\boldsymbol{H}} - \hat{\boldsymbol{H}} = \overline{\boldsymbol{H}}_t - \overline{\boldsymbol{H}}_{t-L} = \overline{\boldsymbol{M}}(\gamma_0 - \ddot{\boldsymbol{\theta}}) \tag{13}$$

The TDE error δ is defined as follows.

$$\delta = \overline{\boldsymbol{M}}^{-1}\left[\overline{\boldsymbol{H}}_t - \hat{\boldsymbol{H}}_t\right] = \gamma_0 - \ddot{\boldsymbol{\theta}} \tag{14}$$

By substituting Equation (8) into Equation (14), the error dynamics of TDC can be expressed as follows.

$$\ddot{e} + \boldsymbol{K}_D\dot{e} + \boldsymbol{K}_P\boldsymbol{e} = \delta \tag{15}$$

which shows the effect of TDE error on tracking error clearly.

Remark 1. *According to Equation* (1), *the friction term* $\boldsymbol{F} \in \boldsymbol{R}^n$ *consists of both static and Coulomb friction forces. These friction forces exhibit rapid changes near* $\dot{\boldsymbol{\theta}} = 0$, *meaning that these fast dynamics can occur within a single sampling period. In such cases, Equation* (10) *cannot be satisfied, resulting in TDE error due to the inaccurate estimation of TDE technology. This, in turn, leads to larger tracking errors, as shown by Equation* (15). *To address this issue, this study proposes a friction compensation method based on parameter identification in Section 3.*

3. Time-Delay Control with Adaptive Fuzzy Logic System

3.1. Derivation of the Proposed Controller

To compensate for TDE error in TDC, we have introduced an adaptive fuzzy logic system (AFLS) as the third element. Therefore, the controller is composed of a TDE element, desired error dynamics, and AFLS. To introduce AFLS, we have used a sliding surface:

$$s = \dot{e} + \boldsymbol{k}_1\boldsymbol{e} \tag{16}$$

where $\boldsymbol{k}_1$ is the $n \times n$ constant matrix to be tuned.

Then, the control input is designed as

$$u = \overline{M}_{\gamma_f} + u_{t-L} - \overline{M}\ddot{\theta}_{t-L} \quad (17)$$

$$\gamma_f = \ddot{\theta}_d + (k_1 + k_2)\dot{e} + k_1 k_2 e + f \quad (18)$$

where $f = [f_1, \ldots f_i, \ldots f_n]$ is the adaptive fuzzy logic system to suppress TDE error; $u_{t-L} - \overline{M}\ddot{\theta}_{t-L}$ is the TDE element to cancel the soft nonlinearities; $\ddot{\theta}_d + (k_1 + k_2)\dot{e} + k_1 k_2 e$ is the error dynamic, which has the small framework with TDC, and k_2 is the designed $n \times n$ constant matrix. Thus, the final form of the controller is proposed by:

$$u = \underbrace{u_{t-L} - \overline{M}\ddot{\theta}_{t-L}}_{\text{Time-Delay Estimation}} + \underbrace{\overline{M}\left[\ddot{\theta}_d + (k_1 + k_2)\dot{e} + k_1 k_2 e\right]}_{\text{The injection of desired error dynamics}} + \underbrace{\overline{M}f}_{\text{AFLS}} \quad (19)$$

Substituting Equation (17) to Equation (2), the equation can be obtained as:

$$\delta = \ddot{e} + (k_1 + k_2)\dot{e} + k_1 k_2 e + f \quad (20)$$

The above equation also shows the influence of f on TDE error. f is designed from two aspects: the fuzzy logic term and adaptive term. Now, the design process is given as follows.

Remark 2. *In order to suppress the influence of TDE error, we designed a sliding-mode surface in this study. Similar to traditional sliding-mode control, the purpose of the sliding-mode surface used here is to guide the system state to a specific trajectory, thereby achieving stable control of the system. Specifically, the sliding-mode surface* $\mathbf{s}$ *used in this study is a hyperplane composed of a state variable* $\dot{\mathbf{e}}$ *and a reference input signal* $\mathbf{k}_1\mathbf{e}$*. If the system state changes and the state point crosses the sliding-mode surface, the controller will adjust the system to guide its state back to the sliding-mode surface, thus achieving stable control of the system.*

3.2. Fuzzy Logic Term

To obtain fuzzy rules, we conducted the following analysis: when $|s_i|$ is large, the expected $|f_i|$ should also be large to ensure the convergence speed of the system. On the other hand, when $|s_i|$ is small, allowing for a smaller $|f_i|$ can avoid oscillation. Moreover, when $|s_i|$ is zero, $|f_i|$ can also be zero. Therefore, we can define the rules as follows:

IF s_i is NB, THEN f_i is NB
IF s_i is NS, THEN f_i is NS
IF s_i is ZE, THEN f_i is ZE
IF s_i is PS, THEN f_i is PS
IF s_i is PB, THEN f_i is PB

In the system, s_i is the input of the fuzzy system, while f_i is its output. Both variables are divided into five fuzzy subsets: positive big (PB), positive small (PS), zero (ZE), negative small (NS), and negative big (NB). These subsets are represented by Gaussian membership functions, which are defined as follows:

$$\mu_A(x_i) = \exp\left[-\left(\frac{x_i - \alpha}{\sigma}\right)^2\right] \quad (21)$$

where the subscript A denotes the fuzzy sets such as $NB, \ldots, PB$; x_i is s_i and f_i; α is the center of A and σ is the width of A.

Choosing the product inference engine, singleton fuzzification, and center average defuzzification, f_i can be written as:

$$f_i = \frac{\sum_{m=1}^{M} \theta_{f_i}^m \mu_A^m(s_i)}{\sum_{m=1}^{M} \mu_A^m(s_i)} = \theta_{f_i}^T \psi_{f_i}(s_i) \quad (22)$$

where $\theta_{f_i} = \left[\theta_{f_i}^1, \ldots, \theta_{f_i}^m \ldots, \theta_{f_i}^M\right]^T$, $\psi_{f_i}(s_i) = \left[\psi_{f_i}^1(s_i), \ldots, \psi_{f_i}^m(s_i) \ldots, \psi_{f_i}^M(s_i)\right]^T$ and $\psi_{f_i}^m(s_i) = \mu_{A^m}(s_i)/\sum_{m=1}^M \mu_{A^m}(s_i).\theta_{f_i}$ is chosen as the parameter to be updated and therefore is called the parameter vector. $\psi_{f_i}(s_i)$ can be regarded as the weight of the parameter vector which is called the function basis vector.

3.3. Adaptive Scheme

Define $\theta_{f_i}^*$ so that $f_i = \theta_{f_i}^{*T} \psi_{f_i}(s_i)$ is the optimal estimation for ε, and there exists the optimal estimation error $w_i > 0$, which satisfying

$$\left|\delta_i - \theta_{f_i}^* \psi_{f_i}(s_i)\right| \leq w_i \tag{23}$$

Define

$$\tilde{\theta}_{f_i} = \theta_{f_i} - \theta_{f_i}^*, \tag{24}$$

Then

$$f_i = \tilde{\theta}_{f_i}^T \psi_{f_i}(s_i) + \theta_{f_i}^* \psi_{f_i}(s_i) \tag{25}$$

After that, choose the adaptive law as

$$\dot{\tilde{\theta}}_{f_i} = \eta_i s_i \psi_{f_i}(s_i) \tag{26}$$

where η_i is a positive constant.

3.4. Stability Analysis

Theorem 1. *Let γ_i be the control gain satisfying $0 < \gamma_i < 1$. Then, the recursive FCC controller defined in Equation (19) ensures that all closed-loop system signals are bounded and achieve asymptotic output tracking, i.e.,*

$$\lim_{t \to \infty} e(t) = 0, \tag{27}$$

where $e(t)$ is the tracking error signal.

In other words, Theorem 1 guarantees the stability of the closed-loop system and the convergence of the tracking error to zero as t approaches infinity. This result is significant as it confirms the effectiveness of the proposed recursive FCC controller in achieving high-precision control of the system.

Proof. The Lyapunov function is chosen as:

$$V = \frac{1}{2}\mathbf{s}^T \mathbf{s} + \frac{1}{2}\sum_{i=1}^{n}\left(\frac{1}{\eta_i}\tilde{\theta}_{f_i}^T \tilde{\theta}_{f_i}\right) \tag{28}$$

From Equation (10), the following equation can be obtained.

$$\dot{\mathbf{s}} = \boldsymbol{\delta} - \mathbf{f} - \mathbf{k}_2 \mathbf{s} \tag{29}$$

Then, the derivative of V can be expressed as

$$\begin{aligned} \dot{V} &= \mathbf{s}^T \dot{\mathbf{s}} + \sum_{i=1}^{n} \frac{1}{\eta_i} \tilde{\theta}_{f_i}^T \dot{\tilde{\theta}}_{f_i} \\ &= \mathbf{s}^T (\boldsymbol{\varepsilon} - \mathbf{f} - \mathbf{k}_2 \mathbf{s}) + \sum_{i=1}^{n} \frac{1}{\eta_i} \tilde{\theta}_{f_i}^{\tilde{\theta}_{\theta_i}} \dot{l}_i \\ &= \sum_{i=1}^{n} s_i(\varepsilon_i - f_i - k_{1i} s_i) + \sum_{i=1}^{n} \frac{1}{\eta_i} \tilde{\theta}_{f_i}^T \dot{\tilde{\theta}}_{f_i} \end{aligned} \tag{30}$$

As $f_i = \tilde{\theta}_{f_i}^T \psi_{f_i}(s_i) + \theta_{f_i}^{*T} \psi_{f_i}(s_i)$, then

$$\begin{aligned}
\dot{V} &= \sum_{i=1}^{n} s_i(\varepsilon_i - f_i - k_{1i}s_i) + \sum_{i=1}^{n} \frac{1}{\eta_i} \tilde{\theta}_{f_i}^T \dot{\tilde{\theta}}_{f_i} \\
&= \sum_{i=1}^{n} s_i \left[\varepsilon_i - \tilde{\theta}_{f_i} \psi_{f_i}(s_i) - \theta_{f_i}^{*T} \psi_{f_i}(s_i) - k_{1i}s_i \right] + \sum_{i=1}^{n} \frac{1}{\eta_i} \tilde{\theta}_{i_i}^T \dot{\tilde{\theta}}_{f_i} \\
&= -\sum_{i=1}^{n} s_i k_{1i} s_i + \sum_{i=1}^{n} s_i \left[\varepsilon_i - \theta_{f_i}^{*} \psi_{f_i}(s_i) \right] + \sum_{i=1}^{n} \left[\frac{1}{\eta_i} \tilde{\theta}_{f_i}^T \dot{\tilde{\theta}}_{f_i} - \tilde{\theta}_{f_i}^T s_i \psi_{f_i}(s_i) \right] \\
&= -\sum_{i=1}^{n} s_i k_{1i} s_i + \sum_{i=1}^{n} s_i \left[\varepsilon_i - \theta_{f_i}^{*} T \psi_{f_i}(s_i) \right] + \sum_{i=1}^{n} \frac{1}{\eta_i} \tilde{\theta}_{f_i}^T \left[\dot{\tilde{\theta}}_{f_i} - \eta_i s_i \psi_{f_i}(s_i) \right]
\end{aligned} \tag{31}$$

As $\dot{\tilde{\theta}}_{\hat{k}_i} = \eta_i s_i \psi_{f_i}(s_i)$, the above equation can be expressed as

$$\dot{V} = -\sum_{i=1}^{n} s_i k_{1i} s_i + \sum_{i=1}^{n} s_i \left[\varepsilon_i - \theta_{f_i}^{*} T \psi_{f_i}(s_i) \right] \tag{32}$$

According to Equation (23), there exists a small positive constant, which satisfies

$$\left| \varepsilon_i - \theta_{f_i}^{*} \psi_{f_i}(s_i) \right| \le w_i \le \gamma_i |s_i| \tag{33}$$

where $0 < \gamma_i < 1$.

Then, Equation (32) can be expressed as:

$$\begin{aligned}
\dot{V} &= -\sum_{i=1}^{n} s_i k_{1i} s_i + \sum_{i=1}^{n} s_i \left[\varepsilon_i - \theta_{f_i}^{*} \psi_{f_i}(s_i) \right] \\
&\le -\sum_{i=1}^{n} s_i k_{1i} s_i + \sum_{i=1}^{n} s_i \gamma_i |s_i| \\
&\le -\sum_{i=1}^{n} s_i k_{1i} s_i + \sum_{i=1}^{n} \gamma_i s_i^2 \\
&= -\sum_{i=1}^{n} (k_{1i} - \gamma_i) s_i^2 \le 0
\end{aligned} \tag{34}$$

□

According to Equations (28) and (34), V is bounded and greater than or equal to zero. Since k_{1i} and γ_i are both positive constants, $\dot{V}$ is negative semi-definite. Thus, $\dot{V}$ converges to zero as t approaches infinity.

Remark 3. *By Barbalat lemma, we can conclude that if a function $f(t)$ is uniformly continuous and bounded, and its derivative converges to zero as t approaches infinity, then $f(t)$ converges to a finite limit as t approaches infinity. In our case, V is bounded and its derivative $\dot{V}$ converges to zero as t approaches infinity. Therefore, we can conclude that $\lim_{t\to\infty} V$ exists.*

When the system reaches a steady state, the first derivative of the error signal (i.e., $\dot{e}$) also tends to zero. Thus, we have $\dot{e} = \dot{\theta} - \dot{\theta}_d$, and:

$$\lim_{t\to\infty} \dot{e} = \lim_{t\to\infty} (\dot{\theta} - \dot{\theta}_d) \quad = \lim_{t\to\infty} \frac{1}{T}(\theta - \theta_d) - \dot{\theta}_d \ = 0 \tag{35}$$

Similarly, we can show that $\lim_{t\to\infty} ke = 0$.

Thus, this controller with the adaptive law in Equation (26) can drive the overall system tracking error to converge to zero, that is

$$\lim_{t\to\infty} s = \lim_{t\to\infty} (\dot{e} + ke) = 0, \tag{36}$$

as

$$\lim_{t\to\infty} \theta = \theta_d \text{ and } \lim_{t\to\infty} \dot{\theta} = \dot{\theta}_d \tag{37}$$

Therefore, it has been proven that with this control method and by applying Equation (19) as the input, the actual joint positions will converge to the desired ones.

4. Numerical Experiments

This study aims to verify the effectiveness of the proposed robot arm control method through simulations conducted on a two-degree-of-freedom robot arm, as illustrated in Figure 1. The experimental objectives of this paper consist of two aspects: first, to investigate the performance of the FCC under different f values, and second, to validate the advantages and disadvantages of the TDE, NTSM, and FCC under different frictional force disturbances. Therefore, we conducted three sets of comparative experiments to evaluate the performance of three algorithms under three different conditions: no friction, normal friction disturbance, and significant friction disturbance.

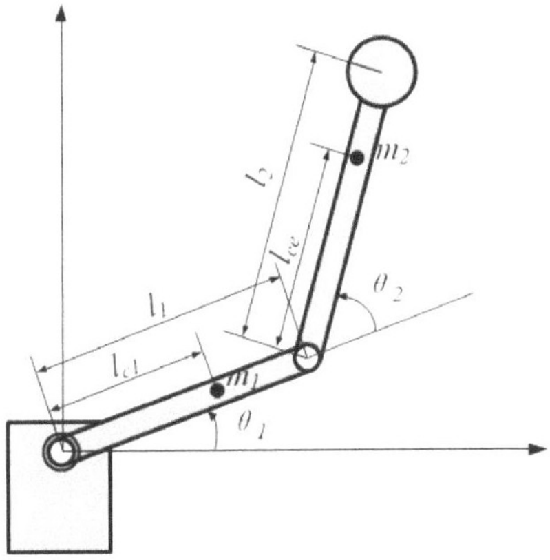

Figure 1. Diagram of two-DOF robot manipulator.

To facilitate the simulation, the dynamic model of the robot arm system is presented in Equation (1). Detailed information on the model is presented below:

$$\boldsymbol{M}(\boldsymbol{\theta}) = \begin{bmatrix} \alpha + 2\delta\cos(\theta_2) + 2\eta\sin(\theta_2) & \beta + \delta\cos(\theta_2) + \eta\sin(\theta_2) \\ \beta + \delta\cos(\theta_2) + \eta\sin(\theta_2) & \beta \end{bmatrix} \tag{38}$$

$$\boldsymbol{C}(\boldsymbol{\theta}, \dot{\boldsymbol{\theta}}) = \begin{bmatrix} [-2\delta\sin(\theta_2) + 2\eta\cos(\theta_2)]\dot{\theta}_2 & [-\delta\sin(\theta_2) + \eta\cos(\theta_2)]\dot{\theta}_2 \\ [\delta\sin(\theta_2) - \eta\cos(\theta_2)]\dot{\theta}_1 & 0 \end{bmatrix} \tag{39}$$

$$\boldsymbol{G}(\boldsymbol{\theta}) = \begin{bmatrix} \delta e_2\cos(\theta_1 + \theta_2) + \eta e_2\sin(\theta_1 + \theta_2) + (\alpha - \beta + e_1)e_2\cos(\theta_1) \\ \delta e_2\cos(\theta_1 + \theta_2) + \eta e_2\sin(\theta_1 + \theta_2) \end{bmatrix} \tag{40}$$

Friction severely affects the control performance of robot systems. Therefore, we select the friction term as follows:

$$\boldsymbol{F}(\boldsymbol{\theta}, \dot{\boldsymbol{\theta}}) = \begin{bmatrix} F_{v1}\dot{\theta}_1 + F_{c1}\,\mathrm{sgn}(\dot{\theta}_1) \\ F_{v2}\dot{\theta}_2 + F_{c2}\,\mathrm{sgn}(\dot{\theta}_2) \end{bmatrix} \tag{41}$$

where $\alpha = I_1 + m_1 l_{c1}^2 + I_e + m_e l_{ce}^2 + m_e l_1^2$, $\beta = I_e + m_e l_{ce}^2, \delta = m_e l_1 l_{ce} \cos \delta_e$, $\eta = m_e l_1 l_{ce} \sin \delta_e$, $e_1 = m_1 l_1 l_{c1} - I_1 - m_1 l_1^2, e_2 = g/l_1; m_1$ denotes the mass of first link; l_{cl} is the distance between the mass center of the first link and the first joint; I_l is the moment of inertia of the first link; m_e is the mass of second link with payload; l_{ce} is the distance between the mass center of second link and the second joint; I_e is the moment of inertia of the second link; δ_e is the angle relative to the original second link. The physical parameters of the robot manipulator are shown in Table 1.

Table 1. The physical parameters of the robot manipulator.

Indices	l_1	l_2	l_{c1}	l_{ce}	I_1	I_2	m_1	m_e	δ_e
Value	1	1.2	0.5	1	0.083	0.4	1	3	0

This study compares the performance of three controllers in experiments:

1. FCC: This is a time-delay estimation controller equipped with AFLS (as given in Equation (20)), which is thoroughly described in Section 3 of this paper. The control gains for this controller are set as follows: $\overline{M}_1 = 0.15$, $\overline{M}_2 = 0.15$, $k_{11} = 10$, $k_{12} = 10$, $k_{21} = 70$, and $k_{22} = 70$.
2. NTSM: This technique is founded on the principles of TDE and sliding-mode control. It achieves the high-precision control of nonlinear dynamic systems by incorporating supplementary terms on the sliding surface. These terms help to mitigate the impacts that traditional sliding-mode control may generate, resulting in a superior level of control.
3. TDC: A widely used control method for systems with a delay that effectively solves delay problems and improves the stability and precision of the control system.

To quantitatively evaluate the control performance of these three controllers, the study uses the maximum value, mean value, and standard deviation of the tracking error as performance indicators, marked as MAV, RMS, and Var, respectively, whose definitions can be found in [29]. In the three experiments, a normal level sinusoidal trajectory with sufficient smoothness is used, defined as $x_{1d}(t) = [2\sin(0.5\pi t); 2\sin(0.5\pi t)]^T rad$.

4.1. Comparative Experiment of Three Controllers under No-Friction Disturbance

This section compares the simulation results of three algorithms at $F_{v1} = 0$, $F_{c1} = 0$, $F_{v2} = 0$, and $F_{c2} = 0$, as presented in Figures 2–5. The FCC achieves maximum tracking errors of approximately 0.01 rad and 0.005 rad for joint 1 and joint 2, respectively. Compared to the TDC algorithm with linear error dynamics, the FCC exhibits smaller tracking errors. Furthermore, Figure 4 demonstrates that there is no jitter in the control inputs of the two joints.

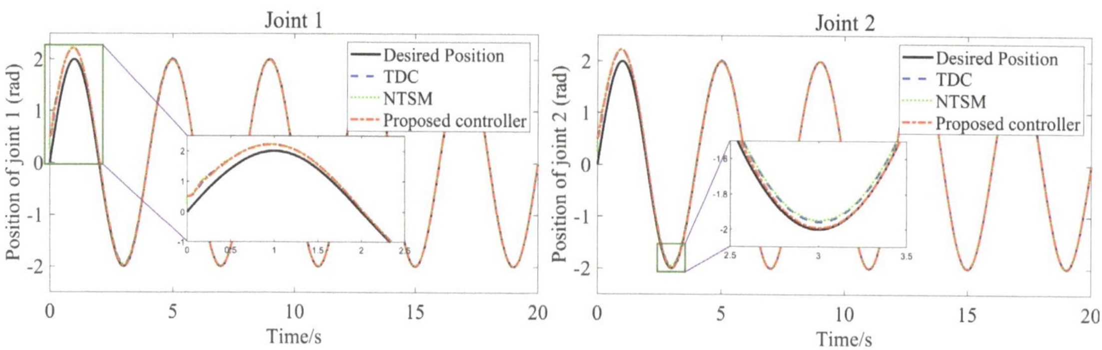

Figure 2. Position-tracking curves for the three controllers under no-friction disturbance.

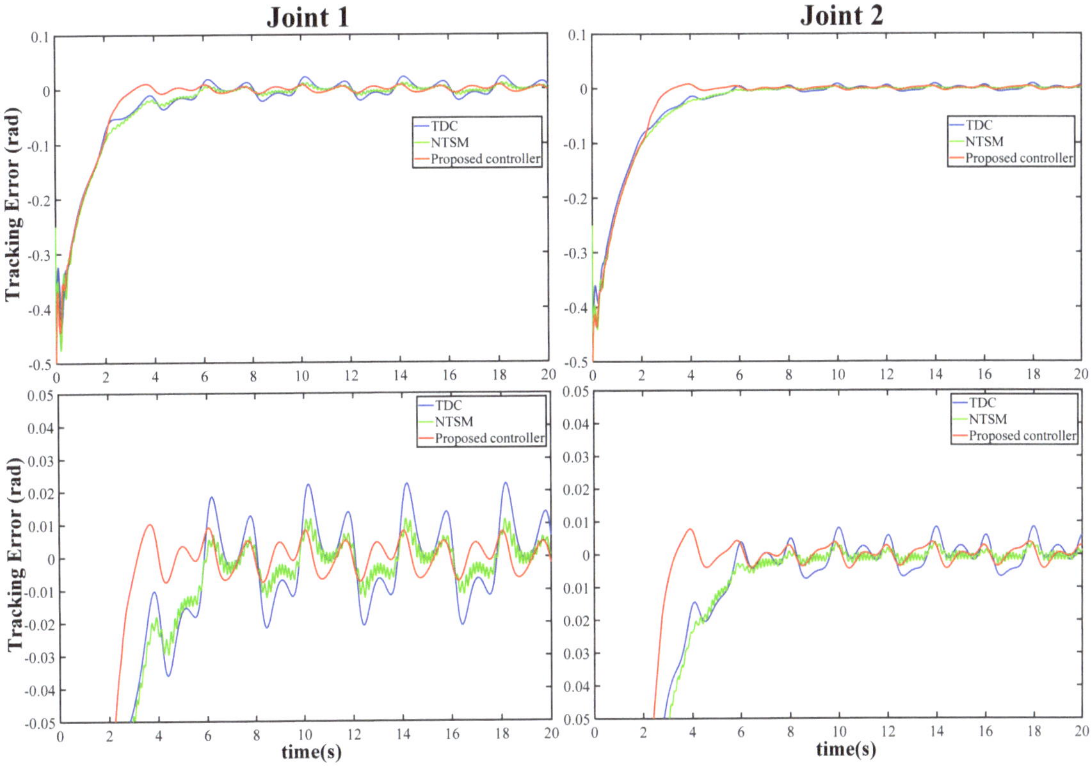

Figure 3. Tracking-error curves of the three controllers under no-friction disturbance.

As illustrated in Figures 2–5, the TDC algorithm effectively eliminates uncertainties and exhibits good tracking performance. Building upon this, the NTSM algorithm achieves high-precision tracking while also avoiding control jitter. Additionally, the FCC is proposed in this paper to achieve high-precision anti-interference control. In contrast to the TDC and NTSM algorithms, the FCC not only eliminates uncertainties but also achieves excellent tracking performance in the presence of frictional force interference.

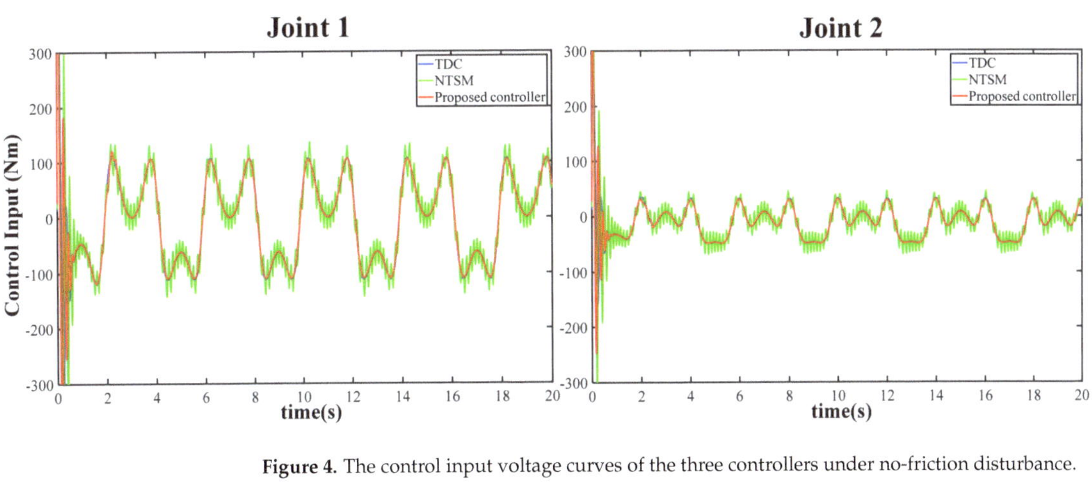

Figure 4. The control input voltage curves of the three controllers under no-friction disturbance.

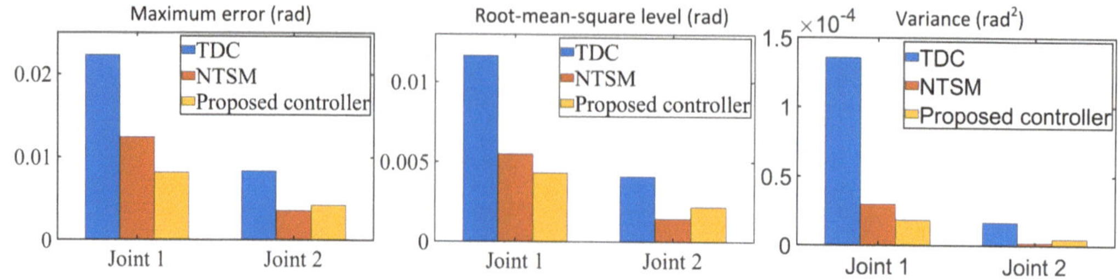

Figure 5. Performance indices evaluation during the last three cycles under no-friction disturbance.

Figure 5 demonstrates the FCC's outstanding tracking performance without frictional interference. The variance data in Figure 5 shows that the TDC controller's tracking performance on joint 1 and joint 2 improved by 86.198% and 71.286%, respectively, after adding AFLS.

Remark 4. *The variance of motion error in a robotic arm is a crucial metric for assessing its performance, enhancing control systems, optimizing motion trajectory planning, and predicting motion trajectories. To enhance the performance of a robotic arm, this study analyzes its errors by calculating variance and utilizes it as a reference point to gauge the extent of performance enhancement. For instance, on joint 1, the variance of TDC is* 1.3575×10^{-4} *, and the variance of FCC is* 1.875×10^{-5}*. As a result of this calculation, it is determined that the performance of joint 1 has been enhanced by 86.198%. It is important to note that all performance improvement ratios in this study are based on variance calculations.*

4.2. Comparative Experiment of Three Controllers under Normal Friction Disturbance

To evaluate the high-precision tracking performance of the proposed algorithm, a certain amount of nonlinear friction disturbance was introduced to the system in this section by setting $F_{v1} = 50$, $F_{c1} = 50$, $F_{v2} = 50$, and $F_{c2} = 50$. Under these conditions, unmodeled nonlinear friction was identified as the primary source of disturbance and was used to test the robustness of the proposed FCC.

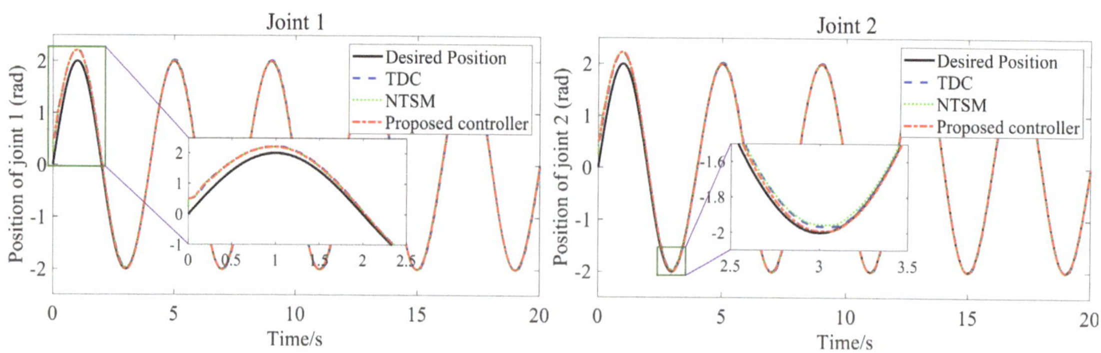

Figure 6. Position-tracking curves for the three controllers under normal friction disturbance.

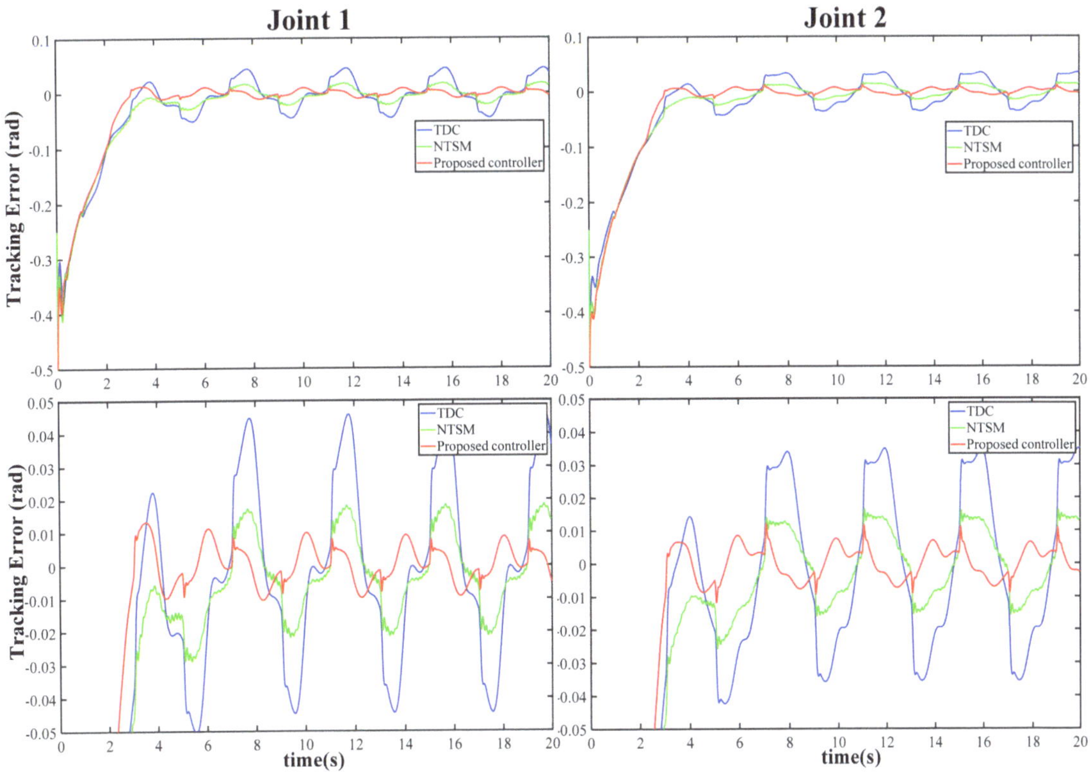

Figure 7. Tracking-error curves of the three controllers under normal friction disturbance.

The simulation results presented in Figure 6 demonstrate that the FCC can accurately track the desired motion trajectory of the load. Figure 7 displays the tracking errors of the three controllers, indicating that the proposed FCC achieved the best tracking performance among the three controllers. The AFLS compensation mechanism resulted in better tracking performance than the NTSM. Figure 8 shows the control input voltage of the FCC, which is both smooth and limited. The performance indicators of the last three cycles are summarized in Figure 9. In the presence of normal friction disturbance, the FCC exhibited excellent anti-interference performance in the three indicators of MAV, RMS, and Var, which is the best among the three controllers. This is attributed to the faster convergence efficiency and high robustness of AFLS. Based on the variance shown in Figure 9, it is evident that the addition of AFLS improved the performance of TDC by 96.105% and 96.376% in joint 1 and joint 2, respectively.

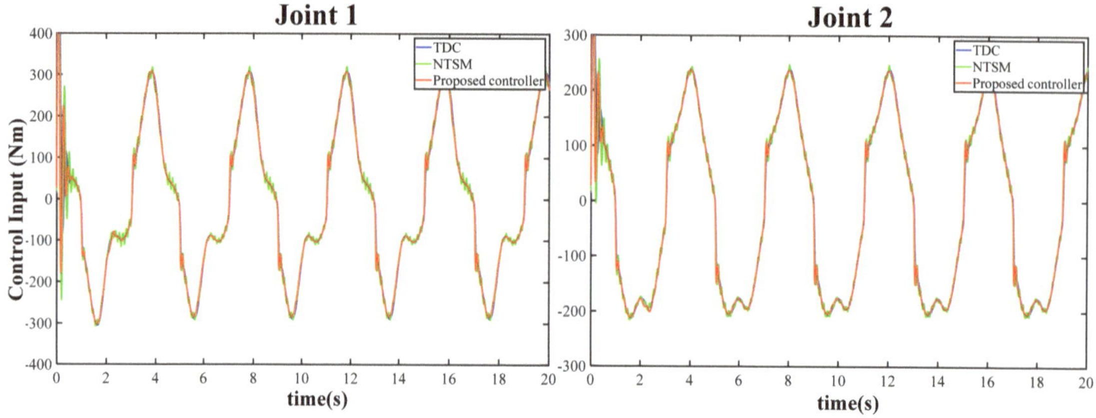

Figure 8. The control input voltage curves of the three controllers under normal friction disturbance.

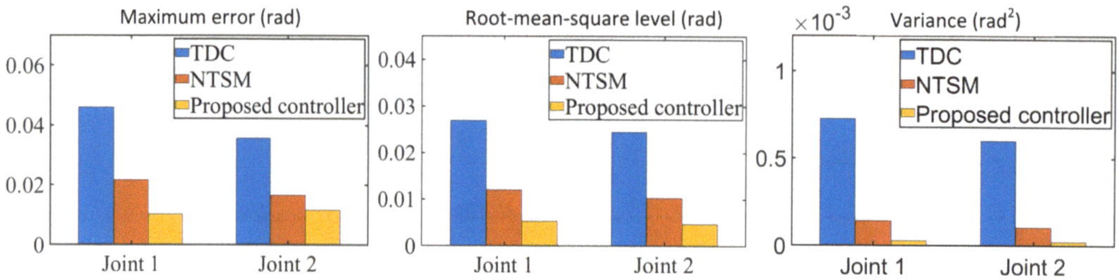

Figure 9. Performance indices evaluation during the last three cycles under normal friction disturbance.

4.3. Comparative Experiment of Three Controllers under Significant Frictional Disturbance

In this section's three comparative experiments, a significant amount of frictional disturbance was introduced by setting $F_{v1} = 100$, $F_{c1} = 100$, $F_{v2} = 100$, and $F_{c2} = 100$.

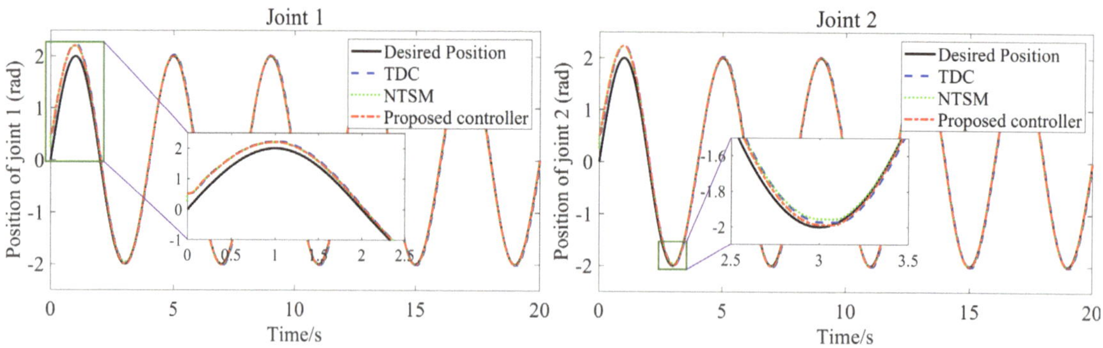

Figure 10. Position-tracking curves for the three controllers under significant frictional disturbance.

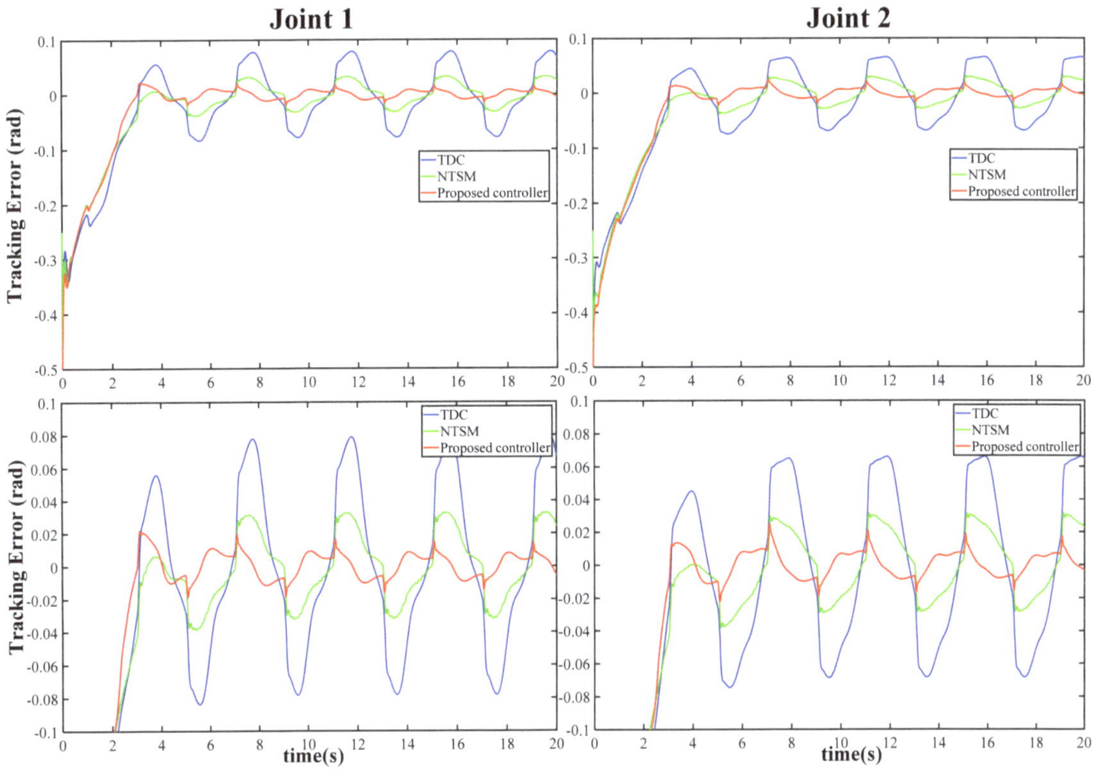

Figure 11. Tracking-error curves of the three controllers under significant frictional disturbance.

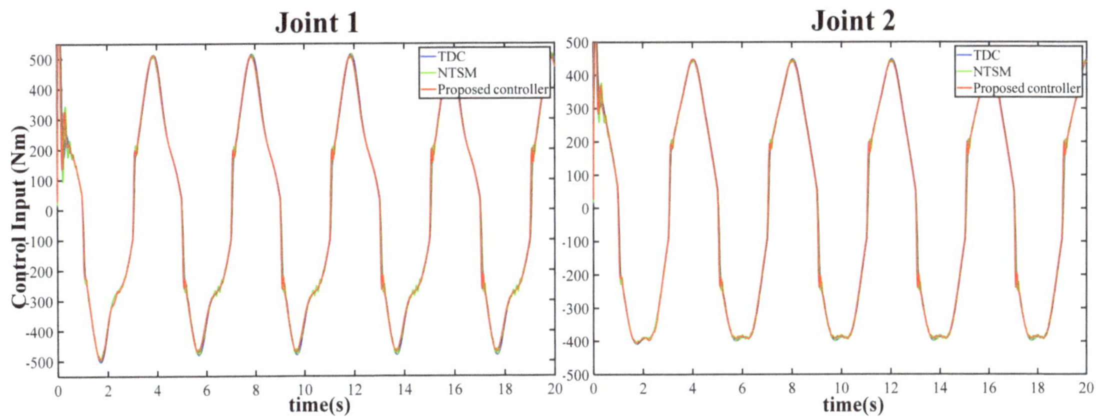

Figure 12. The control input voltage curves of the three controllers under significant frictional disturbance.

The tracking trajectories and tracking errors of the three controllers are presented in Figures 10 and 11. Figure 12 illustrates the control input voltage of the three controllers, indicating that their input voltage remains smooth and constrained. Figure 13 summarizes the performance indicators for the last three cycles, with TDC demonstrating the worst tracking performance due to its low robustness to nonlinear friction disturbance. In contrast, the FCC delivers the best performance in all performance indicators. Specifically, according to the variance shown in Figure 9, the addition of AFLS improves TDC performance by 97.885% and 97.712% for joint 1 and joint 2, respectively. The numerical experimental results reveal that, compared with the other two controllers, the proposed FCC delivers the best tracking performance in both transient and steady-state aspects.

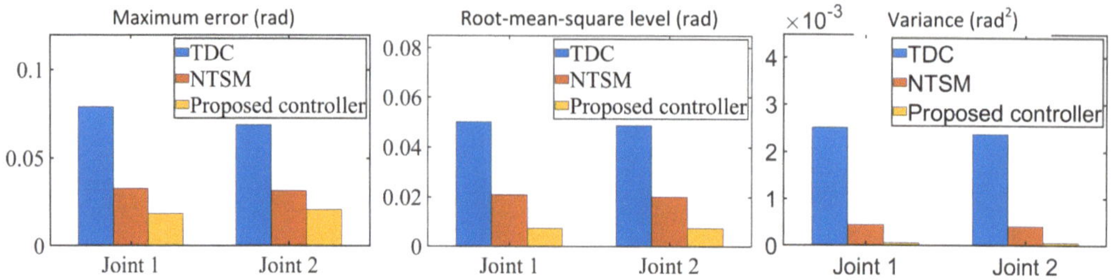

Figure 13. Performance indices evaluation during the last three cycles under significant frictional disturbance.

We conducted a series of comparative experiments to verify the effectiveness and fast convergence of the FCC under various levels of frictional disturbances. The results showed that the proposed control algorithm exhibited excellent position-tracking performance. As the disturbance increased, the performance of FCC became increasingly superior, confirming the superiority and effectiveness of AFLS in terms of disturbance rejection. Therefore, the FCC not only possesses the fast convergence and efficiency of the TDC algorithm but also has the high disturbance-rejection capability and faster convergence rate of the AFLS algorithm. This further confirms the feasibility and effectiveness of the FCC in practical control applications.

4.4. Further Analysis of Error for Three Algorithms

In this study, we analyzed the kinematic errors of three algorithms (TDC, NTSM, and FCC) applied to two joints of a robotic arm, using 20,000 samples. Box plots were used to display the data distribution and evaluate algorithm performance.

Figure 14 illustrates the distribution of kinematic errors in the robotic arm, with the x-axis indicating the joints and the y-axis indicating error values. Each box represents the error distribution of a joint, with the upper and lower boundaries indicating the upper and lower quartiles (Q3 and Q1), the middle line representing the median (Q2), and the internal line of the box indicating the mean. Outliers are shown outside the upper and lower limits of the box.

TDC 1 and TDC 2 represent the TDC algorithm applied to joints 1 and 2, respectively. Similarly, NTSM 1, NTSM 2, FCC 1, and FCC 2 represent the NTSM and FCC algorithms applied to joints 1 and 2 of the robotic arm. From Figure 14, it is apparent that the TDC algorithm has the largest error distribution, with the greatest distance between the upper and lower limits, but there are no outliers. The error distribution of the NTSM algorithm is relatively stable, with a lower height of the box and a smaller distance between the upper and lower limits, except for an outlier in NTSM 2. The error distribution of the FCC algorithm is the most stable, with the lowest average height of the box and the smallest distance between the upper and lower limits, and there are no outliers.

The box plot indicates that the FCC algorithm has smaller and more evenly distributed kinematic errors in each joint of the robotic arm, indicating superior tracking performance. Therefore, the FCC algorithm proposed in this study is an effective algorithm for controlling robotic arms.

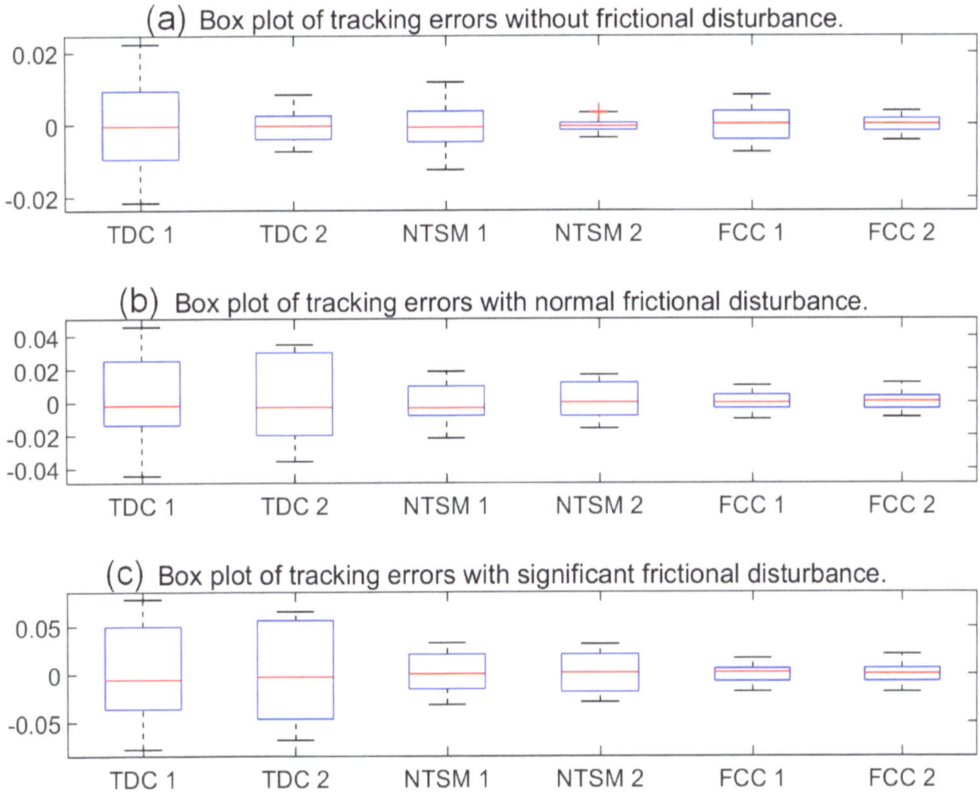

Figure 14. Tracking-error curves of the three controllers under significant frictional disturbance.

5. Conclusions

In this paper, we propose a novel friction compensation controller (FCC) that integrates time delay estimation (TDE) and an adaptive fuzzy logic system (AFLS). The FCC method utilizes the TDE technique to eliminate and estimate the unknown dynamic functions of the system and incorporates an element for injecting expected error dynamics and an AFLS as the third element to handle strong nonlinearity and TDE errors. Additionally, an adaptive rule is designed to update the parameters of the fuzzy logic system online, improving the performance of the controller. Compared to existing TDC methods, the proposed method effectively suppresses TDE errors and provides a faster convergence rate. Moreover, the controller in this paper does not require complex offline parameter identification or precomputed models but can be implemented directly online, making it more suitable for practical systems.

Numerical experiments demonstrate that the proposed friction compensation control algorithm has fast convergence, high efficiency, and strong anti-interference capability, and achieves excellent position-tracking performance even under a large amount of nonlinear interference. Specifically, the addition of AFLS improves the performance of TDC by an average of 93.0623% and 88.125% for joint 1 and joint 2, respectively.

This study provides novel ideas and methods for the development of robot arm controllers and serves as a reference for the design of controllers in other industrial automation fields. The results indicate that the proposed FCC is a promising solution for controlling robotic systems with frictional effects and nonlinearities.

Author Contributions: Conceptualization, Y.S. and X.L.; methodology, X.L.; software, X.L.; validation, Y.S. and X.L.; formal analysis, Y.S. and X.L.; investigation, X.L.; data curation, X.L.; writing—original draft preparation, Y.S. and X.L.; writing—review and editing, Y.S. and X.L.; visualization, Y.S. and X.L.; supervision, Y.W.; project administration, Y.W.; funding acquisition, Y.W. All authors have read and agreed to the published version of the manuscript.

Funding: This work was supported in part by the National Natural Science Foundation of China under Grant 51975336, in part by the Key Research and Development Program of Shandong Province under Grant 2020JMRH0202, in part by Shandong Province New Old Energy Conversion Major Industrial Tackling Projects under Grant 2021-13, in part by Key R&D project of Jining City (2021DZP005), in part by Key R&D Program of Shandong Province (2022CXGC020701), and in part by Education and Teaching Reform Research Projects of Shandong University (2022Y133, 2022Y124, 2022Y312).

Data Availability Statement: The datasets of the current study are available from the corresponding author upon reasonable request.

Conflicts of Interest: The authors declare that they have no conflict of interest.

References

1. Izadbakhsh, A. An observer-based output tracking controller for electrically driven cooperative multiple manipulators with adaptive Bernstein-type approximator. *Robotica* **2022**, *40*, 2295–2319. [CrossRef]
2. Thomasson, O.; Battarra, M.; Erdoğan, G.; Laporte, G. Scheduling twin robots in a palletising problem. *Int. J. Prod. Res.* **2018**, *56*, 518–542. [CrossRef]
3. Sun, Y.; Wan, Y.; Ma, H.; Liang, X. Compensation control of hydraulic manipulator under pressure shock disturbance. *Nonlinear Dyn.* **2023**, 1 –17. [CrossRef]
4. Mir-Nasiri, N. Design, modelling and control of four-axis parallel robotic arm for assembly operations. *Assem. Autom.* **2004**, *24*, 365–369. [CrossRef]
5. Sabet, S.; Poursina, M. Computed torque control of fully-actuated nondeterministic multibody systems. *Multibody Syst. Dyn.* **2017**, *41*, 347–365. [CrossRef]
6. Jin, M.; Kang, S.H.; Chang, P.H. Robust compliant motion control of robot with nonlinear friction using time-delay estimation. *IEEE Trans. Ind. Electron.* **2008**, *55*, 258–269. [CrossRef]
7. Sun, Y.; Wan, Y.; Ma, H.; Liang, X. Workspace Description and Evaluation of Master-Slave Dual Hydraulic Manipulators. *Actuators* **2023**, *12*, 9. [CrossRef]
8. Cio, Y.S.L.K.; Raison, M.; Menard, C.L.; Achiche, S. Proof of concept of an assistive robotic arm control using artificial stereovision and eye-tracking. *IEEE Trans. Neural Syst. Rehabil. Eng.* **2019**, *27*, 2344–2352. [CrossRef]
9. Lee, J.; Deshpande, N.; Caldwell, D.G.; Mattos, L.S. Microscale precision control of a computer-assisted transoral laser microsurgery system. *IEEE/ASME Trans. Mechatron.* **2020**, *25*, 604–615. [CrossRef]
10. Chen, C.; Liu, Z.; Zhang, Y.; Xie, S. Coordinated motion/force control of multiarm robot with unknown sensor nonlinearity and manipulated object's uncertainty. *IEEE Trans. Syst. Man, Cybern. Syst.* **2016**, *47*, 1123–1134. [CrossRef]
11. Zhu, Y.; Qiao, J.; Zhang, Y.; Guo, L. High-precision trajectory tracking control for space manipulator with neutral uncertainty and deadzone nonlinearity. *IEEE Trans. Control Syst. Technol.* **2018**, *27*, 2254–2262. [CrossRef]
12. Zhang, A.; Ma, S.; Li, B.; Wang, M.; Guo, X.; Wang, Y. Adaptive controller design for underwater snake robot with unmatched uncertainties. *Sci. China Inf. Sci.* **2016**, *59*, 1–15. [CrossRef]
13. Hsia, T.S. A new technique for robust control of servo systems. *IEEE Trans. Ind. Electron.* **1989**, *36*, 1–7. [CrossRef]
14. Hsia, T.S.; Lasky, T.A.; Guo, Z. Robust independent joint controller design for industrial robot manipulators. *IEEE Trans. Ind. Electron.* **1991**, *38*, 21–25. [CrossRef]
15. Kali, Y.; Saad, M.; Benjelloun, K.; Khairallah, C. Super-twisting algorithm with time delay estimation for uncertain robot manipulators. *Nonlinear Dyn.* **2018**, *93*, 557–569. [CrossRef]
16. Ahmed, S.; Wang, H.; Tian, Y. Adaptive high-order terminal sliding mode control based on time delay estimation for the robotic manipulators with backlash hysteresis. *IEEE Trans. Syst. Man Cybern. Syst.* **2019**, *51*, 1128–1137. [CrossRef]
17. Van, M.; Ge, S.S.; Ren, H. Finite time fault tolerant control for robot manipulators using time delay estimation and continuous nonsingular fast terminal sliding mode control. *IEEE Trans. Cybern.* **2016**, *47*, 1681–1693. [CrossRef]
18. Jin, M.; Lee, J.; Ahn, K.K. Continuous nonsingular terminal sliding-mode control of shape memory alloy actuators using time delay estimation. *IEEE/ASME Trans. Mechatron.* **2014**, *20*, 899–909. [CrossRef]
19. Lee, J.; Chang, P.H.; Jamisola, R.S. Relative impedance control for dual-arm robots performing asymmetric bimanual tasks. *IEEE Trans. Ind. Electron.* **2013**, *61*, 3786–3796. [CrossRef]
20. Liang, X.; Wan, Y.; Zhang, C.; Kou, Y.; Xin, Q.; Yi, W. Robust position control of hydraulic manipulators using time delay estimation and nonsingular fast terminal sliding mode. *Proc. Inst. Mech. Eng. Part I: J. Syst. Control Eng.* **2018**, *232*, 50–61. [CrossRef]

21. Geng, J.; Sheng, Y.; Liu, X. Time-varying nonsingular terminal sliding mode control for robot manipulators. *Trans. Inst. Meas. Control* **2014**, *36*, 604–617. [CrossRef]
22. Jin, M.; Kang, S.H.; Chang, P.H.; Lee, J. Robust control of robot manipulators using inclusive and enhanced time delay control. *IEEE/ASME Trans. Mechatron.* **2017**, *22*, 2141–2152. [CrossRef]
23. Adhikary, N.; Mahanta, C. Sliding mode control of position commanded robot manipulators. *Control Eng. Pract.* **2018**, *81*, 183–198. [CrossRef]
24. Lee, H.J.; Lee, J.J. Time delay control of a shape memory alloy actuator. *Smart Mater. Struct.* **2004**, *13*, 227. [CrossRef]
25. Cho, S.J.; Jin, M.; Kuc, T.Y.; Lee, J.S. Stability guaranteed auto-tuning algorithm of a time-delay controller using a modified Nussbaum function. *Int. J. Control* **2014**, *87*, 1926–1935. [CrossRef]
26. Jin, F.; Qiu, T. Adaptive time delay estimation based on the maximum correntropy criterion. *Digit. Signal Process.* **2019**, *88*, 23–32. [CrossRef]
27. Bae, H.J.; Jin, M.; Suh, J.; Lee, J.Y.; Chang, P.H.; Ahn, D.s. Control of robot manipulators using time-delay estimation and fuzzy logic systems. *J. Electr. Eng. Technol.* **2017**, *12*, 1271–1279. [CrossRef]
28. Arimoto, S.; Kawamura, S.; Miyazaki, F.; Tamaki, S. Learning control theory for dynamical systems. In Proceedings of the 1985 24th IEEE Conference on Decision and Control, Fort Lauderdale, FL, USA, 11–13 December 1985; pp. 1375–1380.
29. Yao, J.; Deng, W.; Jiao, Z. Adaptive control of hydraulic actuators with LuGre model-based friction compensation. *IEEE Trans. Ind. Electron.* **2015**, *62*, 6469–6477. [CrossRef]

Disclaimer/Publisher's Note: The statements, opinions and data contained in all publications are solely those of the individual author(s) and contributor(s) and not of MDPI and/or the editor(s). MDPI and/or the editor(s) disclaim responsibility for any injury to people or property resulting from any ideas, methods, instructions or products referred to in the content.

actuators MDPI

Article

Research on Fault-Tolerant Control of Distributed-Drive Electric Vehicles Based on Fuzzy Fault Diagnosis

Shaopeng Zhu [1,2], Haojun Li [1], Guodong Wang [3], Chenyang Kuang [1], Huipeng Chen [4,5], Jian Gao [5,6,*] and Wei Xie [7]

1 Power Machinery & Vehicular Engineering Institute, College of Energy Engineering, Zhejiang University, Hangzhou 310058, China; spzhu@zju.edu.cn (S.Z.); hjli26@zju.edu.cn (H.L.); chenyang_kuang@zju.edu.cn (C.K.)
2 Key Laboratory of Clean Energy and Carbon Neutrality of Zhejiang Province, Hangzhou 310013, China
3 Nanjing Research Institute of Electronic Engineering, Nanjing 210007, China; wguodong0606@163.com
4 School of Mechanical Engineering, Hangzhou DianZi University, Hangzhou 310018, China; hpchen@hdu.edu.cn
5 Jiaxing Research Institute, Zhejiang University, Jiaxing 314031, China
6 Polytechnic Institute, Zhejiang University, Hangzhou 310058, China
7 Fujian Provincial Key Laboratory of Intelligent Identification and Control of Complex Dynamic System, Quanzhou 362216, China; wei.xie@fjirsm.ac.cn
* Correspondence: gj_zju2022@163.com; Tel.: +86-139-1691-0595

Abstract: This paper addresses the fault problem in distributed-four-wheel-drive electric vehicle drive systems. First, a fault-factor-based active fault diagnosis strategy is proposed. Second, a fault-tolerant controller is designed to reconstruct motor drive torque based on vehicle stability. This controller ensures that the vehicle maintains stability by providing fault-free motor output torque based on fault diagnosis results. To validate the effectiveness of the fault diagnosis and fault-tolerant control, SIL simulation is conducted using MATLAB/Simulink and CarSim. A hardware-in-the-loop (HIL) simulation platform with the highest confidence level is established based on NI PXI and CarSim RT. Through the HIL simulation experiments, it is shown that the proposed control strategy can accurately diagnose the operating state of the motor, rebuild the motor torque based on stability, and demonstrate robust stability when the drive system fails. Under various fault conditions, the maximum error in the vehicle lateral angular velocity is less than 0.017 rad/s and the maximum deviation in the lateral direction is less than 0.7 m. These findings substantiate the highly robust stability of the proposed method.

Keywords: distributed drive; electric vehicle; fault diagnosis; failure control; torque reconstruction

Citation: Zhu, S.; Li, H.; Wang, G.; Kuang, C.; Chen, H.; Gao, J.; Xie, W. Research on Fault-Tolerant Control of Distributed-Drive Electric Vehicles Based on Fuzzy Fault Diagnosis. *Actuators* **2023**, *12*, 246. https://doi.org/10.3390/act12060246

Academic Editors: Oscar Barambones, Jose Antonio Cortajarena and Patxi Alkorta

Received: 17 May 2023
Revised: 12 June 2023
Accepted: 12 June 2023
Published: 13 June 2023

Copyright: © 2023 by the authors. Licensee MDPI, Basel, Switzerland. This article is an open access article distributed under the terms and conditions of the Creative Commons Attribution (CC BY) license (https://creativecommons.org/licenses/by/4.0/).

1. Introduction

A distributed-drive electric vehicle typically refers to an electric vehicle driven by wheel-side motors or hub motors, offering several advantages over classical centralized single-motor-drive electric vehicles. By eliminating the need for transmission shafts, gearboxes, and other components, the distributed-drive electric vehicle reduces vehicle weight, optimizes the chassis structure, and ensures a more compact design, higher space utilization, and improved transmission efficiency [1]. Moreover, the ability to individually control each driving motor allows for more flexible and precise torque control, enabling the implementation of advanced features such as operational stability, anti-slip driving, and an electronic differential [2,3]. Therefore, it serves as an ideal platform for the application of advanced vehicle dynamics control theories. Furthermore, the redundancy provided by the distributed-drive electric vehicle drive system, with multiple power sources, enables the coordinated control of multiple actuator units to achieve the desired driving torque. This redundancy configuration ensures that even if some of the drive system motors fail, the vehicle can maintain the intended driving state, thus ensuring driving safety [4,5].

However, at the same time, a distributed-drive electric vehicle drive system with multiple actuator redundancy configurations increases the probability of failure. To ensure stable and safe driving, failure control can be employed to take advantage of the redundancy configuration of the drive system actuators. In a distributed-drive electric vehicle, failure control mainly refers to measures such as fault diagnosis, separation, and control of the driving motor in the drive system hardware or software after the occurrence of a failure, but before the vehicle becomes unstable. This reduces the tendency toward instability and improves the safety and reliability of driving [6].

Fault diagnosis provides the theoretical basis for failure control, and precise and reliable identification results are essential for failure control. Reference [7] proposes a fault diagnosis method for distributed-drive electric vehicles based on an online estimator of the tire–road friction coefficient. This method can work well even in the presence of interference and modeling errors and has good robustness. Reference [8] proposes a fault diagnosis method based on actuator redundancy. By comparing the actual motor torque with its nominal value, motor faults can be detected more accurately, and faulty motors can be assessed and located. Reference [9] designs a distributed-drive electric vehicle fault detector based on the Extended Kalman Filter (EKF) wheel hub motor, which can observe hub motor parameter changes caused by faults and identify motor faults. Reference [10] applies differential algebraic methods to diagnose and isolate faults by establishing a motor model. After identifying the faulty motor through fault diagnosis, an effective failure control algorithm needs to be designed to reconstruct torque and fully utilize the advantages of multiple actuator redundancy, ensuring that the vehicle can track the driver's expected trajectory even if the drive system fails. Reference [11] proposes to compensate for the loss of the faulty motor by using an uncompromised one but lacks consideration of actual working conditions. Reference [12] designs a fault control strategy for distributed-four-wheel drive electric vehicles based on sliding mode control and proposes the use of the motor weight matrix to characterize whether it is working properly. Reference [13] applies fault factor limits to the faulty motor and reconstructs the torque of each wheel motor. Reference [14] proposes a distributed-drive electric vehicle drive failure control method based on power and stability to cope with different failure modes and driving conditions. Reference [15] classifies different failure situations into two categories: the deceleration and stopping type and the maintaining speed type. For the latter, the method suggests shutting down the opposite motor of the faulty motor and using fuzzy control to maintain the expected driving trajectory of the vehicle. References [16,17] propose a hierarchical failure control strategy, which consists of signal acquisition, logic recognition, and failure control layers. Reference [18] designs a torque distribution optimization strategy based on direct lateral force control on top of the conventional drive system failure control to ensure optimal overall performance.

At present, in the event of a failure, the control of the distributed-drive electric vehicle drive system mainly consists of two components: fault diagnosis and failure control. Fault diagnosis primarily involves comparing the actual output torque of the drive motor with the expected output torque or a characteristic value that represents the current output torque, which allows for the determination of whether the motor is malfunctioning. Conversely, failure control mostly employs rule-based methods. Based on the failure mode of the drive system, an appropriate control scheme is employed to reconstruct the driving torque of the remaining functional motors, thereby preserving the vehicle's power and driving stability.

This paper presents the design of a fault diagnosis tool that uses a cumulative judgment approach, comparing the actual and expected values of the drive motor. By relying on this fault diagnostic method, misjudgment of motor failure caused by motor torque fluctuation can be avoided. Leveraging a set of rules designed to prioritize driving stability, a failure control algorithm is developed for the remaining functional motors to ensure the seamless operation of the vehicle.

2. Drive System Fault Diagnosis and Failure Control Strategy Design

This paper presents a failure control strategy, encompassing fault diagnosis and torque reconstruction, as illustrated in Figure 1. Based on the torque distribution outcomes from lateral stability control, the ratio between the actual motor torque signal collected from the drive system and the expected output motor torque, along with its rate of change, are used as input variables for fuzzy control. The output is the failure factor k_l of each drive motor. The current failure mode is identified based on the failure factor k_l, which can be categorized into six different modes. The torque reconstruction process follows a set of predefined rules, ensuring optimal driving stability and satisfying the driver's power requirements to the maximum possible extent, resulting in the final torque input to the drive system, T_{is}.

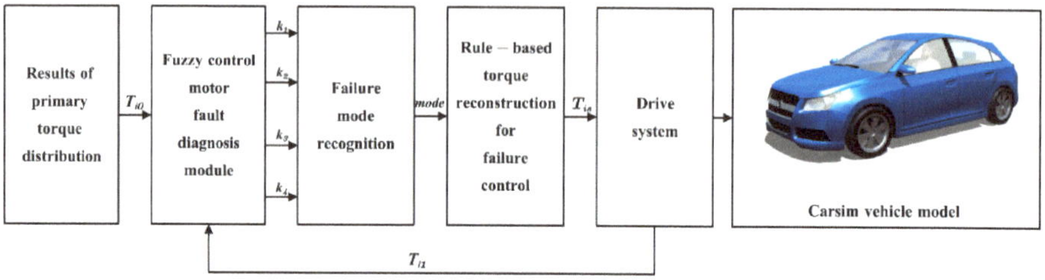

Figure 1. Distributed-Drive Electric Vehicle Drive System Failure Control Framework.

2.1. Active Fault Diagnosis of the Drive System

Based on fuzzy control, a proactive fault diagnosis tool is designed for the drive motor, allowing for timely and accurate monitoring of the motor's working status, proactively detecting malfunctioning motors, isolating them, and taking failure control measures to ensure safe vehicle operation.

In the proactive fault diagnosis tool for the drive motor, the output failure factor k_l has a value of either 0 or 1, with 0 indicating a motor malfunction and 1 indicating normal operation. ρ is defined as:

$$\rho = \frac{T_{i1}}{T_{i0}} \tag{1}$$

In the equation, T_{i0} represents the expected output torque of the drive motor and T_{i1} represents the actual output torque of the drive motor. Both can be obtained by collecting the CAN communication messages of the motor.

A proactive fault diagnosis tool is designed based on fuzzy control, using ρ and its rate of change, $\Delta\rho$, as input variables and σ as the output variable to evaluate the motor's operating status at this moment. First, the variables are fuzzified, and the fuzzy sets for the input and output variables are defined. The range of ρ is [0, 2], divided into five fuzzy sets. The range of $\Delta\rho$ is [−100, 100], divided into three fuzzy sets. The range of the output variable σ is [0.85, 1.05], also divided into three fuzzy sets. The partition of the fuzzy variable fuzzy sets is shown in Table 1. The membership functions of each variable are shown in Figures 2–4.

Table 1. Fuzzy sets of input and output variables for the Fault Diagnosis Fuzzy Controller.

ρ	$\Delta\rho$	σ
MS (Smaller)	N (Negative)	F (Failure)
S (Small)	Z (Zero)	Z (Possible failure)
Z (Zero)	P (Positive)	N (Normal)
B (Large)		
MB (Larger)		

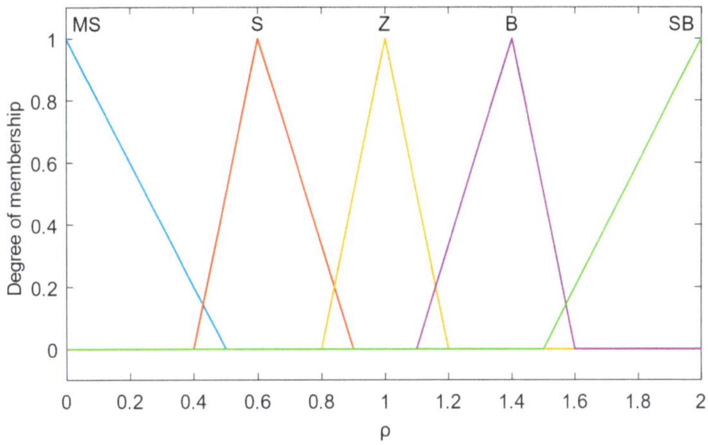

Figure 2. Membership functions of input variable ρ.

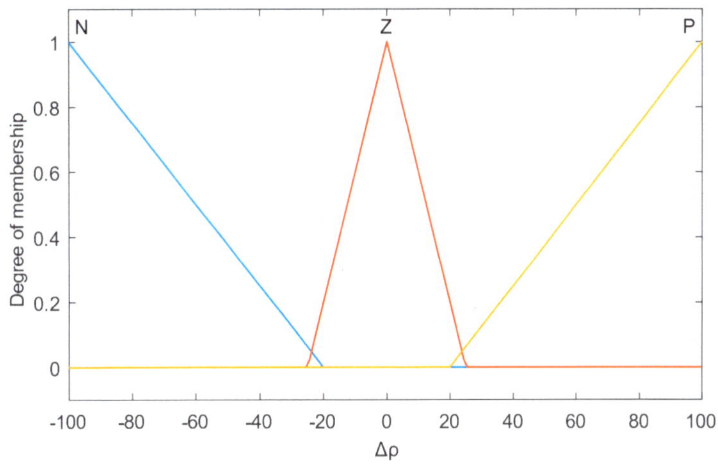

Figure 3. Membership functions of input variable $\Delta\rho$.

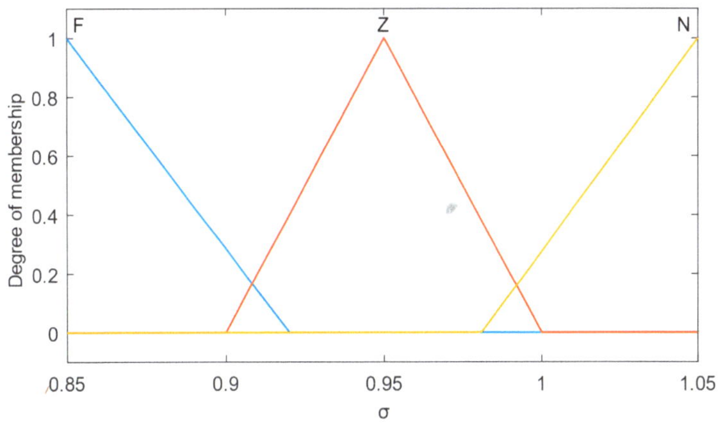

Figure 4. Membership functions of output variable σ.

Subsequently, based on the fuzzy rules stored in the rule base, fuzzy inference determines the corresponding output variable based on the combination of input variables. The design of fuzzy rules influences the control effectiveness of the fuzzy control and they are established by the continuous fine-tuning of experience and model. The fuzzy inference rules established are shown in Table 2.

Table 2. Fuzzy inference rules of the Fault Diagnosis Fuzzy Control.

$\Delta\rho$ \ ρ	MS	S	Z	B	MB
N	F	F	Z	N	F
Z	F	Z	N	Z	F
P	F	N	Z	F	F

The centroid method is adopted in this article for defuzzification, which ensures a smoother output. As the expected output torque may be zero, leading to a zero point in the denominator of ρ and consequently, an infinite value for $\Delta\rho$, a simple filtering process needs to be applied to the expected output torque before feeding it into the fuzzy controller:

$$T_{i0} = \max(0.01, |T_{i0}|) \tag{2}$$

The current motor status is determined based on the output variable σ. If $\sigma \geq 0.95$, the motor is in normal working condition, otherwise, a motor fault is present.

Figure 5 shows a comparison between the expected and actual values of the output torque collected during the on-road test of a distributed-drive vehicle equipped with hub motors on the rear wheels.

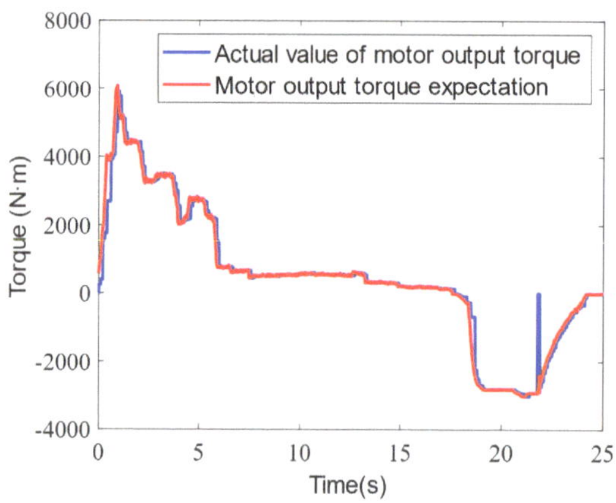

Figure 5. Comparison of the actual and desired output torque of the hub motor.

It can be observed that the actual and expected values of the output torque of the drive motor are not the same, as small deviations exist under normal circumstances. Fuzzy control can ignore these deviations and accurately determine the working status of the motor. Moreover, at approximately 22 s, the motor experiences torque loss due to communication interference and other factors. This situation is brief, and the motor quickly returns to normal operation. To address this issue, the output results of the fuzzy controller need to be cumulatively analyzed. If the number of fault judgments within a given period exceeds the set threshold, the motor is considered to have failed, and it will be deactivated for isolation treatment. The determination period and threshold can be adjusted according to the real-time requirements of the vehicle.

By tracking the motor's working status using two global variables and analyzing them with the set threshold, the failure factor k_l of the motor can be outputted, and $ki = 0$ would lead to the locking of the output results to prevent secondary damage to the motor due to frequent failures.

2.2. Drive System Failure Control Strategy

As shown in Figure 6, the failure mode of the drive motor can be divided into six types: no failure, single-motor failure, double-motor failure on different axles and sides, double-motor failure on the same axle, double-motor failure on the same side, and multiple-motor failure. The analysis of the six failure modes is as follows:

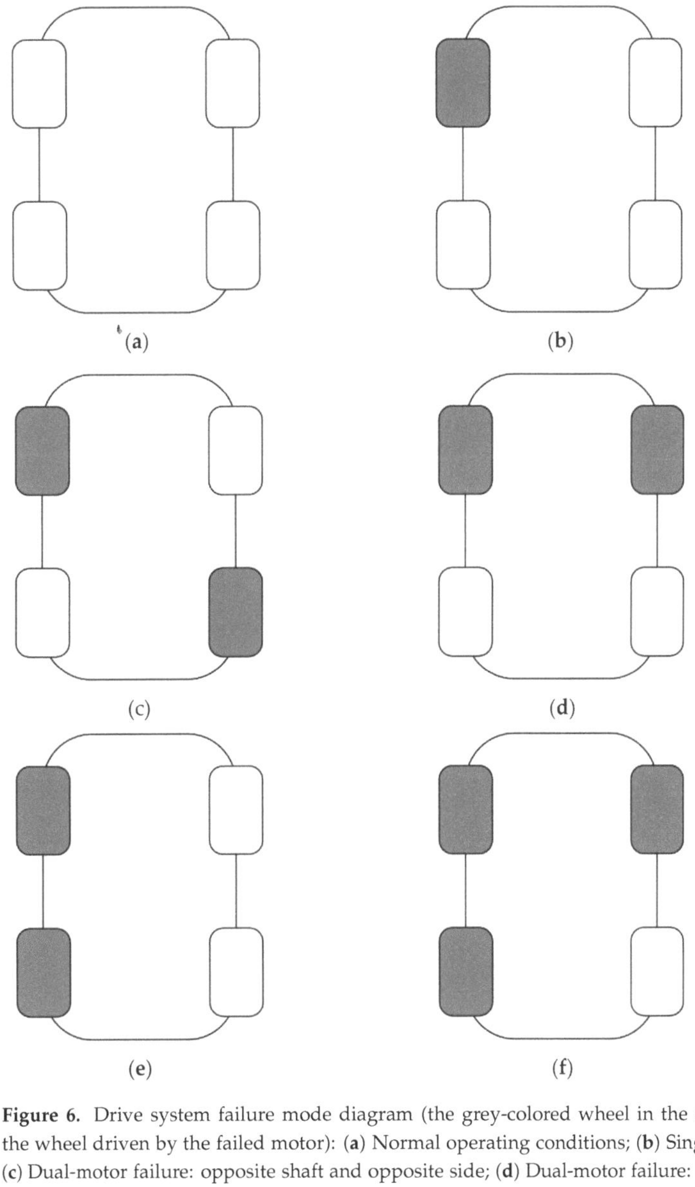

Figure 6. Drive system failure mode diagram (the grey-colored wheel in the figure represents the wheel driven by the failed motor): (**a**) Normal operating conditions; (**b**) Single-motor failure; (**c**) Dual-motor failure: opposite shaft and opposite side; (**d**) Dual-motor failure: coaxial; (**e**) Dual-motor failure: same side; (**f**) Three-motor failure.

(1) No failure: the vehicle operates normally; (2) Single-motor failure: the lost longitudinal force can be compensated for by reconstructing the torque of the normal working

motor, while the lost lateral moment can be compensated for through the driver's steering angle or reconstructing the torque of the motor; (3) Double-motor failure on different axles and sides: the lost longitudinal driving force and lateral moment can be compensated for by reconstructing the torque of the two non-failed motors, prioritizing stability; (4) Double-motor failure on the same axle: in this case, the drive system is considered as a front- or rear-axle drive format; (5) Double-motor failure on the same side: the lost lateral moment cannot be compensated for, and the vehicle needs to be stopped immediately for treatment; (6) Multiple-motor failure: it is impossible to meet driving requirements, and the vehicle needs to be stopped immediately for treatment.

In addition to normal driving and situations that require immediate braking, this article will propose torque reconstruction control strategies based on vehicle stability for failure modes (2)–(4). For this purpose, a seven-degrees-of-freedom distributed-dive vehicle model shown in Figure 7 is considered, which includes the longitudinal and lateral movement of the vehicle, the yaw motion around the *z*-axis, and the rotational motion of the four wheels. In the dynamic equation of the model, W is the wheelbase, a is the distance from the front wheel to the center of gravity, b is the distance from the rear wheel to the center of gravity, r is the rolling radius of the wheel, and δ represents the steering angle of the front wheel. $ij = fl, fr, rl, rr$ respectively represent the left front wheel, right front wheel, left rear wheel, and right rear wheel.

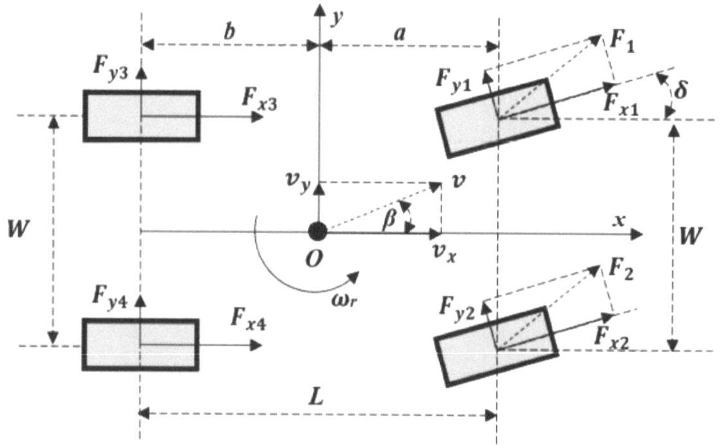

Figure 7. Distributed-drive electric vehicle 7-degrees-of-freedom model.

When a single-motor failure occurs, taking the example of the failure of the left front wheel motor, the motor output should be isolated and shut off, i.e., $T_{fls} = 0$. To improve the real-time performance of the control algorithm, a distribution method that considers the vertical load distribution of the front and rear axles is used, which can make full use of the road attachment rate, at which point the constraint is expressed as follows:

$$\begin{cases} T_{fls} = 0 \\ T_{fls} + T_{frs} + T_{rls} + T_{rrs} = T_d \\ \left(T_{fls} + T_{frs}\right) a\sin\delta + \left(T_{frs}\cos\delta - T_{fls}\cos\delta + T_{rrs} - T_{rls}\right)\frac{W}{2} = \Delta M \cdot r \\ \frac{T_{fls}+T_{frs}}{T_{rls}+T_{rrs}} = k_z \end{cases} \quad (3)$$

where T_d is the total driving torque, ΔM is the additional lateral moment, and k_z is the ratio of the sum of the front wheel vertical loads to the sum of the rear wheel vertical loads, which is defined as:

$$k_z = \frac{F_{zfl} + F_{zfr}}{F_{zrl} + F_{zrr}} \quad (4)$$

where F_z is the vertical load of the wheel.

Solving Equation (3) yields the drive torque of the three motors that have not failed:

$$\begin{cases} T_{fls} = 0 \\ T_{frs} = \frac{k_z}{1+k_z} T_d \\ T_{rls} = \left(\frac{k_z a \sin\delta}{W} + \frac{k_z \cos\delta + 1}{2} \right) \cdot \frac{T_d}{1+k_z} - \frac{\Delta M \cdot r}{W} \\ T_{rrs} = \frac{\Delta M \cdot r}{W} - \left(\frac{k_z a \sin\delta}{W} + \frac{k_z \cos\delta - 1}{2} \right) \cdot \frac{T_d}{1+k_z} \end{cases} \tag{5}$$

Based on the vehicle's lateral stability, when the motor output exceeds the maximum torque limit, sacrificing some power is necessary to ensure vehicle stability. The new constraint condition can be expressed as follows:

$$\begin{cases} T_{fls} = 0 \\ T_{frs} = T_{frm} \\ \left(T_{fls} + T_{frs} \right) a \sin\delta + \left(T_{frs} \cos\delta - T_{fls} \cos\delta + T_{rrs} - T_{rls} \right) \frac{W}{2} = \Delta M \cdot r \\ \frac{T_{fls} + T_{frs}}{T_{rls} + T_{rrs}} = k_z \end{cases} \tag{6}$$

where T_{frm} is the maximum output torque of the right front wheel motor at the current speed. Solving for the reconstructed output torque of each motor yields the following results:

$$\begin{cases} T_{fls} = 0 \\ T_{frs} = T_{frm} \\ T_{rls} = \frac{T_{frm}}{2k_z} + \frac{\Delta M \cdot r - T_{frm} a \sin\delta}{W} - \frac{T_{frm} \cos\delta}{2} \\ T_{rrs} = \frac{T_{frm}}{2k_z} - \frac{\Delta M \cdot r - T_{frm} a \sin\delta}{W} + \frac{T_{frm} \cos\delta}{2} \end{cases} \tag{7}$$

When a dual-motor failure occurs on opposite sides of the vehicle, taking the example of the failure of the left front and right rear wheel motors, these motors should be isolated and their torque output should be shut off, i.e., $T_{fls} = 0$ and $T_{rrs} = 0$. The constraint condition can be expressed as follows:

$$\begin{cases} T_{fls} = T_{rrs} = 0 \\ T_{fls} + T_{frs} + T_{rls} + T_{rrs} = T_d \\ \left(T_{fls} + T_{frs} \right) a \sin\delta + \left(T_{frs} \cos\delta - T_{fls} \cos\delta + T_{rrs} - T_{rls} \right) \frac{W}{2} = \Delta M \cdot r \end{cases} \tag{8}$$

Solving Equation (8) yields the output torque of the two normally operating motors:

$$\begin{cases} T_{fls} = 0 \\ T_{frs} = \frac{2\Delta M \cdot r + W \cdot T_d}{2a \sin\delta + W \cos\delta + W} \\ T_{rls} = \frac{(2a \sin\delta + W \cos\delta) T_d - 2\Delta M \cdot r}{2a \sin\delta + W \cos\delta + W} \\ T_{rrs} = 0 \end{cases} \tag{9}$$

Based on the vehicle's lateral stability and considering the maximum output torque limit of the motors, the constraint condition should select the maximum output torque of the larger motor after reconstructing the normally operating motors. Taking the example of the right front wheel torque being greater than the left rear wheel, the constraint condition is as follows:

$$\begin{cases} T_{fls} = T_{rrs} = 0 \\ T_{frs} = T_{frm} \\ \left(T_{fls} + T_{frs} \right) a \sin\delta + \left(T_{frs} \cos\delta - T_{fls} \cos\delta + T_{rrs} - T_{rls} \right) \frac{W}{2} = \Delta M \cdot r \end{cases} \tag{10}$$

Solving Equation (10) yields the output torque of the two non-failed motors after reconstruction when a dual-motor failure occurs on opposite sides of the vehicle:

$$\begin{cases} T_{fls} = 0 \\ T_{frs} = T_{frm} \\ T_{rls} = \frac{(2a\sin\delta + W\cos\delta)T_{frm} - 2\Delta M \cdot r}{W} \\ T_{rrs} = 0 \end{cases} \tag{11}$$

When a dual-motor failure occurs on the same axis of the vehicle, taking the example of both motors on the rear axle failing, these motors should be isolated and their output torque should be shut off, i.e., T_{rls} = 0 and T_{rrs} = 0. At this time, the distributed-drive system changes from four-wheel drive to two-wheel drive, and the constraint condition can be expressed as follows:

$$\begin{cases} T_{rls} = T_{rrs} = 0 \\ T_{fls} + T_{frs} + T_{rls} + T_{rrs} = T_d \\ \left(T_{fls} + T_{frs}\right)a\sin\delta + \left(T_{frs}\cos\delta - T_{fls}\cos\delta + T_{rrs} - T_{rls}\right)\frac{W}{2} = \Delta M \cdot r \end{cases} \tag{12}$$

Solving Equation (12) yields:

$$\begin{cases} T_{fls} = \frac{(2a\sin\delta + W\cos\delta)T_d - 2\Delta M \cdot r}{2W\cos\delta} \\ T_{frs} = \frac{(W\cos\delta - 2a\sin\delta)T_d + 2\Delta M \cdot r}{2W\cos\delta} \\ T_{rls} = 0 \\ T_{rrs} = 0 \end{cases} \tag{13}$$

Due to the maximum output torque limit of the motors, the constraint condition should select the maximum output torque of the larger motor after reconstructing the normally operating motors. Taking the example of the right front wheel torque being greater than the left front wheel, the constraint condition is as follows:

$$\begin{cases} T_{rls} = T_{rrs} = 0 \\ T_{frs} = T_{frm} \\ \left(T_{fls} + T_{frs}\right)a\sin\delta + \left(T_{frs}\cos\delta - T_{fls}\cos\delta + T_{rrs} - T_{rls}\right)\frac{W}{2} = \Delta M \cdot r \end{cases} \tag{14}$$

Thus, the output torque of the two motors on the front axle after reconstruction can be obtained when both motors on the rear axle fail:

$$\begin{cases} T_{fls} = T_{flm} \\ T_{frs} = \frac{2\Delta M \cdot r - (W\cos\delta - 2a\sin\delta)T_{frm}}{W\cos\delta + 2a\sin\delta} \\ T_{rls} = 0 \\ T_{rrs} = 0 \end{cases} \tag{15}$$

3. Drive System Failure Control Strategy Simulation Verification

To verify the effectiveness of the designed failure control strategy, this study uses both simulated software and hardware-in-the-loop simulation methods. The simulations are based on the most common driving conditions of vehicles such as straight-line driving and single-lane changing. The stability control effect of the vehicle is analyzed under various failure conditions, including complete and partial failures of a single motor, dual motors on the same axis, and dual motors on opposite sides of the vehicle.

3.1. Software-in-the-Loop Simulation Results

The software-in-the-loop simulation verification of the designed fault-tolerant control strategy for the distributed-drive system mainly includes the verification of the active fault diagnostic of the drive system and the verification of the fault-tolerant control strategy after identifying the failure. Since the simultaneous failure of the two motors on the same axis is an automatic change of the vehicle from four-wheel drive to two-wheel drive, it is not further verified here.

The actual output torque and the expected output torque of the drive motor shown in Figure 5 are used as inputs to verify the effectiveness of the fault diagnostic. The results are shown in Figure 8, which shows that in most cases, the value of σ is greater than or equal to 0.95, and occasionally it is less than 0.95. However, due to the short duration, the motor can still operate normally, and the failure factor remains at 1, indicating that the motor is in a normal operating state. This verifies that the fault diagnosis has good robustness and can avoid misjudgments caused by torque losses due to temporary signal fluctuations.

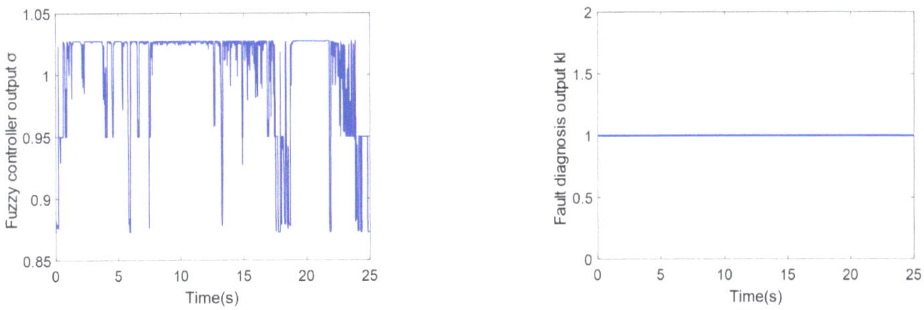

(a) (b)

Figure 8. Robustness simulation results of the drive system fault diagnosis: (**a**) Fuzzy controller output σ; (**b**) Fault diagnosis output k_l.

Based on the actual torque output expectation collected from the wheel hub motor in Figure 5, noise and fault signals are added to simulate the actual torque output of the motor, as shown in Figure 9. At 4 s, the motor fails, and the drive motor is unable to provide the expected output torque for a long time. The fuzzy controller determines that it has failed, and based on the accumulated diagnosis results, analyzes that it has failed. Therefore, at 4.01 s, the fault diagnosis output result becomes 0, indicating that the motor has failed, as shown in Figure 10. At the same time, when the fault diagnostic output indicates that the motor has failed, it is assumed that the motor may continue to fail in the future. To avoid further damage to the motor caused by frequent failures in the future, the diagnostic result is locked and the power output of the motor is turned off.

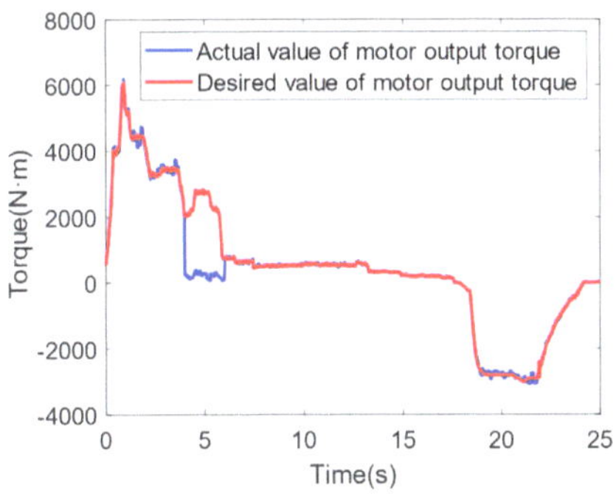

Figure 9. Actual and expected torque values of fault diagnosis motor output.

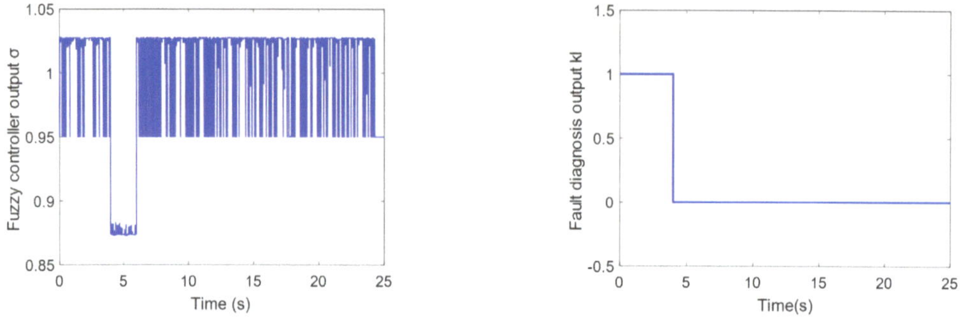

Figure 10. Simulation results of fault diagnosis for the drive system: (**a**) Fuzzy controller output σ; (**b**) Fault diagnosis output k_l.

To verify the control effect when a single motor fails, verification was conducted separately for the straight-line driving condition and the single-lateral-lane-shift driving condition. The simulation control group did not deal with the failure situation. The parameters of the two simulation conditions are shown in Table 3. The left front wheel drive motor failed, and the fault factor is shown in Figure 11. A brief torque loss occurred at 0.5 s, and the motor failed for a long time after 1 s.

Table 3. Simulation condition settings for single-motor failure control.

Operating Conditions	Initial Speed (km/h)	Road Adhesion Coefficient	Throttle Pedal Opening	Simulation End Distance/Time
Straight-line	50	0.85	40%	200 m
Single-lane-shift	50	0.85	15%	10 s

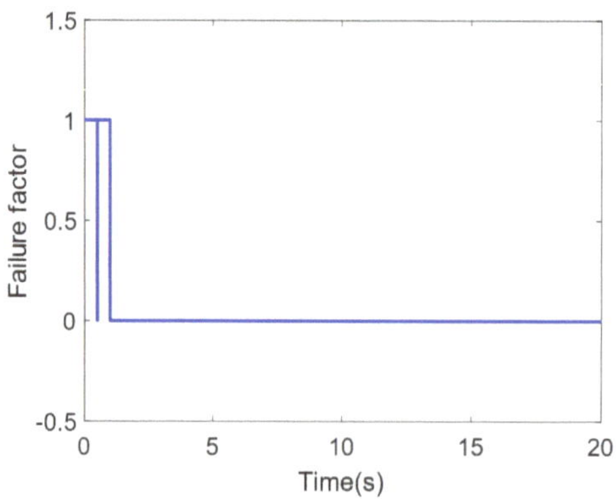

Figure 11. Simulation verification of failure control based on motor fault factor.

In the straight-line driving condition, the steering wheel angle remains unchanged at 0°, and the simulation stops at a longitudinal displacement of 200 m. The simulation results are shown in Figure 12. In Figure 12a, the final speed of the optimized control group at the end of the simulation is 59.21 km/h, while that of the uncontrolled group is 58.38 km/h. By reconstructing the torque of the normal working motor, the driving force lost by the failed motor can be compensated. Therefore, the final speed of the optimized control group is higher, as shown in Figure 12c, while the uncontrolled group in Figure 12d maintains its original torque output. Figure 12b shows that reconstructing the torque can keep the vehicle in a straight line, with a maximum lateral offset of 0.1 m, while the maximum lateral offset of the uncontrolled vehicle has reached 5.98 m, significantly deviating from the expected straight-line trajectory. Figure 12e shows the deviation in the yaw rate between the two groups of vehicles under the current simulation conditions, which is within a small range. The stability performance of both groups of vehicles is good. Since the control strategy proposed in this paper is more focused on tracking the yaw rate within the stable region, the average deviation in the optimized control group's yaw rate is 0.0002 rad/s, while that of the uncontrolled group is 0.0047 rad/s. The optimized control group's tracking of the yaw rate is more accurate.

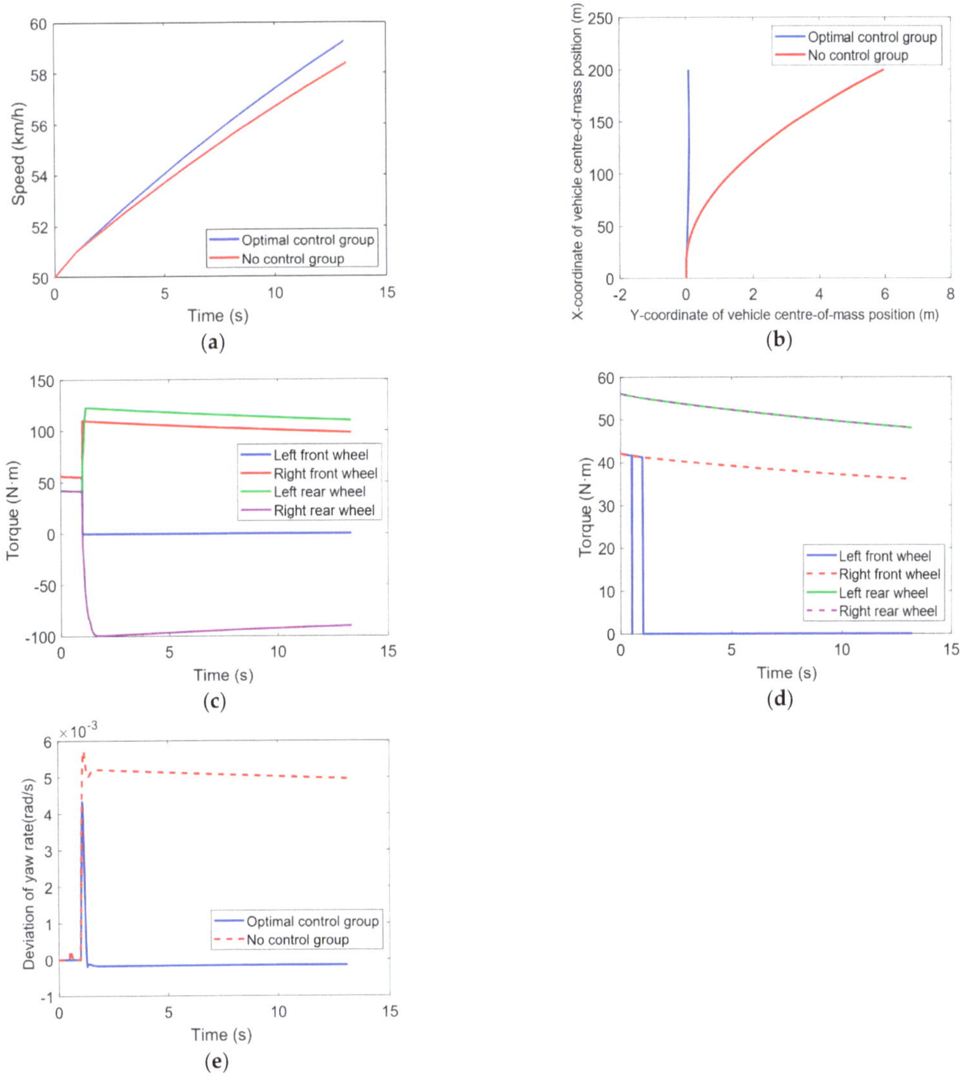

Figure 12. Simulation results for single-motor failure straight-line driving condition: (**a**) vehicle speed; (**b**) vehicle trajectory; (**c**) optimal control of individual wheel torques; (**d**) no control of individual wheel torques; (**e**) deviation in yaw rate.

The steering wheel angle in the single-lane-shift condition is shown in Figure 13:

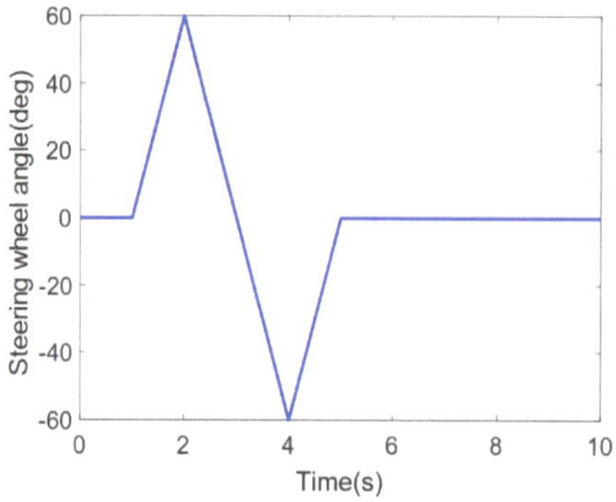

Figure 13. Steering wheel angle input in the single-motor failure single-lane-shift condition.

The simulation results of the single-motor failure single-lane-shift condition are shown in Figure 14. From Figure 14a, it can be seen that the final speed of the optimized control group is 50.55 km/h, while that of the uncontrolled group is 50.27 km/h. Both groups of vehicles can maintain their speed almost unchanged during the driving process. In Figure 14b, the maximum lateral offset of the optimized control group during the single-lane-shift process is 0.67 m compared to the normal driving, which can keep the expected trajectory of the single-lane shift, while that of the uncontrolled group is 1.01 m, and the trajectory has already deviated from the expected single-lane-shift trajectory. The lateral deviation will gradually increase with time. Figure 14c,d show the torque output of each wheel of the two groups of vehicles. The optimized control group detects the failure of the left front motor, isolates and shuts off its power output, and adjusts the torque of the remaining motors to ensure driving stability. Figure 14e shows the deviation in the yaw rate of the two groups of vehicles during driving, with an average value of 0.016 rad/s for the optimized control group and 0.022 rad/s for the uncontrolled group. Adopting failure control can improve the yaw stability of the vehicle during turning.

The effectiveness of the failure control strategy for double-motor failure with different axles and sides is verified, where the failed motors are the left front wheel motor and the right rear wheel motor. The simulation results of the straight-line driving condition are shown in Figure 15. From Figure 15a, it can be seen that the final speed of the optimized control group is 60.31 km/h, while that of the uncontrolled group is 54.63 km/h, because two motors have failed, the uncontrolled group has lost a significant amount of power, and the optimized control group can compensate for the lost driving force by the normal working motors, thus achieving a higher final speed. In Figure 15b, the maximum lateral offset of the optimized control group is 0.02 m, while that of the uncontrolled group is 2.11 m, indicating that reconstructing the torque can keep the vehicle in a straight line, while the uncontrolled vehicle will have a much larger lateral deviation. However, due to the normal working of the remaining motors, a certain lateral force moment will still be generated, resulting in a relatively smaller lateral deviation compared to the case of single-motor failure. Figure 15e shows the deviation in the yaw rate of the two groups of vehicles, with an average deviation of 0.0001 rad/s for the optimized control group and 0.0016 rad/s for the uncontrolled group, similar to the performance with single-motor failure. The optimized control group can better track the expected yaw rate, and both groups of vehicles are within the stable region.

To make the comparison more obvious, the duration of the single-lane-shift condition simulation is set to 20 s, and the simulation results are shown in Figure 16. From Figure 16a, it can be seen that the final speed of the optimized control group is 52.24 km/h, while that of the uncontrolled group is 48.38 km/h. Similar to the straight-line driving condition, the uncontrolled group will lose some power, and the optimized control group can meet the driver's power demand. In Figure 16b, the lateral deviation in the optimized control group compared to normal driving during the driving process is 0.52 m, which can maintain the expected single-lane-shift trajectory, while that of the uncontrolled group is 1.84 m, and the trajectory has already deviated from the expected single-lane-shift trajectory. The lateral deviation will gradually increase with time. Figure 16c,d show the torque output of each wheel of the two groups of vehicles. The optimized control group adjusts the torque of the other two normal working motors on the diagonal to improve driving stability. Figure 16e shows the deviation in the yaw rate of the two groups of vehicles during driving, with an average value of 0.0083 rad/s for the optimized control group and 0.01 rad/s for the uncontrolled group. Adopting failure control can improve the yaw stability of the vehicle during turning.

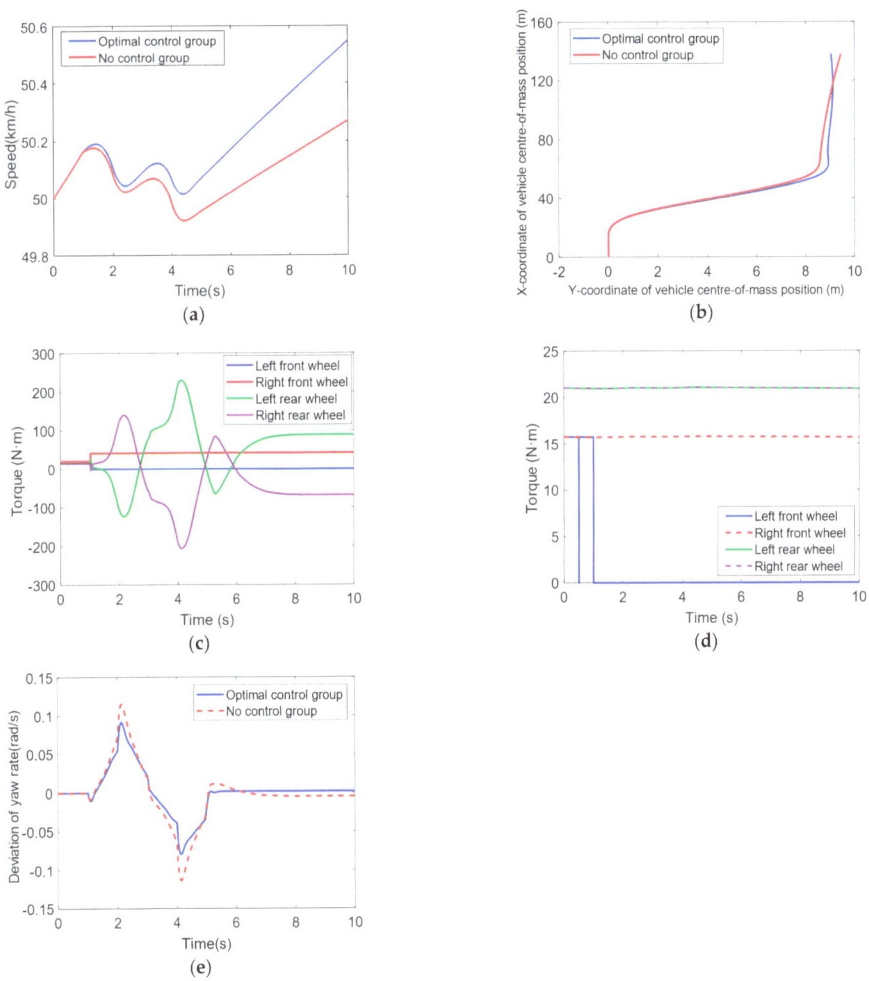

Figure 14. Simulation results for single-motor failure with single-lane shift: (**a**) vehicle speed; (**b**) vehicle trajectory; (**c**) optimal control of individual wheel torques; (**d**) no control of individual wheel torques; (**e**) deviation in yaw rate.

The simulation results for the two failure modes under straight-line driving and single-lane-shift driving conditions are presented in Table 4. This failure control strategy effectively diagnoses the occurrence of failures, exhibits robust performance, and successfully isolates the failed motors while reconstructing the torque of the remaining motors based on the specific failure mode, ensuring both driving stability and power demand.

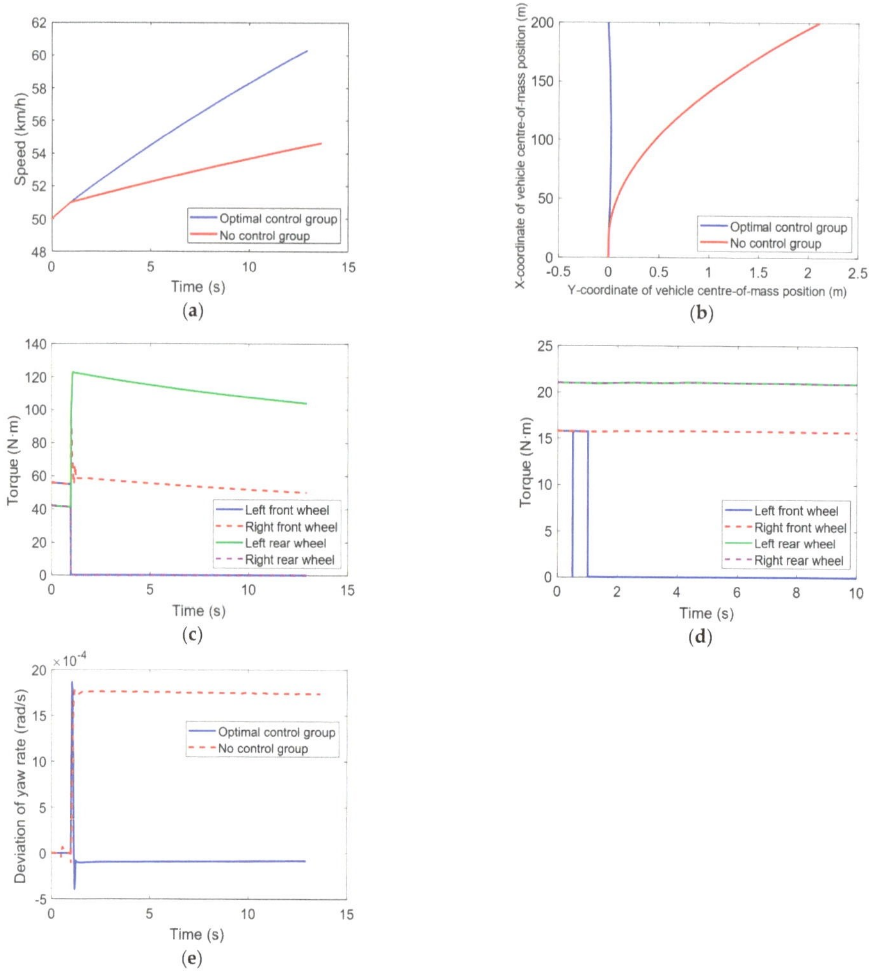

Figure 15. Simulation results of double-motor failure on the opposite side of the opposite axis under the straight-line driving condition: (**a**) vehicle speed; (**b**) vehicle trajectory; (**c**) optimal control of individual wheel torques; (**d**) no control of individual wheel torques; (**e**) deviation in yaw rate.

Table 4. Failure control simulation results.

Failure Mode	Condition	The Mean Value of Angular Velocity Deviation of the Pendulum (rad/s)		Max. Lateral Deflection (m)		Final Speed (km/h)	
		Optimized Control Group	No Control Group	Optimized Control Group	No Control Group	Optimized Control Group	No Control Group
Single-motor failure	Straight-line driving condition	0.0002	0.0047	0.10	5.98	59.21	58.38
	Single-lane-shift condition	0.0163	0.0217	0.67	1.01	50.55	50.27
Double-motor failure on the opposite side of the opposite axis	Straight-line driving condition	0.0001	0.0016	0.02	2.11	60.31	54.63
	Single-lane-shift condition	0.0083	0.0100	0.52	1.84	52.24	48.38

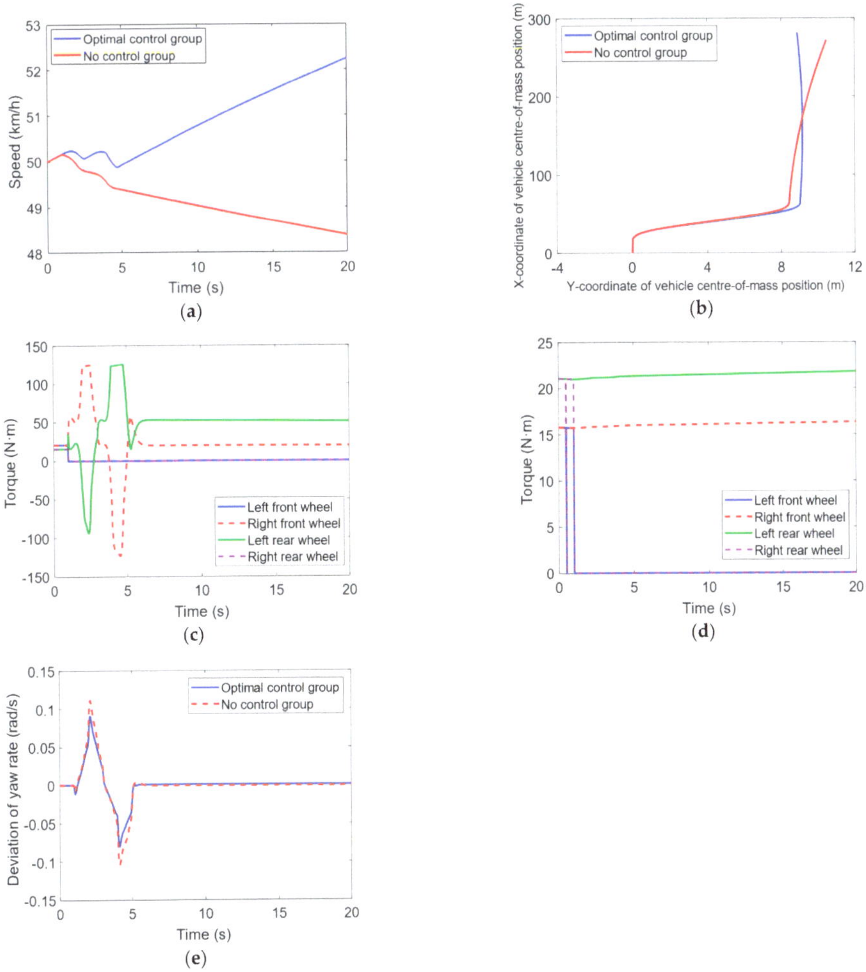

Figure 16. Simulation results of double-motor failure on the opposite side of the opposite axis under single-lane-shift condition: (**a**) vehicle speed; (**b**) vehicle trajectory; (**c**) optimal control of individual wheel torques; (**d**) no control of individual wheel torques; (**e**) deviation in yaw rate.

3.2. Hardware-in-the-Loop Simulation Results

To further verify the effectiveness of the failure control strategy in an actual vehicle control unit (VCU), HIL testing was conducted. In this paper, a hardware-in-the-loop simulation platform based on NI PXI and CARSIM RT was constructed, as shown in Figure 17. The platform is seamlessly integrated with the Simulink control model in SIL for co-simulation. It consists of three main parts: the host computer, the tested VCU hardware, and the industrial PC cabinet equipped with real-time processors and digital-to-analog input and output boards.

Due to the inconsistency between the simulation cycle and the communication cycle of HIL testing, the threshold for fault determination needs to be adjusted during HIL testing to adapt to the actual controller. The minimum sampling period of the constructed HIL testing platform was recorded as 10 ms. Therefore, the judgment period was set to 100 ms. If the number of times when σ was less than 0.95 during the judgment period exceeded eight, it was determined that the drive motor had failed and isolation processing was needed.

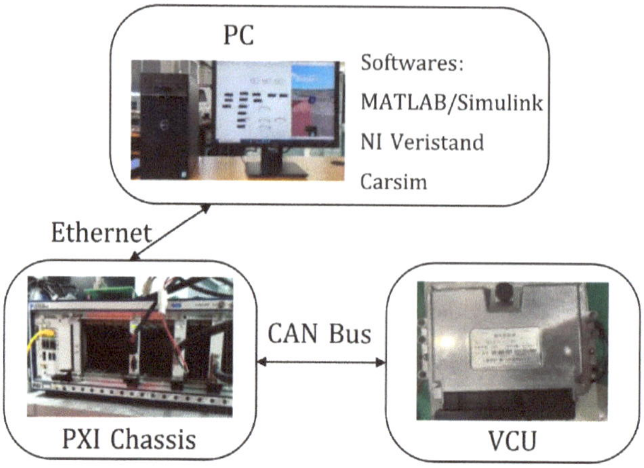

Figure 17. HIL testing platform.

The actual output torque of the drive motor shown in Figure 5 and the expected output torque were used as inputs to the fault diagnostic tool for HIL testing, as shown in Figure 9. The test result is shown in Figure 18. Figure 18a is the output σ of the designed fault diagnostic tool, while Figure 18b always outputs k_l = 1, indicating that the motor is in normal working condition. The HIL test results have demonstrated that the designed fault diagnostic tool has good robustness which can avoid misjudgment caused by torque loss due to transient signal fluctuations. For Figure 18c,d, when the motor fails at 4 s, the drive motor cannot provide the expected output torque for a long time. The fuzzy controller determines that the motor has failed, and based on the cumulative judgment result, the fault diagnostic output results indicate that it has failed at 4.08 s. This is consistent with the simulation results. The HIL test results have demonstrated that the designed fault diagnostic tool can still diagnose whether the motor has failed in a timely and accurate manner in actual controllers. Compared with the simulation results, the influence of signal fluctuations is reduced due to the different sampling frequencies of the controller and the CAN communication frequency, which is the reason why the threshold needs to be adjusted according to the sampling frequency of the controller and the CAN communication frequency.

To verify the consistency between HIL testing and the simulation results, the test was conducted using the operating conditions in Table 3. The HIL test under the condition of a single-motor failure in straight-line driving is shown in Figure 19. The results of the HIL test are consistent with the simulation results; when the lateral angular velocity deviation is in a small range in a stable situation, the average value of the lateral angular velocity deviation optimized by the control group is 0.0005 rad/s, while that of the uncontrolled group is 0.0047 rad/s. The control group can track the lateral angular velocity more accurately. By reconstructing the torque of the non-failed motor, the expected power of the driver can be guaranteed, and the final speed of the optimized control group is higher than that of the uncontrolled group. At the same time, adopting failure control can ensure that the vehicle travels along the expected straight trajectory, with a maximum lateral deviation of 0.01 m, whereas the maximum lateral deviation of the uncontrolled vehicle is 5.99 m.

The simulation of a single-motor failure in a single-lane-shift condition is shown in Figure 20. Both groups of vehicles can maintain a speed of around 50 km/h, but the trajectory of the uncontrolled vehicle gradually deviates from the expected single-lane-shift trajectory, while the optimized control group can continue to travel on the expected trajectory through adjustment of the torque of each wheel. Figure 20e shows the lateral angular velocity deviation of the two groups of vehicles while driving. The average value of the optimized control group is 0.016 rad/s, while that of the uncontrolled group is

0.022 rad/s. Compared with the uncontrolled group, the optimized control group can reduce the deviation from the expected lateral angular velocity and improve the stability during turning, which is consistent with the simulation results.

The simulation of a straight-line driving condition with dual-motor failure on opposite sides is shown in Figure 21. In Figure 21a, the final speed of the optimized control group is 60.32 km/h, while that of the uncontrolled group is 54.64 km/h. The optimized control group can maintain the expected power. It can be seen from Figure 21b that the maximum lateral deviation of the optimized control group is 0.072 m, which is slightly increased compared to the simulation results, but it can still prevent the vehicle from running off the road. The maximum lateral deviation of the uncontrolled group is 2.111 m, and the vehicle gradually deviates from the straight-line-driving trajectory. Both groups of vehicles are stable. The average lateral angular velocity deviation of the optimized control group is 0.0002 rad/s, while that of the uncontrolled group is 0.0016 rad/s. The lateral angular velocity deviation is improved compared to the uncontrolled group.

The simulation of a single-lane-shift working condition with dual motor failure on opposite sides is shown in Figure 22. Similar to the straight-line driving condition in Figure 22a, the uncontrolled group will lose some power and the vehicle speed gradually decreases. The final speed is 48.38 km/h, while the optimized control group can meet the driver's power requirements, and the final speed is 52.27 km/h. In Figure 22b, the optimized control group can still maintain the expected single-lane-shift trajectory, while the trajectory of the uncontrolled group vehicle has deviated. With time, it will gradually deviate from the expected trajectory. Figure 22e shows the lateral angular velocity deviation of the two groups of vehicles while driving. Both groups of vehicles are in a stable operating region. Due to the two normal working motors on both sides, the uncontrolled group can provide some additional lateral torque, which makes it more stable compared to the single motor failure condition. The average lateral angular velocity deviation of the optimized control group is 0.0082 rad/s, while that of the uncontrolled group is 0.01 rad/s. Adopting failure control can improve the driving stability after dual-motor failure on opposite sides, and the lateral angular velocity deviation is smaller, which is consistent with the simulation results.

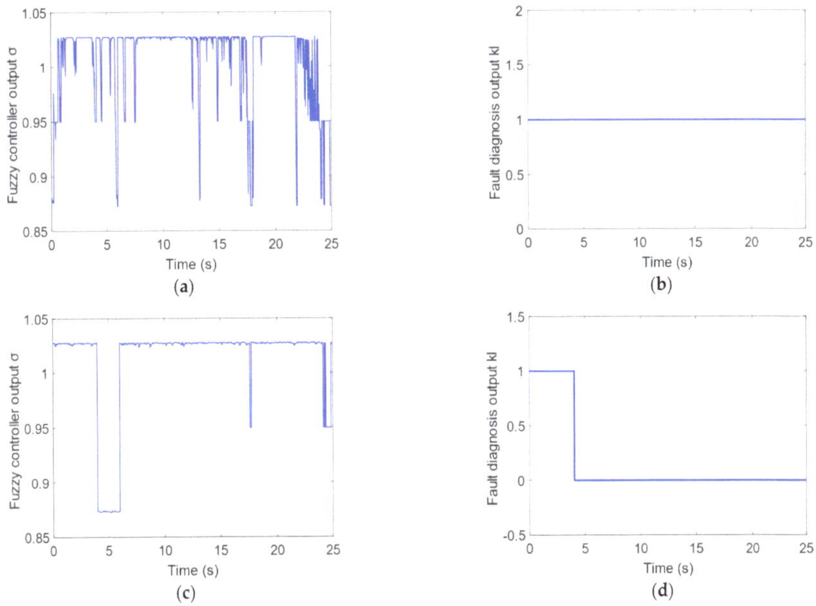

Figure 18. Drive system fault diagnosis HIL test results: (**a**) Fuzzy controller output for fault diagnosis σ; (**b**) Failure factor k_l; (**c**) Fuzzy controller output for fault diagnosis σ; (**d**) Failure factor k_l.

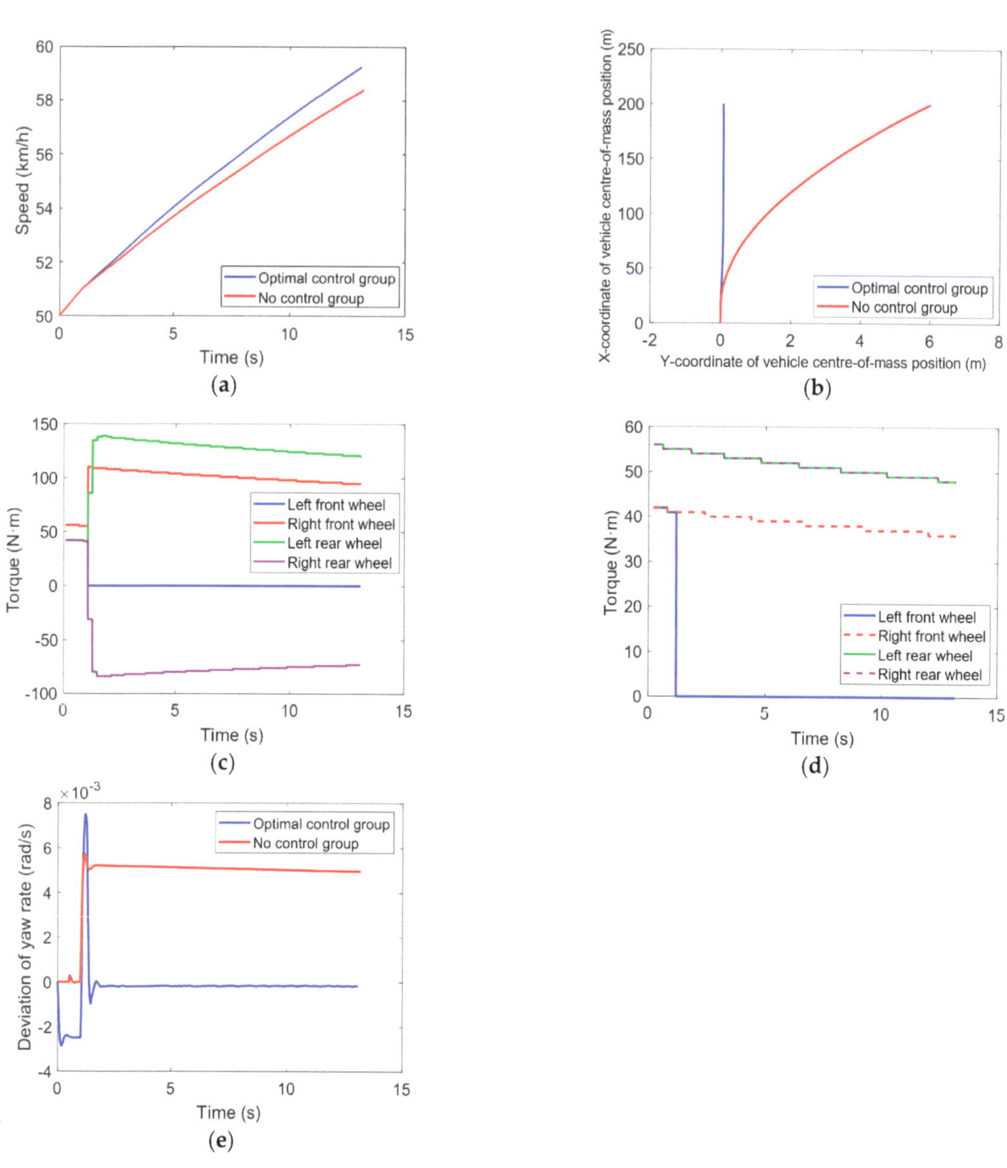

Figure 19. HIL test results for single-motor failure in straight-line operation: (**a**) vehicle speed; (**b**) vehicle trajectory; (**c**) optimal control of individual wheel torques; (**d**) no control of individual wheel torques; (**e**) deviation in yaw rate.

The hardware-in-the-loop (HIL) test results for the two failure modes in the respective working conditions are presented in Table 5. These results are consistent with the simulation findings, validating the efficacy of the proposed failure control strategy for the distributed-four-wheel-drive system. The strategy can effectively diagnose drive-system faults in the actual VCU. It can also isolate the faulty motor according to the failure condition and restructure the torque of the remaining non-faulty motors, thus improving the driving stability of the vehicle, maintaining the expected power, and enabling the vehicle to follow the intended driving trajectory.

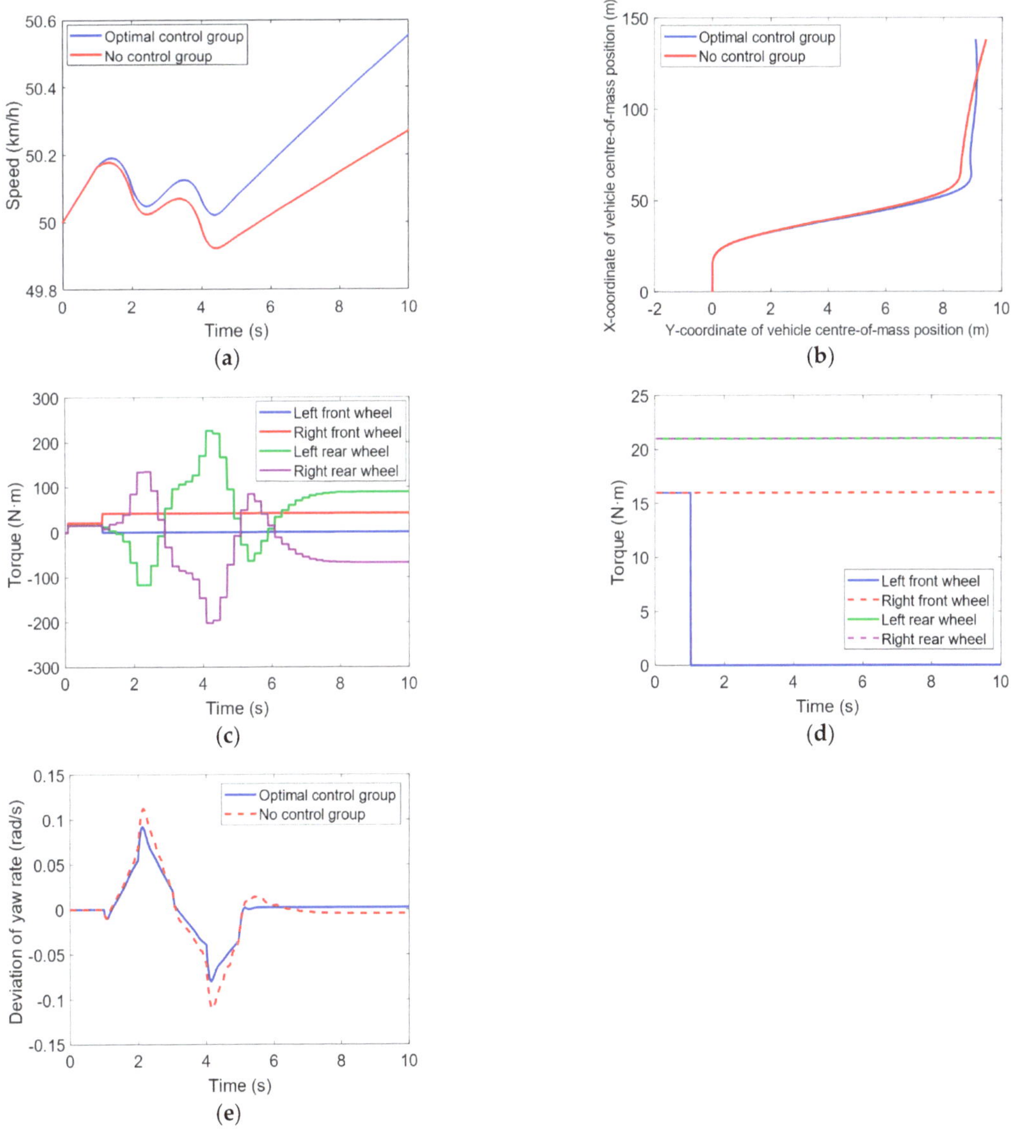

Figure 20. HIL test results for single-motor failure single-lane-shift condition: (**a**) vehicle speed; (**b**) vehicle trajectory; (**c**) optimal control of individual wheel torques; (**d**) no control of individual wheel torques; (**e**) deviation in yaw rate.

Table 5. Failure control HIL test results.

Failure Mode	Condition	The Mean Value of Angular Velocity Deviation of the Pendulum (rad/s)		Max. Lateral Deflection (m)		Final Speed (km/h)	
		Optimized Control Group	No Control Group	Optimized Control Group	No Control Group	Optimized Control Group	No Control Group
Single-motor failure	Straight-line driving condition	0.0005	0.0047	0.01	5.99	59.24	58.39
	Single-lane-shift condition	0.0162	0.0219	0.70	1.01	50.54	50.28

Table 5. *Cont.*

Failure Mode	Condition	The Mean Value of Angular Velocity Deviation of the Pendulum (rad/s)		Max. Lateral Deflection (m)		Final Speed (km/h)	
		Optimized Control Group	No Control Group	Optimized Control Group	No Control Group	Optimized Control Group	No Control Group
Double-motor failure on the opposite side of the opposite axis	Straight-line driving condition	0.0002	0.0016	0.07	2.11	60.32	54.64
	Single-lane-shift condition	0.0082	0.0100	0.53	1.84	52.27	48.38

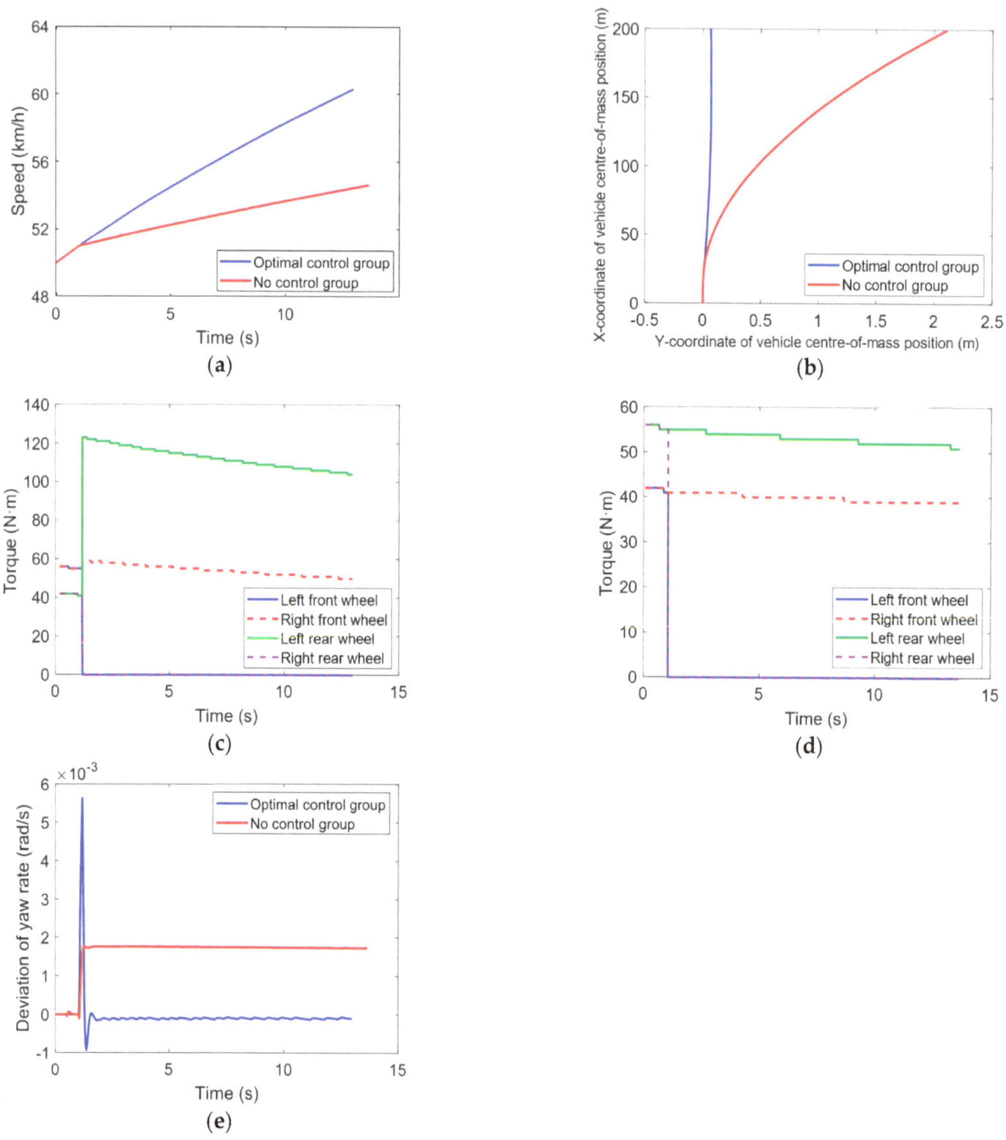

Figure 21. HIL test results for hetero-shaft and hetero-side dual-motor failure in straight-line driving condition: (**a**) vehicle speed; (**b**) vehicle trajectory; (**c**) optimal control of individual wheel torques; (**d**) no control of individual wheel torques; (**e**) deviation in yaw rate.

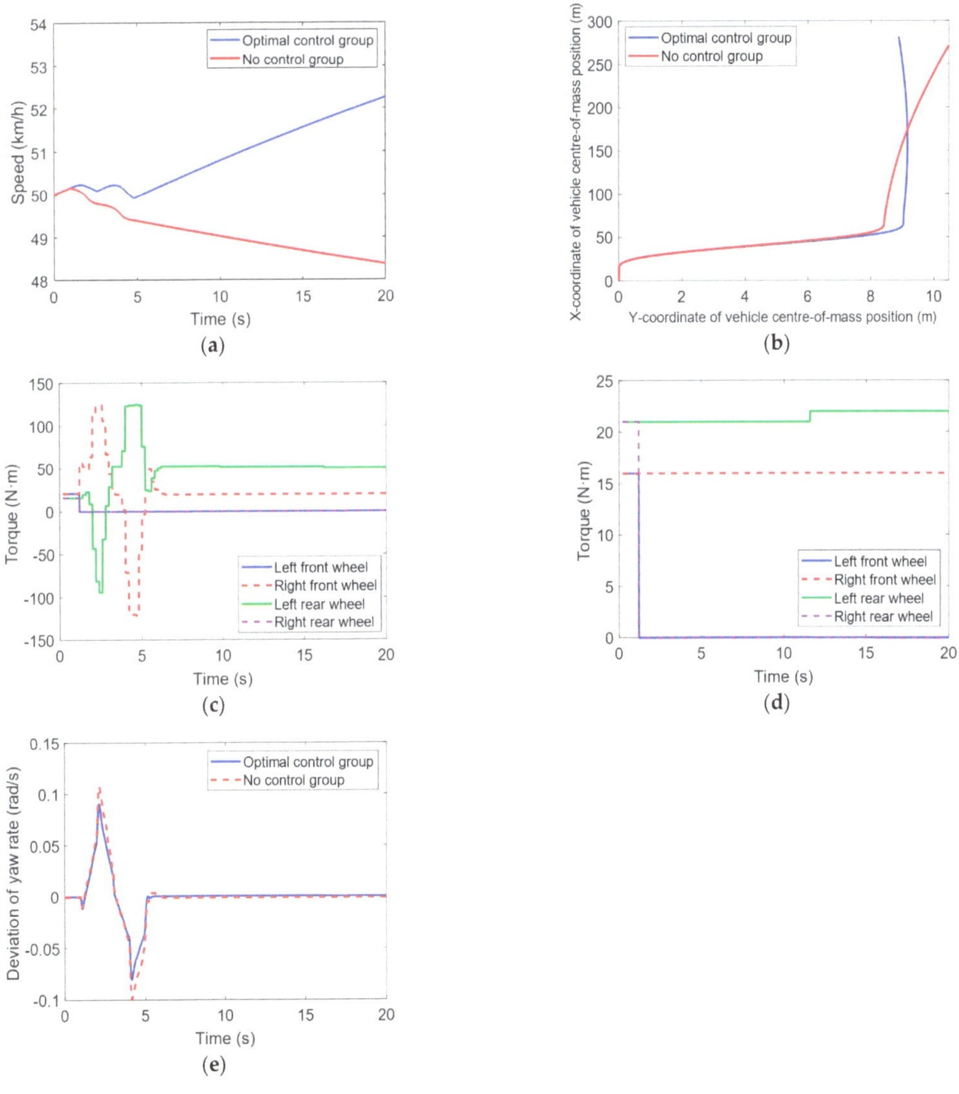

Figure 22. HIL test results for single-lane-shift condition with dual-motor failure on the opposite shaft and opposite side: (**a**) vehicle speed; (**b**) vehicle trajectory; (**c**) optimal control of individual wheel torques; (**d**) no control of individual wheel torques; (**e**) deviation in yaw rate.

4. Conclusions

The fault-tolerant control strategy designed in this study comprises two key components: fault diagnosis and fault-tolerant control. The actual output torque of the drive motor and the ratio of the expected output torque along with its rate of change are used as inputs to design a fuzzy controller, which produces the motor fault diagnosis result. The accumulated diagnosis results within a specified time period are analyzed to determine if the motor has failed. The fault-tolerant controller identifies the failure mode based on the diagnostic results of the four motors and applies fault-mode rules to restructure the torque of the non-failed motors. This fault-tolerant control strategy effectively mitigates the impact of interference signals, accurately identifies the failed motors and their failure mode, and exhibits robust performance. By adopting torque reconstruction, this strategy can maintain driving power and stability. The software simulation and HIL test results show good consistency, confirming the effectiveness and reliability of the designed control strategy.

Author Contributions: Conceptualization, S.Z.; Software, C.K.; Investigation, H.C. and J.G.; Writing—original draft, H.L.; Writing—review & editing, H.L. and H.C.; Supervision, G.W.; Funding acquisition, W.X. All authors have read and agreed to the published version of the manuscript.

Funding: This research was funded by [Control design of new energy vehicle air conditioning compressor based on intelligent multi-objective optimization] grant number [ZDLQ2020002].

Data Availability Statement: No new data were created or analyzed in this study.

Acknowledgments: The Open Project Program of Fujian Provincial Key Laboratory of Intelligent Identification and Control of Complex Dynamic System (2022A0006); Control design of new energy vehicle air conditioning compressor based on intelligent multi-objective optimization (ZDLQ2020002).

Conflicts of Interest: The authors declare no conflict of interest.

References

1. Zhang, X. *Modeling and Dynamics Control for Distributed Drive Electric Vehicles*; Springer Vieweg: Berlin, Germany, 2021. [CrossRef]
2. Zhou, H.; Jia, F.; Jing, H.; Liu, Z.; Guvenc, L. Coordinated Longitudinal and Lateral Motion Control for Four Wheel Independent Motor-Drive Electric Vehicle. *IEEE Trans. Veh. Technol.* **2018**, *67*, 3782–3790. [CrossRef]
3. Chen, T.; Chen, L.; Xu, X.; Cai, Y.; Jiang, H.; Sun, X. Passive fault-tolerant path following control of autonomous distributed drive electric vehicle considering steering system fault. *Mech. Syst. Signal Process.* **2019**, *123*, 298–315. [CrossRef]
4. Un-Noor, F.; Padmanaban, S.; Mihet-Popa, L.; Mollah, M.N.; Hossain, E. A Comprehensive Study of Key Electric Vehicle (EV) Components, Technologies, Challenges, Impacts, and Future Direction of Development. *Energies* **2017**, *10*, 1217. [CrossRef]
5. Ding, N.; Taheri, S. An adaptive integrated algorithm for active front steering and direct yaw moment control based on direct Lyapunov method. *Veh. Syst. Dyn.* **2010**, *48*, 1193–1213. [CrossRef]
6. Wang, Z.; Shao, J.; He, Z. Fault Tolerant Sensorless Control Strategy with Multi-States Switching Method for In-Wheel Electric Vehicle. *IEEE Access* **2021**, *9*, 61150–61158. [CrossRef]
7. Wang, R.; Wang, J. Fault-Tolerant Control for Electric Ground Vehicles with Independently-Actuated In-Wheel Motors. *J. Dyn. Syst. Meas. Control* **2012**, *134*, 021014. [CrossRef]
8. Wang, R.; Wang, J. Actuator-Redundancy-Based Fault Diagnosis for Four-Wheel Independently Actuated Electric Vehicles. *IEEE Trans. Intell. Transp. Syst.* **2013**, *15*, 239–249. [CrossRef]
9. Li, C.; Chen, G.; Zong, C.; Liu, W. Fault-tolerant control for 4WID/4WIS electric vehicle based on EKF and SMC. *SAE Int. J. Passeng. Cars Electron. Electr. Syst.* **2015**, *9*, 1–8. [CrossRef]
10. Rkhissi-Kammoun, Y.; Ghommam, J.; Boukhnifer, M.; Mnif, F. New fault-tolerant induction motor control architecture for current sensor fault in electric vehicle. In Proceedings of the 2018 15th International Multi-Conference on Systems, Signals & Devices (SSD), IEEE, Yasmine Hammamet, Tunisia, 19–22 March 2018; pp. 65–70.
11. Kamen, D.; Morrell, J.B.; Robinson, D.W.; Reich, R.K.; Heinzmann, J.D.; LeMay, P.; Meyer, S.R.; Sachs, J.M.; Field, J.D. Method and System for Fail-Safe Motor Operation. U.S. Patent 6,965,206, 15 November 2005.
12. Lopes, A.; Araújo, R.E. Fault-tolerant control based on sliding mode for over-actuated electric vehicles. In Proceedings of the 2014 IEEE International Electric Vehicle Conference (IEVC), IEEE, Florence, Italy, 17–19 December 2014; pp. 1–6.
13. Wang, J.; Gao, S.; Wang, K.; Wang, Y.; Wang, Q. Wheel torque distribution optimization of four-wheel independent-drive electric vehicle for energy efficient driving. *Control Eng. Pract.* **2021**, *110*, 104779. [CrossRef]
14. Xin, X.; Zheng, H.; Xu, H.; Qin, G. Control strategies for four in-wheel drove electric vehicles when the motor drive systems fail. In Proceedings of the 2014 American Control Conference, IEEE, Portland, OR, USA, 4–6 June 2014; pp. 885–890.
15. Sun, X.; Shi, Z.; Cai, Y.; Lei, G.; Guo, Y.; Zhu, J. Driving-Cycle-Oriented Design Optimization of a Permanent Magnet Hub Motor Drive System for a Four-Wheel-Drive Electric Vehicle. *IEEE Trans. Transp. Electrif.* **2020**, *6*, 1115–1125. [CrossRef]
16. Wanner, D.; Wallmark, O.; Jonasson, M.; Drugge, L.; Trigell, A.S. Control allocation strategies for an electric vehicle with a wheel hub motor failure. *Int. J. Veh. Syst. Model. Test.* **2015**, *10*, 263. [CrossRef]
17. Lee, K.; Lee, M. Fault-Tolerant Stability Control for Independent Four-Wheel Drive Electric Vehicle Under Actuator Fault Conditions. *IEEE Access* **2020**, *8*, 91368–91378. [CrossRef]
18. Liu, L.; Shi, K.; Yuan, X.; Li, Q. Multiple model-based fault-tolerant control system for distributed drive electric vehicle. *J. Braz. Soc. Mech. Sci. Eng.* **2019**, *41*, 531. [CrossRef]

Disclaimer/Publisher's Note: The statements, opinions and data contained in all publications are solely those of the individual author(s) and contributor(s) and not of MDPI and/or the editor(s). MDPI and/or the editor(s) disclaim responsibility for any injury to people or property resulting from any ideas, methods, instructions or products referred to in the content.

actuators

Article

Balancing Control of an Absolute Pressure Piston Manometer Based on Fuzzy Linear Active Disturbance Rejection Control

Hongda Wu, Xianyi Zhai, Teng Gao *, Nan Wang, Zongsheng Zhao and Guibing Pang

School of Mechanical Engineering and Automation, Dalian Polytechnic University, Dalian 116034, China; whd0604@foxmail.com (H.W.); zhaixianyi9421@foxmail.com (X.Z.); 15582307677@163.com (N.W.); zzsheng2021@163.com (Z.Z.); pangguibingsx@163.com (G.P.)

* Correspondence: ddpmobile_smt@126.com

Abstract: As an international standard pressure-measuring instrument, the absolute pressure piston manometer's working medium is gas, so the actual working process will be affected by many internal uncertainties and external disturbances, leading to its long stability time and poor performance. In this paper, a fuzzy linear active disturbance rejection control strategy (FLADRC) for absolute pressure piston manometers is proposed to address these problems. First, the characteristics of the main components are analyzed according to the actual working principle of the system to establish a theoretical model of the controlled system. Second, the corresponding linear active disturbance rejection controller (LADRC) is designed according to the model. The principle of fuzzy control is introduced to adaptively adjust the controller parameters of the LADRC in real time, which improves the disadvantages of the LADRC parameters, which are difficult to rectify and have poor immunity to disturbances due to fixed parameters, and the stability of the control method is subsequently demonstrated. Finally, a simulation model is built in the Simulink environment in MATLAB, and three different pressure operating points are selected for the corresponding experiments to make a comparative analysis with Kp, PID, and LADRC. The results show that FLADRC enables the absolute pressure piston manometer to achieve better stability and greater immunity to disturbances. This also verifies the effectiveness and feasibility of the control strategy in practical engineering applications.

Keywords: absolute pressure piston manometer; system theoretical model; linear disturbance rejection control; fuzzy control; parameter tuning

Citation: Wu, H.; Zhai, X.; Gao, T.; Wang, N.; Zhao, Z.; Pang, G. Balancing Control of an Absolute Pressure Piston Manometer Based on Fuzzy Linear Active Disturbance Rejection Control. *Actuators* **2023**, *12*, 275. https://doi.org/10.3390/act12070275

Academic Editors: Oscar Barambones, Jose Antonio Cortajarena and Patxi Alkorta

Received: 9 June 2023
Revised: 3 July 2023
Accepted: 4 July 2023
Published: 6 July 2023

Copyright: © 2023 by the authors. Licensee MDPI, Basel, Switzerland. This article is an open access article distributed under the terms and conditions of the Creative Commons Attribution (CC BY) license (https://creativecommons.org/licenses/by/4.0/).

1. Introduction

In recent years, absolute pressure measurement technology has become one of the important symbols to measure the degree of industrial technology development of a country and is also an important evaluation indicator of industrial safety [1–3]. As an international standard pressure measurement instrument with high accuracy and stability, the absolute pressure piston manometer is based on the hydrostatic equilibrium principle and Pascal's law for pressure measurement. As the working medium is a gas, the adiabatic manometer is subject to many nonlinear and time-varying factors, which results in relatively long stabilization times and unsatisfactory stability performance. Therefore, it is crucial to design a high-quality control system to address these issues.

Relatively few scholars at home and abroad have studied piston manometer balancing control methods, but for quadrotors [4,5], inverted pendulums [6–8], balancing vehicles [9,10], permanent-magnet synchronous motors [11,12], and other such similar balancing systems, there are many modern control algorithms currently applied, such as proportional-integral-derivative (PID) control [13–15], linear quadratic regulator control (LQR) [16,17], robust control [18,19], fuzzy control [20–23], adaptive control [24,25], sliding mode control [26–28], etc. Theoretically, the control methods applied to these equilibrium systems can also be applied to piston manometers. The authors in [29] introduced the relaxation factor into model predictive control (MPC), on which it was combined with hybrid PID control theory for applied control to improve the control performance of smart car path tracking across the board. The authors in [30] proposed a Udwadia–Kalaba theory-based adaptive robust control (UKBARC), applied to a permanent-magnet linear synchronous motor, which can both transform the control task with the desired trajectory as the desired constraint and deal effectively with system uncertainty. The authors in [31] proposed a sliding mode control method incorporating an adaptive strategy to achieve altitude tracking of a quadrotor under strong external disturbances, but the method requires high accuracy of the model and a relatively complex controller structure. The authors in [32] proposed an improved fuzzy logic controller based on Lyapunov's criterion for the real-time oscillation and stability control of a coupled two-arm inverted pendulum, which improved the transient and steady-state response speed of the system to a certain extent, but the amount of operations in this controller is large and more difficult to implement in practical engineering. The authors in [33] combined the quantum particle swarm algorithm (QPSO) with the LQR control method. It used the former to search for the optimal values of the Q matrix as well as the weighting coefficients in the controller, improving the overall control performance of the balancing vehicle, but the time for the algorithm to perform the optimal search still needs to be improved.

Active disturbance rejection control (ADRC) is an "observation + compensation" nonlinear control method proposed by Professor Jingqing Han et al. [34] to preserve the advantages of PID and overcome its shortcomings. However, the internal parameters are too many, and the rectification process is tedious. Gao [35] then introduced the concepts of linearization and bandwidth based on the advantages of active disturbance suppression techniques and proposed linear active disturbance rejection control (LADRC) for the first time, making the controller parameters more intuitive and simpler to rectify. In addition, to further improve the control performance of LADRC, many researchers have made further improvements. An improved third-order LADRC controller was proposed by the authors in [36], introducing the total disturbance differential signal in the LESO and applying a series of first-order inertial links, which improves the system's ability to suppress high-frequency noise. The authors in [37] combined robust sliding mode control (SMC) theory with LADRC control to overcome the bandwidth limitation of LADRC itself and improve the control accuracy for the system. The authors in [38] proposed an improved linear active disturbance suppression control (MLADC) method to compensate the system model into a linear state observer (LESO) to improve its observation accuracy, reduce its state error and enhance the robustness of the system.

However, it is easy to see that most improvements have a common limitation, namely that the controller parameters can only be adjusted manually by trial and error, which is complex and time-consuming, and the fixed parameters also lead to a less robust system, which cannot be applied to specific practical engineering problems. As a nonlinear system, absolute pressure piston manometers cannot be modeled with absolute accuracy. Moreover, when implementing control, uncertainties within the system and external disturbances are inevitable, which can reduce the efficiency of the system [39], and the stability performance of the system needs to be improved. Fuzzy control is considered to be a control scheme to improve the robustness and adaptability of the system and has been widely used in industry. It can dynamically adjust the parameters of the controller according to the output of the system, thus enabling the controller to track the input signal faster. Therefore, to improve the anti-disturbance capability of the controlled system and make it achieve equilibrium quickly and stably, this paper proposes a FLADRC-based equilibrium control strategy for the adiabatic piston manometer, which introduces the idea of fuzzy control for adaptive parameter adjustment based on LADRC. The feasibility and effectiveness of the control scheme are verified by simulations in MATLAB.

The other sections of the article are organized as follows: In Section 2, the theoretical modeling of the adiabatic piston manometer is presented. In Section 3, the specific design and stability analysis of the FLADRC controller is carried out for the controlled system model. In Section 4, the simulation results of the four control schemes, FLADRC, LADRC, PID, and Kp, are compared and analyzed to verify the performance advantages of the proposed control scheme. Finally, the summary of the research work in this paper is discussed in Section 5.

2. Theoretical Modeling of an Absolute Pressure Piston Manometer

2.1. The Working Principle of the Absolute Pressure Piston Manometer

The mechanical structure of the absolute pressure piston manometer as a nonlinear time-varying system is shown in Figure 1. Before the system starts to work, the corresponding weights are automatically configured according to the measured pressure. The pure gas in the cylinder, which is the working medium for pressurizing the system, is first added quickly to the originally vacuumed piston cylinder through the inlet valve until the weights and the piston is jacked up. Immediately afterward, the piston is subjected to constant changes in height due to the pressure in the piston cylinder and other uncertainties. The gas flow from the inlet and outlet valves is then adjusted in turn according to the actual situation so that the piston is finally stabilized at the preset height. At this point, the weights and piston are then mechanically balanced as a whole, thus completing the measurement and calibration of the pressure.

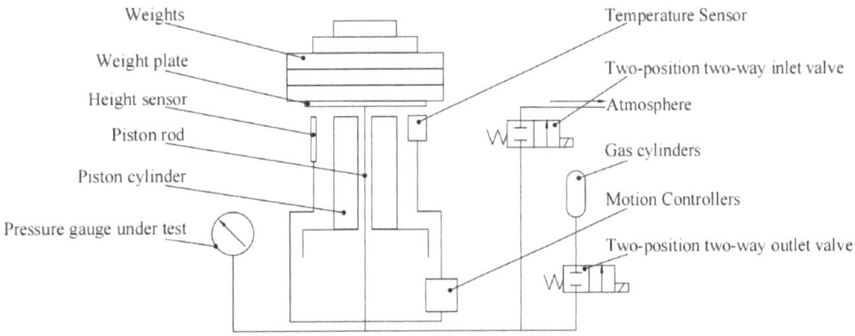

Figure 1. Schematic diagram of the structure of an absolute pressure piston manometer.

Based on the above working principle, it can be seen that the most important part of the system is the piston, followed by the electromagnetic switching valve. Therefore, to build a theoretical model for an absolute pressure piston manometer, it is necessary to start with these two parts and carry out the corresponding characteristic analysis.

2.2. Kinetic Analysis of the Weight Combination Section

Throughout the actual working of the system, the weight, the weight plate, and the check piston can be seen as a whole, collectively referred to as the weight combination part. In the force analysis of this whole, the weight combination is subjected to its own gravity G, the pressure F_v resulting from the change in mass of the gas in the piston cylinder (F_i and F_o depending on the inlet and outlet processes), the resistance F_f of the gas preventing the change in the height of the piston, the pressure F_q lost by the gas leakage and the elastic force F_z resulting from the compression of the gas in the cylinder. Take vertical upwards as the positive direction.

As can be seen from Figure 2, the weight combination part is subjected to different magnitudes and directions of partial forces in the two different working processes of rising and falling, which have to be analyzed separately, thus listing the kinetic equilibrium equations of the weight combination part in in the two states of motion according to Newton's second law, as in Equation (1).

$$\begin{cases} F_i - G - F_f - F_q + F_z = m_w\ddot{h} \\ -F_o - G + F_f - F_q + F_z = -m_w\ddot{h} \end{cases} \quad (1)$$

where m_w is the total mass of the combined portion of weights and h is the real-time height at which the piston rises. The process by which the gas is compressed and thus expands is similar to that of a spring, and the process by which the gas in the gap in the side of the piston prevents the piston from rising is similar to that of a damping system, so the resistance F_f can be calculated using Equation (2) and the elastic force F_z can be calculated using Equation (3).

$$F_f = c\dot{h} \quad (2)$$

$$F_z = kx \quad (3)$$

where c is the dynamic viscosity of the gas medium, k is the coefficient of elasticity after the compression and expansion of the gas in the cylinder is equivalent to a spring system, and x is the actual compression of the gas in the cylinder. Taking the process of gas intake as an example, the process of calculating the elasticity coefficient after equivalence is as follows: the instantaneous moment when the intake valve has been fed once is analyzed and calculated, at which time the gas volume is compressed and reduced by ΔV, and the pressure is increased by ΔP. According to the basic principles of elasticity, there is Equation (4).

$$\Delta P = -E(\Delta V/V_0) \quad (4)$$

where V_0 is the cylinder volume corresponding to the primary gas intake and is a fixed value that can be calculated from the flow rate of the intake valve, and E is the modulus of elasticity of the gas media. At this point, according to the calculation Equation of the elastic force generated by the compression of the gas (5), the calculation Equation of the piston cylinder volume (6), and Equation (3), the calculation formula of the elasticity coefficient (7) can be deduced.

$$F = -\Delta P \cdot A = E(\Delta V/V_0)A \quad (5)$$

$$\Delta V = x \cdot A \quad (6)$$

$$k = (EA^2)/V_0 \quad (7)$$

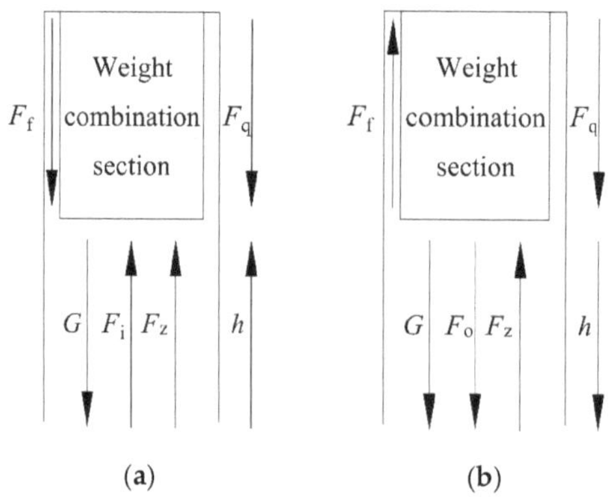

Figure 2. Analysis of the forces on the combined weight section under different working conditions. (**a**) Piston rise; (**b**) Piston drop.

The relationship between the gas compression and the actual rise of the piston is $x = h_n - h$, where h_n is the ideal height of the piston rise after n gas inlets without taking into account the gas compression, which needs to be determined according to the actual mass of the intake gas.

In summary, the elastic force F_z in Equation (1) can be expressed by Equation (8).

$$F_z = (EA^2)(h_n - h)/V_0 \tag{8}$$

The pressure F_v in Equation (1) is the driving force for the entire combined part of the piston and can be calculated using the actual gas state Equation (9) as well as the pressure Equation (10) in the closed vessel together.

$$PV_2 = ZnRT \tag{9}$$

$$F_v = P \cdot A = (ZnRTA)/V_2 \tag{10}$$

where R is the molar constant of the gas media, T is the thermodynamic temperature, Z is the compression coefficient of the gas media, P is the real-time pressure in the piston cylinder, n is the number of moles of gas in the cylinder, and the formula is $n = m_t/M$. Taking the inlet process as an example, m_t is the mass of gas entering the piston cylinder through the inlet valve, and M is the molar mass of the gas media. V_2 is the volume in the cylinder after the piston has risen to an expected height and is a fixed value.

2.3. Flow Analysis of Switching Valves

Solenoid switching valves, as actuators in the controlled system, can be divided according to their actual function into inlet and outlet valves, the most important characteristic of which is the real-time gas flow through the valve body. The switching valve can be seen as a throttle plate with a regularly varying orifice diameter, each switch corresponding to a maximum opening and a zero opening. The equation for the gas mass flow rate of the solenoid switch valve is obtained as follows [40].

$$m_{in} = C_d A_1 \sqrt{2\Delta P_1 \rho} \tag{11}$$

$$m_{out} = C_d A_1 \sqrt{2\Delta P_2 \rho} \tag{12}$$

where m_{in} is the gas inlet mass of the single switch of the inlet valve body, m_{out} is the outlet mass of the single switch of the gas outlet valve body, C_d is the flow coefficient, A_1 is the flow area of the valve opening, ΔP_1 is the pressure difference between the left and right of the inlet valve body, ΔP_2 is the pressure difference between the left and right of the inlet valve body, and ρ is the density of the gas medium in the gas cylinder.

Remark 1. *The nonlinear factors in the controlled system are linearized, assuming that the control command u and the actual mass flow rate in and out of the valve are equal, i.e., the actuator can fully satisfy the control force command. However, in practical engineering, due to the limitations of the physical mechanism, the frequency of the number of times the solenoid valve can be switched on and off is limited, and therefore the inlet and outlet flow rates per unit of time are also bounded, so they are converted into an inequality form constraint, which is reflected in Equation (13).*

$$\begin{cases} m_{1t} \le m_1 = 15m_{in} & \text{Gas in} \\ m_{2t} \le m_2 = 15m_{out} & \text{Gas out} \end{cases} \tag{13}$$

2.4. Analytical Correction of Other Uncertainties

In the process of theoretical modeling, the internal uncertainties of the controlled system should be considered to be comprehensively as possible to obtain a higher accuracy of the system modeling. Compensating these factors into the controller can also, to some extent, improve the control performance of the system [41].

2.4.1. Analytical Correction of Piston Effective Area

During the operation of the manometer, as the temperature, as well as the pressure, continues to rise, the piston will produce a certain elastic deformation, and its effective working area will then change. Assuming that the influence of external factors is not taken into account, the effective area of the piston can be calculated by Equation (14).

$$A = \pi r^2 + \pi r \delta \tag{14}$$

where r is the piston rod radius, and δ is the clearance between the piston rod and the piston cylinder.

However, in the actual operation of the manometer, the effective area of the piston is affected by the temperature and pressure when the effective area of the piston is calculated by Equation (15):

$$A_0 = A[1 + (\alpha_c + \alpha_e)(\theta - 20)](1 + \lambda P) \tag{15}$$

where a_c is the thermal deformation coefficient of the piston cylinder, a_e is the thermal deformation coefficient of the piston, θ is the temperature of the piston system during actual operation, λ is the elastic deformation coefficient of the piston at the bottom of the piston, and P is the working pressure at the bottom of the piston rod.

2.4.2. Analytical Correction of Gas Leakage Volumes

During the operation of the system, there is a gap between the checking piston and the cylinder. As the system pressure rises, some of the gas media will leak through the gap, resulting in pressure loss within the piston cylinder and affecting the rise height of the piston. According to engineering thermodynamics, knowledge can be known between the piston rod and the piston cylinder, in which the concentric annular gap flow can be approximated as a flat slit flow [42]; at this time, the formation of differential pressure flow leakage formula for Equation (16):

$$Q_t = \frac{\pi d \delta^3 \Delta P_0}{12 \mu L} \tag{16}$$

where Q_t is the real-time gas leakage, d is the diameter of the checking piston, ΔP_0 the pressure difference between the two ends of the piston gap leakage surface, L is the length of the flow path, and δ the width of the annular gap.

The value of the pressure loss due to gas leakage can be calculated according to Poiseuille's law, as shown in Equation (17).

$$F_q = \frac{8 \mu L Q_t}{\pi r^4} \tag{17}$$

2.5. Establishment of the Differential Equilibrium Equations of the System

Combined with the characterization of the two core components of the system and the analytical correction of some of the uncertainties inherent in the actual operation of the system, this leads to a specific mathematical theoretical model of the adiabatic piston manometer, i.e., the differential equilibrium equations of the system. Taking the kinetic equilibrium Equation (1) as the main body, Equations (2), (8), (10), (12), (13), (15) and (17) are substituted into it to supplement it, and finally, the overall differential equilibrium Equation (18) of the system is obtained.

$$\begin{cases} \frac{m_{1t} ZRTA_0}{MV_2} - a - m_w g - c\dot{h} - \frac{8\mu L Q_t}{\pi r^4} + (EA_0{}^2)(\frac{m_{1t}}{\rho A_0} - h)/V_0 = m_w \ddot{h} \\ b - \frac{m_{2t} ZRTA_0}{MV_2} - m_w g + c\dot{h} - \frac{8\mu L Q_t}{\pi r^4} + (EA_0{}^2)(h - \frac{m_{1t}}{\rho A_0})/V_0 = -m_w \ddot{h} \end{cases} \tag{18}$$

where m_{1t} is the real-time inlet mass flow rate of the inlet valve, and m_{2t} is the real-time inlet mass flow rate of the outlet valve, both of which are input signals to the system and the output signal is the real-time rise height h of the piston, the others are fixed system parameters.

According to the actual working principle of the system, it can be determined that the real-time height of the piston needs to be compared with the set value when realizing the system control, and based on the result, the switching of the working state equation is selected. Equation (18) from top to bottom is the differential balance equation for the inlet and outlet states of the system, respectively. Each switch is made based on the previous working state, so we add the outgoing gas factor a and incoming gas factor b to Equation (18). a is the effect of the total outgoing air volume on the driving pressure before the system is switched to the inlet state, and b is the effect of the total air intake on the driving pressure before the system switches to the outgoing air state, both of which are constants when in the differential equations of their respective states.

To make the differential equilibrium equation of the system more intuitive and concise and to facilitate the design of the corresponding controller, the author rearranges and simplifies Equation (18) to obtain Equation (19).

$$\begin{cases} K_1 m_{1t} = K_2 \ddot{h} + K_3 \dot{h} + K_4 h + K_{51} \\ K_1 m_{2t} = K_2 \ddot{h} + K_3 \dot{h} + K_4 h + K_{52} \end{cases}$$
$$K_1 = \frac{ZRTA_0}{MV_2} + \frac{EA_0}{\rho V_0}$$
$$K_2 = m_w$$
$$K_3 = c \tag{19}$$
$$K_4 = \frac{EA_0{}^2}{V_0}$$
$$K_{51} = a + m_w g + \frac{8\mu L Q_t}{\pi r^4}$$
$$K_{52} = b - m_w g - \frac{8\mu L Q_t}{\pi r^4}$$

As can be seen from Equation (19), the equilibrium equations for both the inlet and outlet conditions are very similar, with the same constant coefficients for all terms except for the constant term. In the design and simulation of the controller, the constant term only affects the initial point of the system and does not affect the control performance of the system or the simulation results, so the constant term can be ignored. To make the controller structure more simple and clear, combined with Remark 1, the two separate actuator outputs can be regarded as a positive and negative control quantity, with a positive value representing the inlet valve working alone and a negative value representing the outlet valve working alone, so that the final differential equilibrium equation of the system becomes Equation (20).

$$\begin{cases} K_1 m_t = K_2 \ddot{h} + K_3 \dot{h} + K_4 h \\ m_1 < m_t < m_2 \end{cases} \tag{20}$$

3. Design of the Controller

As the absolute pressure piston manometer is subject to many internal uncertainties during actual operation, the LADRC controller can effectively estimate the internal and external disturbances of the system in real time while compensating accordingly. The LADRC was therefore chosen for balanced control of the system, but given the fixed parameters, the controller lacked adaptability, poor control performance, and weak resistance to interference. Therefore, this section introduces the idea of fuzzy control to improve the traditional LADRC method and proposes a fuzzy linear self-anti-disturbance-based absolute pressure piston manometer controller. The details are as follows. Numbered lists can be added as follows:

3.1. Design of the LADRC

3.1.1. Structure of the LADRC

LADRC, as a modern control method, enables the system to maintain good dynamics and a steady state under disturbances such as noise, load disturbances, and changes in the mathematical model and process parameters. In this subsection, the corresponding LADRC controller is designed based on the theoretical model of the system developed, as shown in Figure 3. The components include a linear differential tracker (LTD), a linearly expansive state observer (LESO), and a linear error feedback control rate (LESF), with the specific design of each part as follows.

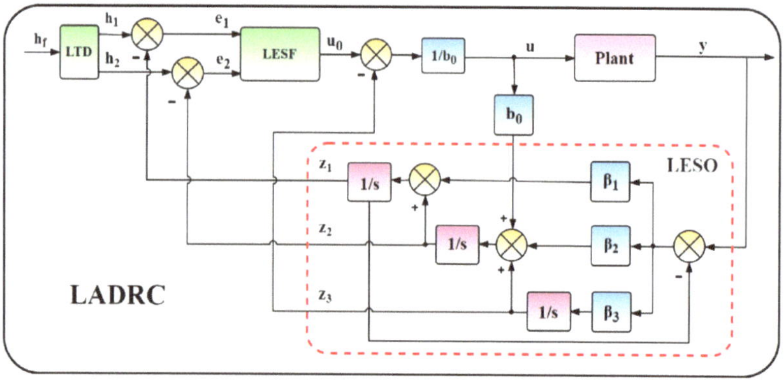

Figure 3. Schematic diagram of the second-order LADRC structure.

3.1.2. Design of the LTD

The LTD arranges the transition process for the input height signal and selects the appropriate parameters, which can make the output fast and overshoot-free, eliminating the contradiction between the amount of overshoot and the speed of control and also preventing sudden changes in the input signal, obtaining the differential signal and improving the robustness of the control system. The LTD design for the piston's ideal height is as follows:

$$\begin{cases} \dot{h}_1 = h_2 \\ \dot{h}_2 = -2r^2(h_1 - h_f) - 2rh_2 \end{cases} \tag{21}$$

where: h_f is the given value input value, h_1 is the softened height given value, and h_2 is the differential signal of h_1, which are used later in the design of the state error feedback controller. r is the tracking speed factor.

3.1.3. Design of the LESO

LESO is the central part of the self-anti-disturbance control, which is capable of both tracking and estimating the height of the piston rise in the manometer and its differential signal in finite time when the system has an input signal, and estimating the total disturbance to the system in real time [43]. It can also be seen from Equation (20) that the adiabatic piston manometer is a second-order single-input, single-output system and, according to the operating principle of the LADRC, the system equation can be expressed as:

$$\begin{cases} \ddot{h}(t) = f(h(t), \dot{h}(t), t) + b_0 u(t) \\ y(t) = h(t) \end{cases} \tag{22}$$

where $u(t)$ is the input variable; $y(t)$ is the output variable; b_0 is the system gain; $f\left(h(t),\dot{h}(t),t\right)$ is the generalized perturbation of the system, abbreviated as $f()$ for neatness, expanding $f()$ to a new state variable. The resulting state variables have been chosen as follows.

$$\begin{cases} x_1(t) = y(t) = h \\ x_2(t) = \dot{y}(t) = \dot{h} \\ x_3(t) = f() \end{cases} \tag{23}$$

The system equations can then be expressed in the form of an equation of state as follows:

$$\begin{cases} \dot{x}_1 = x_2 \\ \dot{x}_2 = x_3 + bu \\ \dot{x}_3 = \dot{f} \end{cases} \Rightarrow \begin{cases} \dot{x} = Ax + Bu + C\dot{f} \\ y = Dx \end{cases} \tag{24}$$

Of which $x = \begin{bmatrix} x_1 & x_2 & x_3 \end{bmatrix}^T$, $A = \begin{bmatrix} 0 & 1 & 0 \\ 0 & 0 & 1 \\ 0 & 0 & 0 \end{bmatrix}$, $B = \begin{bmatrix} 0 & b & 0 \end{bmatrix}^T$, $C = \begin{bmatrix} 0 & 1 & 0 \end{bmatrix}^T$, $D = \begin{bmatrix} 1 & 0 & 0 \end{bmatrix}$.

As a result, the corresponding LESO is designed as:

$$\begin{cases} \dot{z} & = Az + Bu + \beta(y - \hat{y}) \\ & = Az + Bu + \beta(y - Dz) \\ \hat{y} & = Dz \end{cases} \Rightarrow \begin{cases} \dot{z}_1 = z_2 + \beta_1(y - z_1) \\ \dot{z}_2 = z_3 + \beta_2(y - z_1) + b_0 u \\ \dot{z}_3 = \beta_3(y - z_1) \end{cases} \tag{25}$$

where $z_i = \begin{bmatrix} z_1 & z_2 & z_3 \end{bmatrix}^T$ is the matrix of observer output estimates, z_1 tracks the estimated height of the piston rise in the actual manometer, z_2 tracks the estimated speed of the piston rise, and z_3 is the estimate of the total perturbation $f()$; and $\beta_i = \begin{bmatrix} \beta_1 & \beta_2 & \beta_3 \end{bmatrix}^T$ is the observer gain matrix. The bandwidth method was proposed to rectify the observer gain parameters [35], let the bandwidth of the observer be ω_0 and try to configure all the eigenvalues of the observer as $-\omega_0$, i.e.,

$$\lambda_0(s) = \left| SI - (A - \beta D) \right| = s^3 + \beta_1 s^2 + \beta_2 s + \beta_3 = (s + \omega_0)^3 \tag{26}$$

At this point, the above equation satisfies the Hurwitz condition, and β_1, β_2, and β_3 all become functions with respect to ω_0, i.e.,

$$\begin{cases} \beta_1 = 3\omega_0 \\ \beta_2 = 3\omega_0^2 \\ \beta_3 = \omega_0^3 \end{cases} \tag{27}$$

Remark 2. *Observability judgments are made according to Equation (24). Let N = (D, DA)^T, since rank(N) = 3, from which it follows that the expansion system is fully observable.*

3.1.4. Design of the LESF

The LESF takes the error between the TD's output and the estimated system state by linearly combining them to form the initial control quantity and then adding compensation for the total disturbance estimated by the LESO to obtain the final control quantity as in Equation (28).

$$\begin{cases} e_1 = h_1 - z_{i1} \\ e_2 = h_2 - z_{i2} \\ u_0 = k_1 e_1 + k_2 e_2 \\ u = (u_0 - z_3)/b_0 \end{cases} \tag{28}$$

where: e_1 is the error between the LESO's estimate of the piston rise height, and the preset height input value, e_2 is the error between the LESO's estimate of the piston rise speed and the differential signal of the input value; k_1 and k_2 are the feedback control parameters; u_0 is the state error feedback control quantity. For the consideration of the dynamic performance of the LESF link, the same choice of the pole configuration method is used to rectify the two feedback control parameters so that the closed-loop transfer function has a pole of $-\omega_c$ and is solved as.

$$\begin{cases} k_1 = \omega_c^2 \\ k_2 = 2\omega_c \end{cases} \tag{29}$$

where ω_c is the controller bandwidth, for most common engineering objects, ω_c and ω_0 have a multiplicative relationship, i.e., ω_0 = (3~5)ω_c [44], in this paper ω_0 = 4ω_c is taken.

3.2. Design of the FLADRC

3.2.1. Structure of the FLADRC

The application of LADRC to the control of an equilibrium system does reduce the pressure on the adjustment of the control parameters, but the fixed parameters do affect the stability and adaptability of the system to internal and external disturbances. In this paper, fuzzy control is introduced to adjust the parameters $\Delta\omega_c$ and Δb_0 in the LADRC online and in real time so that the two parameters meet the requirements of the errors e_1 and e_2 at different times, thus improving the system's immunity to disturbances and stability performance. The specific control structure of the fuzzy LADRC is shown in Figure 4.

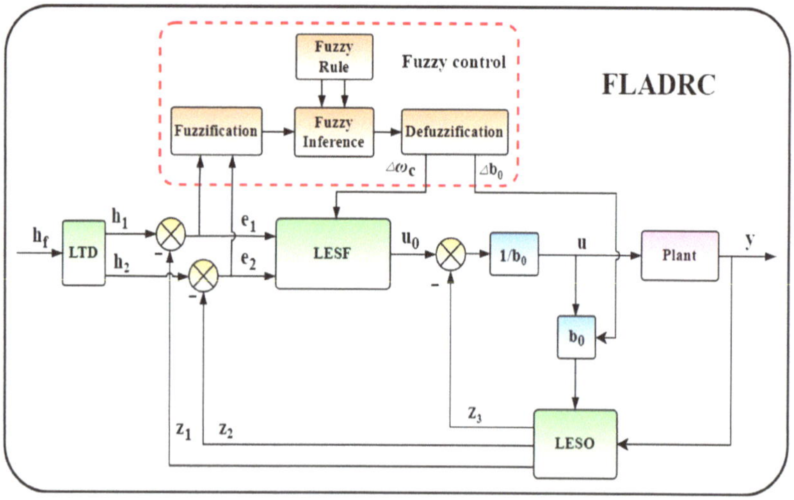

Figure 4. Schematic diagram of the FLADRC structure.

3.2.2. Design of the Fuzzy Controller

The overall process of fuzzy control can be divided into four modules, namely fuzzy quantization processing, which converts precise inputs into fuzzy linguistic values; establishing fuzzy control rules, which establishes specific fuzzy control rules based on expert experience and the actual equilibrium requirements of a specific system; fuzzy inference, which inferred the output values of fuzzy control based on the input–output relationships embedded in the fuzzy inputs and rule base; and defuzzification processing The output values are defuzzified to obtain the actual parameters. The detailed design is as follows.

The fuzzy input linguistic variables selected in this paper are the error e_1 between h_1 and z_1 and its corresponding rate of change of deviation e_2, and the theoretical domains of both are selected to be $[-2 \times 10^{-3}, 2 \times 10^{-3}]$ and $[-1 \times 10^{-3}, 1 \times 10^{-3}]$, respectively. The outputs of the fuzzy controller are chosen to be $\Delta\omega_c$ and Δb_0, where the theoretical domains of the two are assumed to be Range1 and Range2, respectively, the specific values of which are shown in the subsection on Experimental Parameter Settings later on. Seven subsets of the fuzzy language are defined on the respective theoretical domains of the inputs and outputs: {negative big (NB), negative medium (PM), negative small (NS), zero (ZO), positive small (PS), positive medium (NM), positive big (PB)}. Gaussian-type affiliation functions with smooth transitions are chosen for the inputs and high-sensitivity triangular affiliation functions for the outputs.

The traditional Mamdani inference method [45] was used for fuzzy reasoning. Based on the controllable performance of the adiabatic piston manometer and the parameter adjustment method of the LADRC, a table of fuzzy rules for $\Delta\omega_c$, Δb_0 was developed as in in Tables 1 and 2.

Table 1. Fuzzy rules for $\Delta\omega_c$.

e_1	e_2						
	NB	**NM**	**NS**	**ZO**	**PS**	**PM**	**PB**
NB	PB	PB	PM	PM	PS	PS	ZO
NM	PB	PB	PM	PM	PS	ZO	ZO
NS	PM	PM	PM	PS	ZO	NS	NM
ZO	PM	PS	PS	ZO	NS	NM	NM
PS	PS	PS	ZO	NS	NS	NM	NM
PM	ZO	ZO	NS	NM	NM	NM	NB
PB	ZO	NS	NS	NM	NM	NB	NB

Table 2. Fuzzy rules for Δb_0.

e_1	e_2						
	NB	**NM**	**NS**	**ZO**	**PS**	**PM**	**PB**
NB	NS	NS	ZO	ZO	PS	ZO	NB
NM	PS	PS	PS	PS	PS	ZO	PS
NS	PB	PB	PM	PS	PS	ZO	NS
ZO	PB	PM	PM	PS	ZO	ZO	NS
PS	PB	PM	PS	PS	NS	ZO	NS
PM	PM	PS	PS	PS	NS	ZO	NS
PB	NS	ZO	ZO	ZO	NS	ZO	NB

Based on the above control rules, the final control parameter equation for LADRC is derived using the center of gravity method for defuzzification as follows.

$$\begin{cases} \omega_c' = \omega_c + \Delta\omega_c \\ b_0' = b_0 + \Delta b_0 \end{cases} \tag{30}$$

where ω_c and b_0 are the initial control parameters of the LADRC obtained by genetic algorithm search.

3.3. Stability Analysis

As shown in Figure 2, the linear self-anti-disturbance control algorithm is a closed loop and has stability issues. Therefore, this section has been chosen to analyze and demonstrate the stability of LADRC. According to the control rate Equation (28), the closed-loop system consisting of the controlled object (22) is as follows:

$$\ddot{h} = f - z_3 + k_1(h - z_1) + k_2(\dot{h} - z_2) \tag{31}$$

Let $h = h_1, \dot{h} = h_2, \ddot{h} = h_2$, and combine with the equation $e_i = h_i - y_i$. The following equation can be obtained.

$$\begin{cases} \dot{e}_1 = \dot{h}_1 - \dot{y}_1 = h_2 - y_2 \\ \dot{e}_2 = \dot{h}_2 - \dot{y}_2 = h_3 - \ddot{y} = -k_1 e_1 - k_2 e_2 - k_1 \varepsilon_1 - k_2 \varepsilon_2 - \varepsilon_3 \end{cases} \tag{32}$$

where $\varepsilon_i = y_i - z_i$ $(i = 1, 2, 3)$ is the estimation error of LESO. Writing Equation (30) in the form of a state matrix gives Equation (33).

$$\dot{e} = A_e e + A_\varepsilon \varepsilon \tag{33}$$

where $A_e = \begin{bmatrix} 0 & 1 \\ -k_1 & -k_2 \end{bmatrix}$, $A_\varepsilon = \begin{bmatrix} 0 & 0 & 0 \\ -k_1 & -k_2 & 1 \end{bmatrix}$.

Theorem 1. *Assuming that the estimation error of LESO, $\lim\limits_{t\to\infty} \|\varepsilon\|_2 = 0$, there exist controller parameters $k_1 > 0$, $k_2 > 0$, such that the tracking error of the closed-loop system (33) tends to 0.*

Proof of Theorem 1. Solving for Equation (33) yields

$$e(t) = e^{A_e t} e(0) + \int_0^t e^{A_e(t-\tau)} A_\varepsilon \varepsilon d\tau \tag{34}$$

Knowing the matrix Ae, one can choose suitable parameters so that it has two different eigenvalues λ_1 and λ_2, and diagonalize the matrix, then one obtains

$$A_e = p diag\{-\lambda_1, -\lambda_2\} p^{-1} \tag{35}$$

For any $t > 0$, we obtain

$$\|e^{A_e t}\|_2 \leq \|p\|_2 \|p^{-1}\|_2 e^{-\lambda_1 t} \tag{36}$$

where p is determined by the eigenvalue of A_e, and when the eigenvalue has been determined, $\|p\|_2 \|p^{-1}\|_2$ is a constant here, denoted by C. When $t \to \infty$, $e^{-\lambda_1 t} \to 0$, then $\|e^{A_e t}\|_2 \to 0$.

Since the estimation error of LESO $\lim\limits_{t\to\infty} \|\varepsilon\|_2 = 0$, then $\|\varepsilon\|_2$ there is an upper bound, denoted by U, and for any positive number $a > 0$, there exists a time t_a, when $t > t_a$, $\|\varepsilon\|_2 < a$. Using Equation (36), then there is

$$\begin{aligned} \left\| \int_0^t e^{A_e(t-\tau)} A_\varepsilon \varepsilon d\tau \right\|_2 &= \left\| \int_0^{t_d} e^{A_e(t-\tau)} A_\varepsilon \varepsilon d\tau \right\|_2 + \left\| \int_{t_d}^t e^{A_e(t-\tau)} A_\varepsilon \varepsilon d\tau \right\|_2 \\ &\leq C\|A_\varepsilon\|_2 U \int_0^{t_d} e^{\lambda_1 \tau} d\tau e^{-\lambda_1 t} + C\|A_\varepsilon\|_2 e^{-\lambda_1 t} a \frac{e^{\lambda_1 (t-t_d)}}{\lambda_1} \\ &\leq C\|A_\varepsilon\|_2 U \int_0^{t_d} e^{\lambda_1 \tau} d\tau e^{-\lambda_1 t} + C\|A_\varepsilon\|_2 \frac{a}{\lambda_1} \\ &\leq D_1 e^{-\lambda_1 t} + D_2 a \end{aligned} \tag{37}$$

where $D_1 = C\|A_\varepsilon\|_2 U \int_0^{t_d} e^{\lambda_1 \tau} d\tau$, $D_2 = \frac{C\|A_\varepsilon\|_2}{\lambda_1}$, the first term on the right-hand side of Equation (37), has a limit of 0 when $t \to \infty$. The second term, due to the arbitrariness of a, gives

$$\lim_{t\to\infty} \left\| \int_0^t e^{A_e(t-\tau)} A_\varepsilon \varepsilon d\tau \right\|_2 = 0 \tag{38}$$

From $\lim\limits_{t\to\infty} \|e^{A_e t}\|_2 = 0$ and Equation (38), it can be shown that $\lim\limits_{t\to\infty} \|e(t)\|_2 = 0$. Therefore, it is proved that LADRC can make the tracking error of the closed-loop system converge to zero. Following the proof procedure of Theorem 1, it is easy to conclude the following. Assuming that there exists a bounded $f()$ that satisfies the inequality $|f()| \leq D, D > 0$, since there are controller parameters $k_1 > 0$, and $k_2 > 0$, and $\varepsilon_i = y_i - z_i$, then Equation (33) is bounded. This also means that for a bounded input h, the output of the system is also bounded, and the system satisfies BIBO stability. □

4. Simulation Results and Analysis

To verify the effectiveness and feasibility of the FLADRC control strategy proposed in this paper for absolute pressure piston manometers, it was decided to carry out simulation experiments using the Simulink module of MATLAB. Based on the theoretical model, four controllers, Kp, PID, LADRC, and FLADRC, are designed, and their corresponding control performance is compared.

4.1. Experimental Parameter Settings

As the working range of the controlled object selected in this paper is 0–6 MPa, which is a continuous range of values, to further demonstrate the Universal adaptability of FLADRC applied to the absolute pressure piston manometers, three working pressure points in the range are selected for the corresponding simulation analysis, which are low pressure (0.1 MPa), medium pressure (3 MPa) and high pressure (6 MPa). The main parameters of the absolute pressure piston manometer are shown in Tables 3 and 4. The specific parameters of the controller are shown in Table 5, where the optimal parameters for Kp and PID are determined using the Ziegler–Nichols engineering correction method, and the initial optimal parameters for LADRC are determined using a genetic algorithm.

Table 3. System fixed parameters.

Parameters	Value	Parameters	Value
C_d	0.9	a_c	4.5×10^{-5} °C^{-1}
ρ	1.25 kg/m^3	a_e	4.5×10^{-5} °C^{-1}
λ	7.1×10^{-7} MPa^{-1}	θ	21 °C
δ	6×10^{-7} m	T	294 k
R	296.8 J/(kg·K)	μ	1.741×10^{-2} N·s·m^{-2}
Z	0.292	A_1	7.85×10^{-9} m^2
V_0	3.7×10^{-7} m^3	A_0	5×10^{-5} m^2

Table 4. System variable parameters.

Parameters	Value		
	0.1 MPa	**3 MPa**	**6 MPa**
m_1	2.94×10^{-9} kg	3.53×10^{-10} kg	1.12×10^{-9} kg
m_2	2.23×10^{-9} kg	1.93×10^{-9} kg	2.73×10^{-9} kg
m_w	0.5 kg	16 kg	32 kg

Table 5. Controller parameters.

Controllers	Value		
	0.1 MPa	**3 MPa**	**6 MPa**
Kp	55	65	90
PID	$K_p = 26$, $K_i = 0.55$, $K_{1d} = 7.5$	$K_{1p} = 30$, $K_{1i} = 0.5$, $K_{1d} = 7$	$K_{1p} = 40$, $K_{1i} = 0.9$, $K_{1d} = 8$
LADRC	$r = 4$, $\omega_c = 46$, $b_0 = 0.22$	$r = 4$, $\omega_c = 77$, $b_0 = 0.37$	$r = 4$, $\omega_c = 65$, $b_0 = 0.25$
FLADRC	Range1 = [−4, 4] Range2 = [−0.022, 0.022]	Range1 = [−7, 7] Range2 = [−0.037, 0.037]	Range1 = [−6, 6] Range2 = [−0.025, 0.025]

4.2. Experimental Analysis of Stability Performance

Considering that in the actual working process of an absolute pressure piston manometer, the piston has to rise as quickly as possible and be stabilized in the desired position, the desired value of the piston height of the system is taken as the input signal for the simulation, which is uniformly set to a 2 mm step signal, the output signal is the real-time piston height of the piston, the simulation time is set to 35 s, and the error band is taken to be ±2% of the desired height. To facilitate visual analysis, the stable convergence time t_s, the integral IAE of the absolute value of the actual error, and the maximum overshoot rate 6% are chosen as performance indicators in this paper. The real-time height profile of the piston is shown in Figure 5, and the experimental results of the selected performance metrics are shown in Figure 6 and Table 6.

As shown in Figure 5 the FLADRC-controlled pistons all rise at a rate less than Kp, PID, and LADRC, but it is clear from Table 6 that the FLADRC-controlled pistons have the shortest stabilization time, the LADRC has the second shortest stabilization time and the Kp and PID have a longer stabilization time. As can be seen from Figure 6a, the IAE of the FLADRC is not very different from the LADRC overall, with the smallest IAE for the FLADRC at the 0.1 MPa and 3 MPa operating points. At the 6 MPa pressure operating point, the IAE of FLADRC is slightly greater than that of LADRC, but in Figure 6b, it can be seen that FLADRC is the best at suppressing overshoot, with a maximum overshoot of zero at the 0.1 MPa operating point and far less at the 3 MPa and 6 MPa pressure operating points than the other three control strategies. It can be concluded that the FLADRC control strategy proposed in this paper has the best stability performance in the equilibrium control of absolute pressure piston manometers.

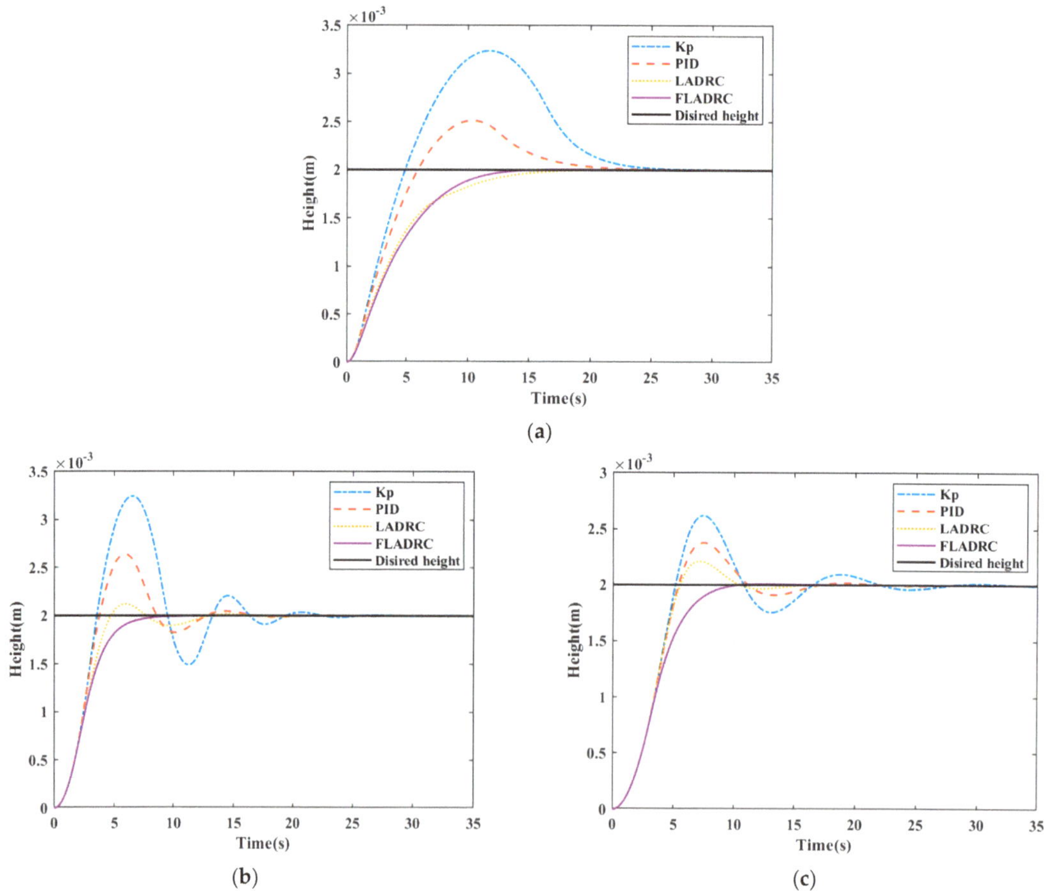

Figure 5. Real-time piston height curves at three different pressure operating points. (**a**) 0.1 MPa; (**b**) 3 MPa; (**c**) 6 MPa.

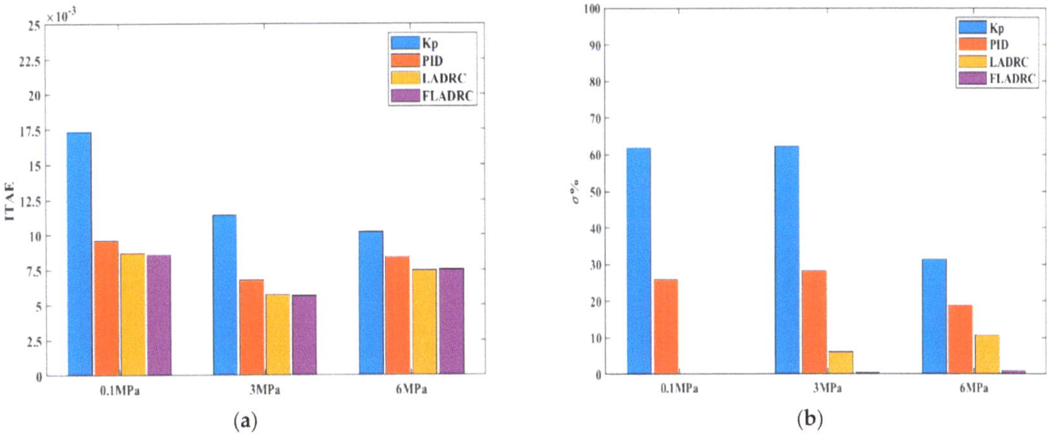

Figure 6. Comparison of the two performance indicators. (**a**) IAE; (**b**) $\sigma\%$.

Table 6. Stable convergence time t_s(s).

Conditions	Kp	PID	LADRC	FLADRC
0.1 MPa	23.36	19.58	14.87	12.37
3 MPa	18.72	14.76	11.84	8.54
6 MPa	20.98	15.79	12.04	8.67

4.3. Experimental Analysis of Interference Immunity Performance

Absolute pressure piston manometers can be subject to many external disturbances during actual operation, such as temperature and humidity, atmospheric pressure, and electromagnetic fluctuations in the environment. Therefore, in the system interference immunity experiments, the system parameters, as well as the controller parameters, are kept constant, and the actual disturbance is assumed to be a time-varying white noise signal, which is added to the input flow of the system model, and its amplitude is set according to the gas incoming and outgoing flow limits at the specific pressure operating point. The corresponding simulation results are shown in Figure 7. The stability time $t_{s(n)}$ and the integral $IAE_{(n)}$ of the absolute value of the steady-state error are taken as the steady-state performance indicators so that they are compared with the experimental results in the previous section, and the corresponding rates of change $\zeta\%$ and $\delta\%$ of the two are derived, which are the system interference immunity performance indicators chosen in this paper. The specific results for $t_{s(n)}$ and $IAE_{(n)}$ are shown in Table 7 and Figure 8, and the corresponding rates of change are calculated in Tables 8 and 9.

It can be seen from Figure 7 that all control strategies stabilize within a certain time after the addition of time-varying disturbances. From Table 7 and Figure 8, it can be easily seen that the stabilization time and IAE(n) of all the control strategies increase to a greater or lesser extent due to the disturbance, but FLADRC still has the shortest stabilization time and the smallest IAE(n). This proves that the steady-state performance of FLADRC is still the best, even when disturbances are added.

Table 7. Stable convergence time $t_{s(n)}$ (s).

Conditions	Kp	PID	LADRC	FLADRC
0.1 MPa	27.62	22.48	17.09	13.26
3 MPa	21.67	16.28	14.85	9.87
6 MPa	25.33	22.03	13.07	10.41

Table 8. Rate of change of t_s ($\zeta\%$).

Conditions	Kp	PID	LADRC	FLADRC
0.1 MPa	18.23%	18.95%	14.92%	7.19%
3 MPa	15.76%	12.06%	11.74%	10.77%
6 MPa	21.97%	13.29%	8.33%	6.22%

Table 9. The rate of change of IAE ($\delta\%$).

Conditions	Kp	PID	LADRC	FLADRC
0.1 MPa	19.21%	10.44%	3.35%	1.51%
3 MPa	15.34%	9.16%	7.63%	2.68%
6 MPa	16.47%	9.79%	4.44%	1.24%

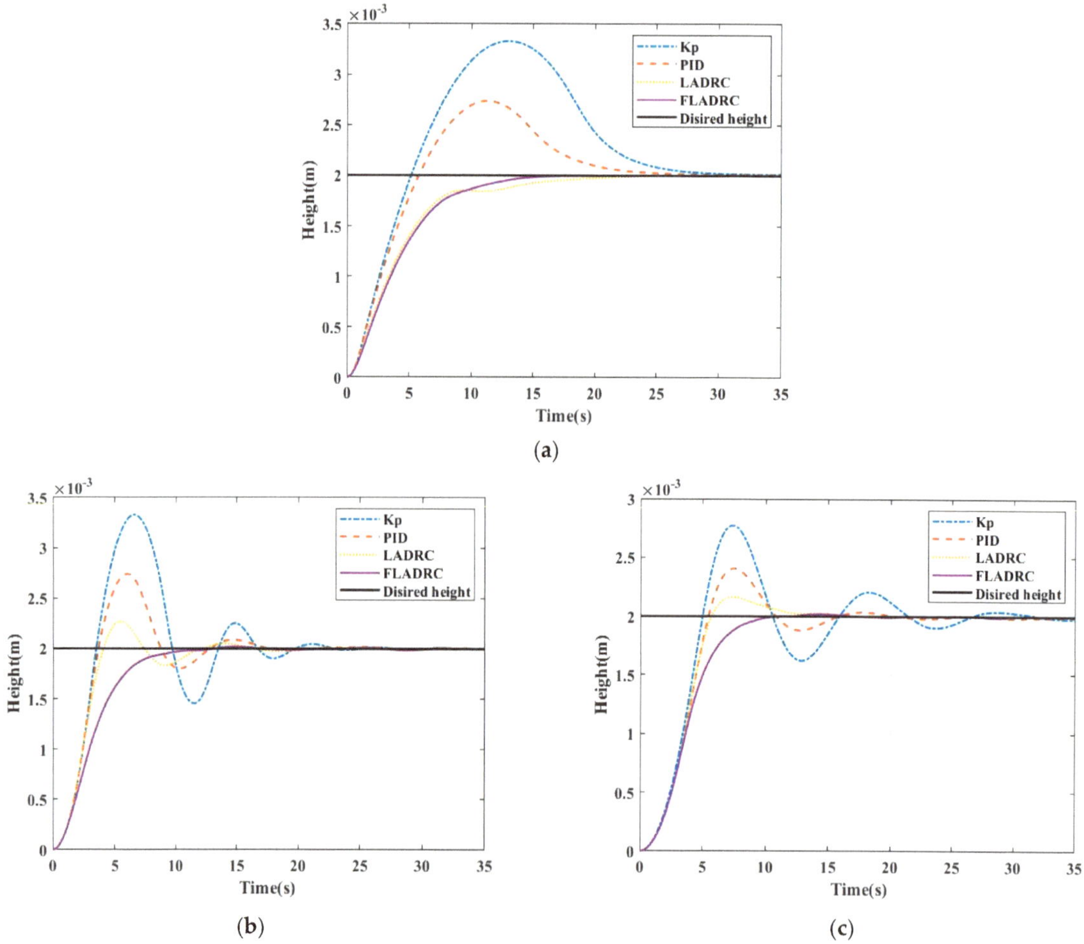

Figure 7. Real-time piston height curves after disturbance at three different pressure operating points. (**a**) 0.1 MPa; (**b**) 3 MPa; (**c**) 6 MPa.

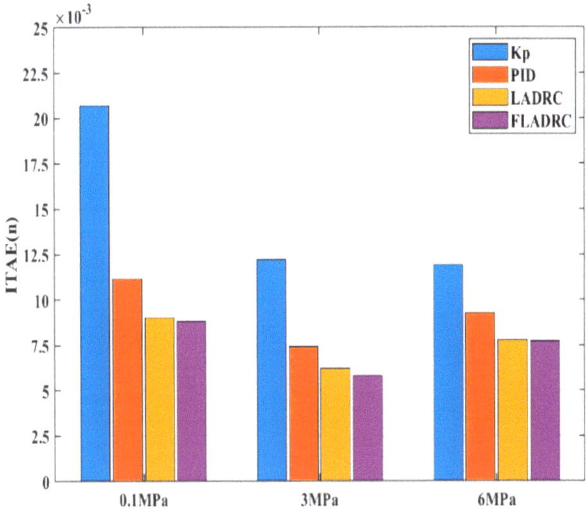

Figure 8. Comparison chart for performance indicator $IAE_{(n)}$.

The immunity performance index of each control strategy was obtained by calculation. As shown in Table 8, the rate of change of the steady-state time ζ is the smallest for FLADRC at all three different pressure operating points, followed by LADRC, while the steady-state times for PID and Kp are relatively more variable. As shown in Table 9, it is also evident that the FLADRC has the smallest rate of change of IAE, δ, followed by the LADRC, PID, and Kp, from largest to smallest. In summary, the steady-state performance of the FLADRC is least affected by time-varying disturbances, which means that the FLADRC has the relatively best interference immunity performance.

Remark 3. *The immunity performance indicators ζ and δ are calculated as follows:*

$$\begin{cases} \zeta = (t_{s(n)} - t_s)/t_s \\ \delta = (IAE_{(n)} - IAE)/IAE \end{cases} \tag{39}$$

4.4. Experimental Analysis of Engineering Energy Consumption

In the actual operation of the system, the inlet and outlet valves are the two actuators, respectively, and the corresponding instantaneous gas mass flow rate is the control quantity of the system. These two control quantities have been abstracted into a positive and negative input signal in Section 2 of the article, with a positive value representing the inlet valve working alone and vice versa for the outlet valve working alone. The control quantity input curves for the system in the steady-state performance experiments are shown in Figure 9.

As shown in Figure 9 at different pressure operating points, the FLADRC used in this paper has the smallest oscillation of the control quantity signal compared to the other three control strategies, which means that the switching frequency of the switching valve is lower. The number of switches between the two actuators, the inlet and outlet valves, is less. This reduces the energy consumption of the actuator operation, reduces the wear of the valve stems in the switching valves, increases the overall actuator life, and reduces the possibility of mechanical errors after long periods of system operation.

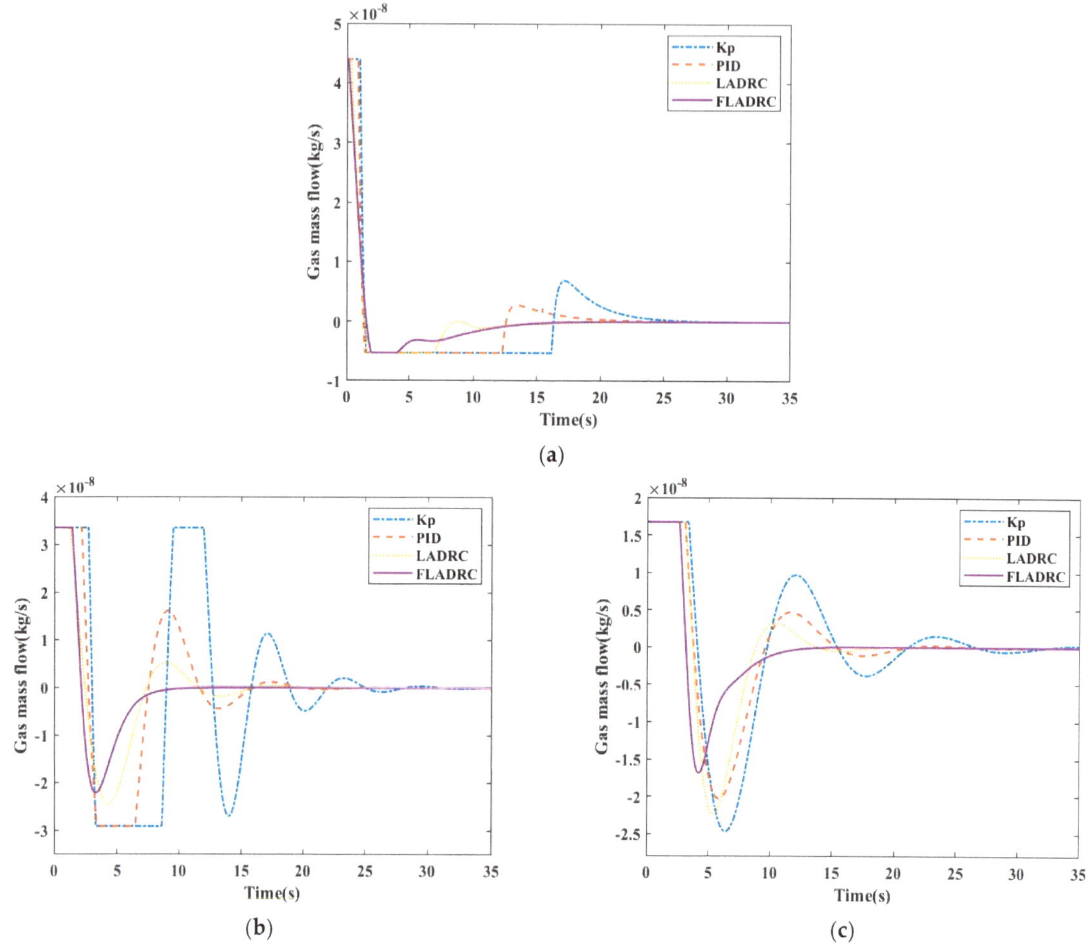

Figure 9. Control volume input curves at three different pressure operating points. (**a**) 0.1 MPa; (**b**) 3 MPa; (**c**) 6 MPa.

5. Conclusions

This paper takes into account the fact that absolute pressure piston manometers are subject to many internal uncertainties and non-linearities during actual operation and that the control performance of the system equilibrium needs to be improved urgently. Therefore, this paper proposes a FLADRC-based equilibrium control method for absolute pressure piston gauges. The control method combines the advantages of LADRC and fuzzy control to effectively estimate and compensate for real-time internal and external disturbances in the system and to achieve adaptive online adjustment of the control parameters. In the contemporary field of control, very little research has been carried out on the controlled systems mentioned in this paper, so a linearized theoretical model of the absolute pressure piston manometer was first developed. Furthermore, the FLADRC controller was designed according to the model, and the initialization parameters were determined by a search for merit. In addition, the stability of the control system was also analyzed. Finally, the simulation is verified in MATLAB's Simulink environment, and its experimental results are compared and analyzed with Kp, PID, and LADRC. The results show that the FLADRC control strategy proposed in this paper has the advantages of short stability time, small overshoot, strong anti-interference capability, and low input energy consumption, verifying that it has important engineering application value for absolute pressure piston manometers.

In future work, it is intended to further commercialize the absolute pressure piston manometer by designing an automatic matching system for the pressure operating point and the corresponding control parameters. An attempt is also made to apply the FLADRC strategy to a piston manometer with a liquid as the working medium and to carry out the corresponding performance verification.

Author Contributions: T.G. designed the project and coordinated the work. H.W. checked and discussed the results and the whole manuscript. X.Z. contributed to the discussion of this study. N.W. and Z.Z. were involved in the data collection for the article. G.P. provided financial support for the article. All authors have read and agreed to the published version of the manuscript.

Funding: Liaoning Provincial Education Department Scientific Research Funding Project NO.LJKZ0510.

Data Availability Statement: Not applicable.

Conflicts of Interest: The authors declare no conflict of interest.

References

1. Yang, Y.; Driver, R.G.; Quintavalle, J.S.; Scherschligt, J.; Schlatter, K.; Ricker, J.E.; Strouse, G.F.; Olson, D.A.; Hendricks, J.H. An integrated and automated calibration system for pneumatic piston gauges. *Measurement* **2018**, *134*, 1–5. [CrossRef]
2. Schmidt, J.W.; A Tison, S.; Ehrlich, C.D. Model for drag forces in the crevice of piston gauges in the viscous-flow and molecular-flow regimes. *Metrologia* **1999**, *36*, 565–570. [CrossRef]
3. Welch, B.; Edsinger, R.; Bean, V.; Ehrlich, C. The reduction of uncertainties for absolute piston gauge pressure measurements in the atmospheric pressure range. *J. Res. Natl. Inst. Stand. Technol.* **1989**, *94*, 343–346. [CrossRef] [PubMed]
4. Li, C.; Jing, H.; Bao, J.; Sun, S.; Wang, R. Robust H_∞ fault tolerant control for quadrotor attitude regulation. *Proc. Inst. Mech. Eng. Part I J. Syst. Control Eng.* **2018**, *232*, 1302–1313. [CrossRef]
5. Romero, A.; Sun, S.; Foehn, P.; Scaramuzza, D. Model Predictive Contouring Control for Time-Optimal Quadrotor Flight. *IEEE Trans. Robot.* **2022**, *38*, 3340–3356. [CrossRef]
6. Wang, X.; Dong, H.; Sun, X.; Yao, X. PD control of inverted pendulum based on adaptive fuzzy compensation. *J. Intell. Fuzzy Syst.* **2016**, *31*, 3013–3019. [CrossRef]
7. Sharma, A.K.; Bhushan, B. Comparison of various fuzzy sliding mode based controller on single link inverted pendulum. *J. Intell. Fuzzy Syst.* **2022**, *42*, 679–688. [CrossRef]
8. Semenov, M.E.; Solovyov, A.M.; Popov, M.A.; Meleshenko, P.A. Coupled inverted pendulums: Stabilization problem. *Arch. Appl. Mech.* **2017**, *88*, 517–524. [CrossRef]
9. Liu, Y.; Huang, X.; Wang, T.; Zhang, Y.; Li, X. Nonlinear dynamics modeling and simulation of two-wheeled self-balancing vehicle. *Int. J. Adv. Robot. Syst.* **2016**, *13*, 156–171. [CrossRef]
10. Huang, C.-N. The Development of Self-Balancing Controller for One-Wheeled Vehicles. *Engineering* **2010**, *2*, 212–219. [CrossRef]
11. Qu, L.; Qiao, W.; Qu, L. Active-Disturbance-Rejection-Based Sliding-Mode Current Control for Permanent-Magnet Synchronous Motors. *IEEE Trans. Power Electron.* **2020**, *36*, 751–760. [CrossRef]
12. Tarczewski, T.; Grzesiak, L.M. Constrained State Feedback Speed Control of PMSM Based on Model Predictive Approach. *IEEE Trans. Ind. Electron.* **2015**, *63*, 3867–3875. [CrossRef]
13. Lee, D.; Cheon, Y.; Ryu, J.-H.; Lee, I.-B. An MCFC operation optimization strategy based on PID auto-tuning control. *Int. J. Hydrogen Energy* **2017**, *42*, 25518–25530. [CrossRef]
14. Rakhtala, S.; Roudbari, E.S. Fuzzy PID control of a stand-alone system based on PEM fuel cell. *Int. J. Electr. Power Energy Syst.* **2016**, *78*, 576–590. [CrossRef]
15. Gün, A. Attitude control of a quadrotor using PID controller based on differential evolution algorithm. *Expert Syst. Appl.* **2023**, *229*, 412–424. [CrossRef]
16. Nocoń, Ł.; Koruba, Z. Modified linear-quadratic regulator used for controlling anti-tank guided missile in vertical plane. *J. Theor. Appl. Mech.* **2020**, *58*, 723–732. [CrossRef]
17. Borrelli, F.; Keviczky, T. Distributed LQR Design for Identical Dynamically Decoupled Systems. *IEEE Trans. Autom. Control* **2008**, *53*, 1901–1912. [CrossRef]
18. Tingting, Y.; Aijun, L.; Taimoor, M.; Amin, R.U. High AOA short landing robust control for an aircraft. *Aircr. Eng. Aerosp. Technol.* **2018**, *91*, 38–49. [CrossRef]
19. Pan, S. Robust control of gyro stabilized platform driven by ultrasonic motor. *Sens. Actuators A Phys.* **2017**, *261*, 280–287. [CrossRef]
20. Abbasi, S.M.M.; Jalali, A. Fuzzy tracking control of fuzzy linear dynamical systems. *ISA Trans.* **2020**, *97*, 102–115. [CrossRef]
21. Yoon, J.H.; Bak, G.-M.; Bae, Y. Fuzzy Control for Chaotic Confliction Model. *Int. J. Fuzzy Syst.* **2020**, *22*, 1961–1971. [CrossRef]

22. Lu, Y. Adaptive-fuzzy control compensation design for direct adaptive fuzzy control. *IEEE Trans. Fuzzy Syst.* **2018**, *26*, 3222–3231. [CrossRef]
23. Canciello, G.; Cavallo, A.; Cucuzzella, M.; Ferrara, A. Fuzzy scheduling of robust controllers for islanded DC microgrids applications. *Int. J. Dyn. Control.* **2019**, *7*, 690–700. [CrossRef]
24. Zhao, Z.; Liu, Z.; He, W.; Hong, K.-S.; Li, H.-X. Boundary adaptive fault-tolerant control for a flexible Timoshenko arm with backlash-like hysteresis. *Automatica* **2021**, *130*, 109690. [CrossRef]
25. Sang, H.; Zhou, Y.; Sun, X.; Yang, S. Heading tracking control with an adaptive hybrid control for under actuated underwater glider. *ISA Trans.* **2018**, *80*, 554–563. [CrossRef] [PubMed]
26. Muñoz-Vázquez, A.J.; Fernández-Anaya, G.; Meléndez-Vázquez, F.; Torres, J.D.S. Generalised conformable sliding mode control. *Math. Methods Appl. Sci.* **2021**, *45*, 1687–1699. [CrossRef]
27. Ma, H.; Wu, J.; Xiong, Z. Discrete-time sliding-mode control with improved quasi-sliding-mode domain. *IEEE Trans. Ind. Electron.* **2016**, *63*, 6292–6304. [CrossRef]
28. Canciello, G.; Cavallo, A.; Schiavo, A.L.; Russo, A. Multi-objective adaptive sliding manifold control for More Electric Aircraft. *ISA Trans.* **2020**, *107*, 316–328. [CrossRef]
29. Nichols, G.P.; Fontenot, J.D.; Gibbons, J.P.; Sanders, M. Evaluation of volumetric modulated Arc therapy for postmastectomy treatment. *Radiat. Oncol.* **2014**, *9*, 66. [CrossRef]
30. Wang, F.; Chen, K.; Zhen, S.; Zheng, H.; Chen, X.; Chen, F. Optimal design of Udwadia–Kalaba theory-based adaptive robust control for permanent magnet linear synchronous motor. *J. Vib. Control* **2023**, 10775463221149086. [CrossRef]
31. Nadda, S.; Swarup, A. On adaptive sliding mode control for improved quadrotor tracking. *J. Vib. Control* **2017**, *24*, 3219–3230. [CrossRef]
32. Jain, A.; Sharma, A.; Jately, V.; Azzopardi, B.; Choudhury, S. Real-Time Swing-Up Control of Non-Linear Inverted Pendulum Using Lyapunov Based Optimized Fuzzy Logic Control. *IEEE Access* **2021**, *9*, 50715–50726. [CrossRef]
33. Luo, Z.; Xu, L. Research on LQR controller of control torque gyro self-balancing vehicle based on QPSO optimization. *J. Phys. Conf. Ser.* **2022**, *2258*, 012071. [CrossRef]
34. Han, J. From PID to Active Disturbance Rejection Control. *IEEE Trans. Ind. Electron.* **2009**, *56*, 900–906. [CrossRef]
35. Gao, Z. Scaling and bandwidth-parameterization based controller tuning. In Proceedings of the 2003 American Control Conference, Denver, CO, USA, 4–6 June 2003; pp. 4989–4996. [CrossRef]
36. Yuan, C.; Zhou, X.; Ma, Y.; Gao, Z.; Zhou, Y.; Wang, C. Improved Application of Third-Order LADRC in Wind Power Inverter. *Energies* **2020**, *13*, 4412. [CrossRef]
37. Wang, Z.; Zhao, T. Based on robust sliding mode and linear active disturbance rejection control for attitude of quadrotor load UAV. *Nonlinear Dyn.* **2022**, *108*, 3485–3503. [CrossRef]
38. Zeng, J.; Huang, Z.; Huang, Y.; Qiu, G.; Li, Z.; Yang, L.; Yu, T.; Yang, B. Modified linear active disturbance rejection control for microgrid inverters: Design, analysis, and hardware implementation. *Int. Trans. Electr. Energy Syst.* **2019**, *29*, e12060. [CrossRef]
39. Ramasamy, K.; Rathinasamy, S.; Kwon, O.M.; Palanisamy, S. Robust tracking control design for fractional-order interval type-2 fuzzy systems. *Nonlinear Dyn.* **2021**, *107*, 3611–3628. [CrossRef]
40. Qiu, T.; Dai, H.; Lei, Y.; Liu, Y. Effects of valve needle speed on flow characteristics in control valve for unit pump fuel system. *Adv. Mech. Eng.* **2018**, *10*, 378–389. [CrossRef]
41. Kavikumar, R.; Kwon, O.-M.; Kaviarasan, B.; Sakthivel, R. Antidisturbance Control Design for Interval Type-2 Fuzzy Stochastic Systems with Input Quantization. *IEEE Trans. Fuzzy Syst.* **2022**, *31*, 1806–1818. [CrossRef]
42. Bobovnik, G.; Kutin, J. Correlation of the leakage flow rate with pressure changes in a clearance-sealed piston prover. *Flow Meas. Instrum.* **2020**, *74*, 101785. [CrossRef]
43. Yu, B.; Shen, A.; Chen, B.; Luo, X.; Tang, Q.; Xu, J.; Zhu, M. A Compensation Strategy of Flux Linkage Observer in SPMSM Sensorless Drives Based on Linear Extended State Observer. *IEEE Trans. Energy Convers.* **2021**, *37*, 824–831. [CrossRef]
44. Liang, Q.; Wang, C.; Pan, J.; Wei, Y.; Wang, Y. Identification of linear self-adjoint control parameters b0 and parameter tuning laws. *Control Decis. Mak.* **2015**, *30*, 1691–1695.
45. Ren, Q.; Bigras, P. A Highly Accurate Model-Free Motion Control System with a Mamdani Fuzzy Feedback Controller combined with a TSK Fuzzy Feed-forward Controller. *J. Intell. Robot. Syst.* **2017**, *86*, 367–379. [CrossRef]

Disclaimer/Publisher's Note: The statements, opinions and data contained in all publications are solely those of the individual author(s) and contributor(s) and not of MDPI and/or the editor(s). MDPI and/or the editor(s) disclaim responsibility for any injury to people or property resulting from any ideas, methods, instructions or products referred to in the content.

actuators MDPI

Article

Research on Friction Compensation Method of Electromechanical Actuator Based on Improved Active Disturbance Rejection Control

Pan Zhang *, Zhaoyao Shi and Bo Yu

Beijing Engineering Research Center of Precision Measurement Technology and Instruments, Beijing University of Technology, No. 100, Pingleyuan, Chaoyang District, Beijing 100124, China; shizhaoyao@bjut.edu.cn (Z.S.)
* Correspondence: zhangpan@bjut.edu.cn

Abstract: The friction factor of harmonic reducers affects the transmission accuracy in electromechanical actuators (EMAs). In this study, we proposed a friction feedforward compensation method based on improved active disturbance rejection control (IADRC). A mathematical model of EMA was developed. The relationship between friction torque and torque current was derived. Furthermore, the compound ADRC control method of second-order speed loop and position loop was studied, and an IADRC control method was proposed. A real EMA was developed, and the working principles of the EMA driving circuit and current sampling were analyzed. The three methods—PI, ADRC, and IADRC—were verified by conducting speed step experiments and sinusoidal tracking experiments. The integral values of time multiplied by the absolute error of the three control modes under the step speed mode were approximately 47.7, 32.1, and 15.5, respectively. Disregarding the inertia of the reducer and assuming that the torque during no-load operation equals the friction torque during constant motion, the findings indicate that, under a load purely driven by inertia, the IADRC control method enhances tracking accuracy.

Keywords: ADRC; friction compensation; integrated electromechanical actuator; PMSM

Citation: Zhang, P.; Shi, Z.; Yu, B. Research on Friction Compensation Method of Electromechanical Actuator Based on Improved Active Disturbance Rejection Control. *Actuators* **2023**, *12*, 445. https://doi.org/10.3390/act12120445

Academic Editors: Oscar Barambones, Jose Antonio Cortajarena and Patxi Alkorta

Received: 20 October 2023
Revised: 25 November 2023
Accepted: 26 November 2023
Published: 30 November 2023

Copyright: © 2023 by the authors. Licensee MDPI, Basel, Switzerland. This article is an open access article distributed under the terms and conditions of the Creative Commons Attribution (CC BY) license (https://creativecommons.org/licenses/by/4.0/).

1. Introduction

An integrated electromechanical actuator mainly comprises a motor, a reducer, a driver, a controller, and a position sensor. Actuators are mainly used in aviation, aerospace, robotics, guided weapons, medical devices, precision instruments, and other fields [1–5]. Reducers used in EMAs mainly include the parallel shaft gear reducer, planetary gear reducer, harmonic reducer, and rotary vector reducer. Harmonic reducers are commonly employed for actuators with high reduction ratios and medium power. The main factors that affect the transmission accuracy of harmonic reducers are clearance, friction, and stiffness. Spong et al. [6] proposed a dynamic modeling method for flexible joints. Based on the friction characteristics of harmonic reducers, Gandhi [7] associated friction with speed and position in the transmission system and used friction identification and nonlinear compensation methods to improve transmission accuracy. Taghirad et al. [8] established a dynamic model of a harmonic reducer, modeled friction losses at high and low speeds, and studied the characteristics of the model through simulation analysis. In the literature [9–12], the influence of temperature and load on friction has been deeply studied. Maré J. C. [9] proposed a generic framework for introducing load and temperature effects in the system-level friction model. Studies [10–12] have analyzed the effects of temperature and load on the friction torque of the harmonic reducer.

High-performance permanent magnet synchronous motors (PMSMs) are used in EMAs. Commonly employed in the control process of PMSM are PID control and state feedback control methods. However, PID control has the drawbacks of slow response speed and weak disturbance rejection. Advanced intelligent algorithms have been incorporated into PID controllers to improve PID control effects, such as genetic algorithm PID, self-tuning PID, artificial intelligence algorithm, and neural network PID [13–16]. Intelligent control algorithms have complex algorithms, high computational complexity, and pose challenges in engineering applications. Active disturbance rejection control (ADRC) operates independently of precise mathematical modeling of the controlled object. Contrary to traditional methods, it accounts for uncertain and complex factors, including unmodeled system components, external disturbances, nonlinear factors, and time-varying elements, by classifying these as the "total disturbance" of the system. ADRC utilizes a constructed extended state observer to estimate this "total disturbance" online and employs a control law for compensation [17]. Applications of ADRC in motor control have shown varying degrees of improvement in motor control efficacy [18–20]. Jin et al. [21] implemented a novel type of linear ADRC, replacing the PID controller, to effectively control a hydraulic cylinder servo system, acknowledging the characteristics of high-order coupling in the electrohydraulic system. Hu et al. [22] established an ADRC control method based on LuGre friction compensation to study the effect of nonlinear friction on the transmission accuracy of the photoelectric stabilization platform. Sira-Ramírez et al. [23] employed ADRC based on high gain generalized proportional integral observers for PMSM large disturbance trajectory tracking systems. Li et al. [24] used second-order ADRC to improve the disturbance rejection and transmission accuracy in the PMSM position control process. Research has been conducted on built-in PMSM control by using ADRC for position sensorless control [25].

To mitigate the impact of nonlinear friction on the precision of EMA transmission, this study proposes an improved ADRC method based on the magnetic field-oriented control (FOC) method. The EMA friction model was added to the IADRC through feedforward compensation to improve the transmission accuracy. First, a mathematical model of the PMSM is presented. Based on this model, combined with the ADRC control principle, an EMA speed loop/position loop composite second-order ADRC is constructed. In the nonlinear error feedback link, a fuzzy control algorithm is incorporated to achieve the adaptive functionality of the EMA control algorithm. The relationship between the no-load friction torque and torque current is derived based on the transmission model of a harmonic reducer. The friction model was established through experimental methods and added to the ADRC control model through feedforward compensation. Based on the above research, an EMA drive control system was developed using STM32F4 as the main control chip, and the aforementioned control strategies were experimentally verified. The experimental results were evaluated and analyzed using the integral of time absolute error (ITAE) and the root mean square error (RMSE).

2. Mathematical Model of the PMSM Established Using the FOC Method

The PMSM is frequently utilized as a torque source in high-performance EMA applications. The PMSM mathematical model mainly includes the voltage equation, magnetic linkage equation, torque equation, and mechanical equation. To simplify the analysis without affecting the control, the winding current is assumed to be a symmetrical three-phase sinusoidal current, motor core saturation is ignored, and the eddy current and hysteresis losses of the motor are not considered. The PMSM adopts the FOC method, which offers the advantages of fast dynamic response, smooth torque, and stable low-speed control. By using FOC, the voltage equation in the d-q coordinate system is

$$\left\{ \begin{array}{l} u_d = R_s i_d + \frac{d\psi_d}{dt} - \omega\psi_q \\ u_q = R_s i_q + \frac{d\psi_q}{dt} + \omega\psi_d \end{array} \right. \tag{1}$$

The d-q axis magnetic linkage equation is

$$\left\{ \begin{array}{l} \psi_d = L_d i_d + \psi_f \\ \psi_q = L_q i_q \end{array} \right. \tag{2}$$

The electromagnetic torque equation is

$$T_e = 1.5p_n[\psi_f i_q + (L_d - L_q)i_d i_q] \tag{3}$$

The second Newton law applied to the motor rotor is

$$J\frac{d\omega_r}{dt} = T_e - T_L - B\omega_r \tag{4}$$

In Equations (1)–(4), Rs is the phase resistance, i_d and i_q are the d-axis and q-axis currents, u_d and u_q are the d-axis and q-axis voltages, L_d and L_q are the d-axis and q-axis inductances, ψ_d and ψ_q are, respectively, the d-axis and q-axis magnetic linkages, p_n is the number of pole pairs, T_e is the electromagnetic torque, J is the rotational inertia, T_L is the load torque, B is the damping coefficient, ψ_f is the permanent magnet magnetic flux, ω is to the electrical angular velocity, and ω_r is the mechanical angular velocity.

$$\omega = p_n\omega_r \tag{5}$$

For surface-mounted PMSM, $L_d = L_q$. When $i_d = 0$ or $L_d = L_q$, Equation (3) can be simplified as

$$T_e = 1.5p_n\psi_f i_q \tag{6}$$

3. Transmission Model of the Harmonic Reducer

The harmonic reducer comprises a circular spine, a flexspline, and a wave generator, as shown in Figure 1. In the EMA, the circular spine is fixed and connects the rotor to the wave generator, whereas the flexspline is connected to the load end. During EMA operation, the wave generator acts as an active component; when the wave generator rotates, the flexspline generates controllable elastic deformation to transmit power. Approximately 30% of the teeth of the flexspline's outer ring and the circular spline's inner ring are in mesh, providing benefits such as a high transmission ratio and substantial load-bearing capacity.

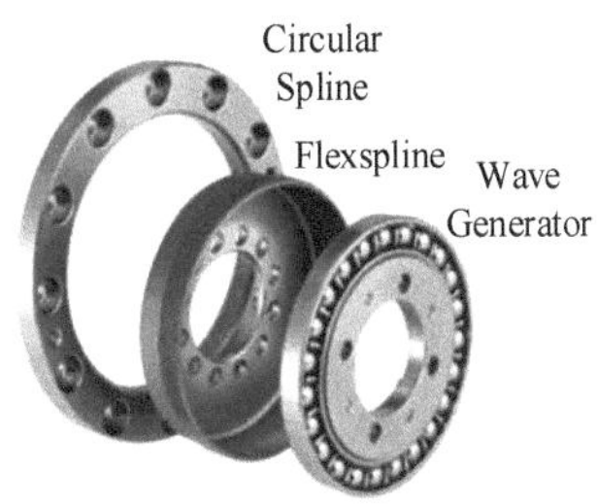

Figure 1. Structure of the harmonic reducer [26].

During the operation of the harmonic reducer, friction arises between the tooth surfaces of the flexspline and the circular spine, between the balls of the flexible bearing and the inner and outer rings, and between the wave generator and the contact surface of the flexspline. When the EMA reciprocates motion, the friction torque affects the transmission accuracy of the system. Friction disturbances in harmonic reducers cannot be ignored in high-performance control processes. High-precision control situations rely on friction compensation. Considering the flexspline as a torsion spring structure; considering the friction between the wave generator, flexspline, and circular spine; and considering the friction between the flexspline and the load, we established a nonlinear friction transmission model of the harmonic reducer based on the friction links in the transmission process, as shown in Figure 2.

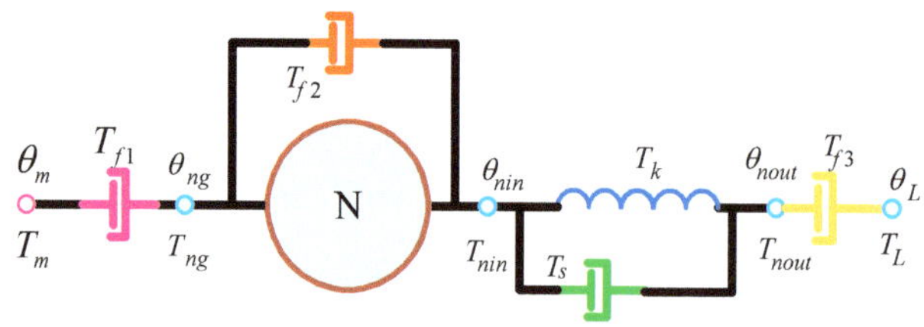

Figure 2. Harmonic drive friction model.

In Figure 2, T_{f1} is the friction generated by the wave generator, T_{f2} refers to the friction between the flexspline and the circular spine, T_{f3} refers to the friction generated by the flexspline, θ_m and T_m are, respectively, the rotor position and torque, θ_{ng} and T_{ng} are, respectively, the output positions and moments of the wave generator, θ_{nin} and T_{nin} are, respectively, the input angle and input torque of the flexspline torsion spring model, θ_{nout} and T_{nout} are, respectively, the displacement and output torque of the flexspline, Tk and Ts are, respectively, the torsion spring force and damping force of the flexspline torsion spring model, T_L is the output torque of the flexspline, and θ_L is the position of the flexspline.

The equilibrium equation of angular displacement and frictional torque between the wave generator and the flexspline is

$$\begin{cases} T_m = T_{ng} + T_{f1} \\ \theta_m = \theta_{ng} \end{cases} \tag{7}$$

The equilibrium equation for the angular displacement and friction moment between the flexspline and the circular spine is

$$\begin{cases} T_{ng} = \frac{1}{N} T_{nin} + T_{f2} \\ \theta_{ng} = N\theta_{nin} \end{cases} \begin{cases} T_k = K_L * \Delta\theta \\ T_{nin} = T_k + T_s \\ T_{nin} = T_{nout} \end{cases} \tag{8}$$

The friction torque T_f acting on the harmonic reducer is $T_f = T_{f1} + T_{f2} + T_{f3}$. The friction torque T_{f3} acting on the load end under low speeds and heavy loads is much smaller than the friction torque T_{f1} acting on the motor and wave generator end under high speeds and light loads and can be ignored; that is, $T_{f3} \approx 0$. $T_L = T_{nout} = T_k + T_s$, and $T_{nout} = f(\Delta\theta, K_L)$. $f(\Delta\theta, K_L) = T_k + T_s = T_L$. k is the stiffness coefficient of the harmonic drive. Thus, the relationship between the input torque, friction torque, and output torque of the harmonic reducer can be expressed as follows:

$$\begin{cases} T_L = f(\Delta\theta, K_L) \\ T_m = \frac{f(\Delta\theta, K_L)}{N} + T_f(t, B, \omega_r, \ldots) \\ \Delta\theta = \frac{\theta_m}{N} - \theta_L \\ f(\Delta\theta, K_L) = K_L(\frac{\theta_m}{N} - \theta_L) \end{cases} \tag{9}$$

where K_L is the equivalent stiffness coefficient of the harmonic reducer, neglecting the rotational inertia of the reducer, J is the rotational inertia of the motor rotor, J_L is the rotational inertia of the load end, and $f(\Delta\theta, K_L) = T_L$. The relationship between the motor torque *Te* and the wave generator torque T_m is

$$\begin{cases} T_e = J\frac{d^2\theta_m}{dt^2} + T_m \\ \\ T_L = J_L\frac{d^2\theta_L}{dt^2} = f(\theta, K_L) \end{cases} \tag{10}$$

The nonlinear friction torque obtained from Equations (9) can be expressed as

$$T_f(t, B, \omega_r, \ldots) = T_m - \frac{T_L}{N} \quad (11)$$

Here, it is assumed that the load is purely inertial. The nonlinear friction force T_f of the harmonic reducer can be expressed as

$$\begin{aligned} T_f(t, B, \omega_r, \ldots) &= T_m - \tfrac{1}{N} T_L \\ &= 1.5 p_n \psi_f i_q - J \tfrac{d^2\theta_m}{dt^2} - \tfrac{1}{N} J_L \tfrac{d^2\theta_L}{dt^2} \end{aligned} \quad (12)$$

When unloaded and running at a constant speed, Equation (12) can be simplified as

$$T_f(t, B, \omega_r, \ldots) = 1.5 p_n \psi_f i_q \quad (13)$$

As can be seen from Equation (13), the friction torque is related to the torque current i_q. During no-load and constant speed operation, the change rule of friction torque can be obtained by measuring the torque current i_q at different speeds and fitting the i_q change curve.

4. Improved ADRC Control Principle

4.1. Composite Second-Order ADRC Control Principle

ADRC does not rely on the precise mathematical model of the controlled object and can perform real-time estimation and compensation for internal and external disturbances in the system [27].

ADRC mainly includes a tracking differentiator (TD), an extended state observer (ESO), and nonlinear state error feedback (NLSEF). The TD executes rapid, non-overshoot tracking of target signals. The ESO monitors the system output and both internal and external disturbances. It isolates interference signals from the controlled output and incorporates compensation for these signals into the control law, thereby enhancing the system's disturbance rejection capabilities [28–30]. NLSEF combines the output of TD and the state variable observation estimation output of ESO in a nonlinear manner and then acts on the controlled object after combining it with the "total disturbance" estimation of the system by ESO. The second-order ADRC principle is illustrated in Figure 3.

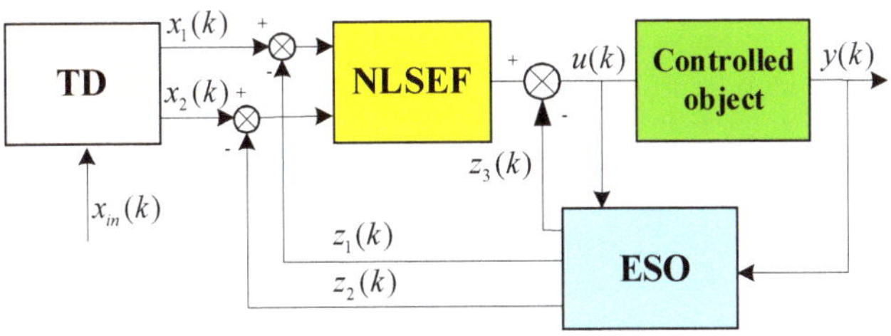

Figure 3. Block diagram of second-order ADRC control.

The TD discrete expression is

$$\begin{cases} x_1(k+1) = x_1(k) + T x_2(k) \\ x_2(k+1) = x_2(k) + T fhan(x_1(k) - x_{in}(k), x_2(k), r, h_1) \end{cases} \quad (14)$$

The TD ensures that x_1 converges to the input signal x_{in}, x_2 is the derivative of the input signal, r is the speed factor and determines the tracking speed, T is the sampling time, and h_1 is the filtering factor. *fhan*$(\cdot)$ is the fastest synthesis function, represented as

$$fhan(\cdot) = \begin{cases} d = rh_1; \ d_0 = h_1 d \\ y = x_1 + h_1 x_2 - x_{in}(k); \ a_0 = \sqrt{d^2 + 8r|y|} \\ a = \begin{cases} x_2 + \frac{(a_0 - d)}{2}, |y| > d_0 \\ x_2 + \frac{y}{h_1}, |y| \leq d_0 \end{cases} \\ fst = \begin{cases} -\frac{ra}{d}, |a| \leq d \\ -rsign(a), |a| > d \end{cases} \end{cases} \tag{15}$$

The ESO discrete expression is

$$\begin{cases} e(k) = z_1(k) - y(k) \\ z_1(k+1) = z_1(k) + T(z_2(k) - \beta_{01} e(k)) \\ z_2(k+1) = z_2(k) + T(z_3(k) - \beta_{02} fal(e(k), \alpha_{01}, \delta) + bu(k)) \\ z_3(k+1) = z_3(k) - T\beta_{03} fal(e(k), \alpha_{02}, \delta) \end{cases} \tag{16}$$

where $e(k)$ is the error signal, $y(k)$ is the system output, $z_1(k)$ is the tracking signal of $y(k)$, $z_2(k)$ is the tracking signal of $z_1(k)$, $z_3(k)$ is the total disturbance of the system, $z_3(k)$ is feed back to the control variable $u(k)$ for compensation, and b is the compensation factor. β_{01}, β_{02}, and β_{03} are the output error correction gains, α_{01} and α_{02} are the nonlinear factors, and δ is the filtering factor.

The NLSEF discrete expression is

$$\begin{cases} e_1(k) = x_1(k) - z_1(k) \\ e_2(k) = x_2(k) - z_2(k) \\ u_0(k) = \beta_1 fal(e_1(k), \alpha_1, \delta) + \beta_2 fal(e_2(k), \alpha_2, \delta) \\ u(k) = u_0(k) - \frac{z_3(k)}{b_0} \end{cases} \tag{17}$$

where $e_1(k)$ and $e_2(k)$ are error signals, and β_1 and β_2 are, respectively, the error gain and differential gain. When $0 < \alpha < 1$, $fal(\cdot)$ achieves a mathematical fitting of "small error with large gain, large error with a small gain." Fuzzy control, variable gain PID, and intelligent control are based on the control concept of "small error with a large gain, large error with small gain" to adjust the output. $fal(\cdot)$ is a nonlinear feedback function and can be expressed as follows:

$$fal(\cdot) = \begin{cases} |e|^{\alpha} \mathrm{sgn}(e), |e| > \delta \\ \frac{e}{\delta^{(1-\alpha)}}, |e| \leq \delta \end{cases} \tag{18}$$

In the EMA control process, a commonly employed strategy is the cascade PI three-loop control, where the current loop constitutes the innermost loop, the speed loop serves as the middle loop, and the position loop functions as the outermost loop. The composite second-order ADRC combines the original speed loop and position loop PI(D) controllers into a single ADRC controller, thus improving the system response speed and reducing overshoot. The principle of the PMSM composite second-order ADRC structure established using the FOC method is shown in Figure 4. w_θ is the total disturbance in position mode, and w_ω is the total disturbance in velocity mode. θ is the measured angle of the rotor, and ω is the speed of the rotor.

In the composite second-order ADRC control mode, the control structure of the speed loop is the same as that of the position loop, except for the different input variables of the controller and the control parameters of the ADRC. By adjusting the input of TD and the control parameters of ADRC, speed-mode and position-mode operation can be achieved. The parameters that must be adjusted in second-order ADRC mainly include r and h in TD; β_{01}, β_{02}, and β_{03} in ESO; and β_1 and β_2 in NLSEF. Although there are many parameters that must be adjusted, the three stages have their own engineering significance, and the principle of separate directional adjustment can be used to adjust the parameters of each stage.

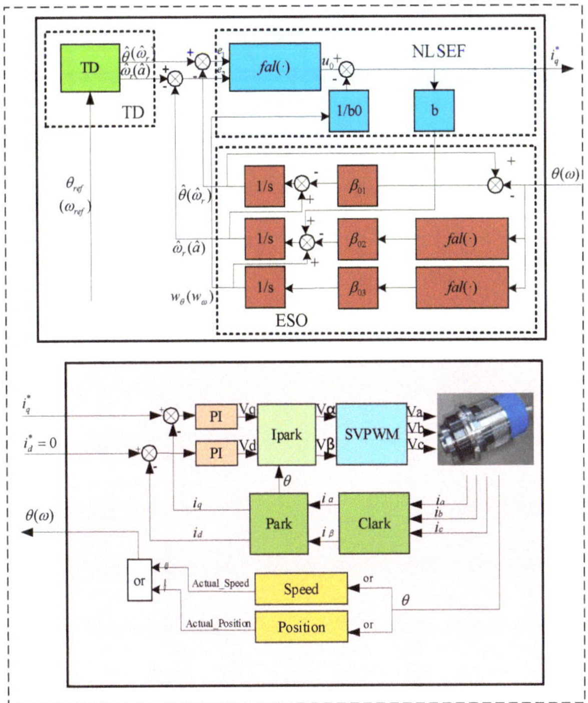

Figure 4. Block diagram of composite second−order ADRC.

4.2. Fuzzy ADRC Control Principle

During the operation of the EMA, the load is variable, and long-term service may cause changes in lubrication conditions and contact surface wear, resulting in parameter drift in the Stribeck friction model. In addition, parameters such as inductance, resistance, and magnetic linkage may drift with temperature changes. To adapt to the time-varying characteristics of the model, the control parameters of the controller must be modified adaptively.

However, ADRC does not possess parameter self-correction capability. To impart adaptive capability to ADRC, fuzzy logic control is integrated. Online adjustment of ADRC parameters is facilitated through the application of fuzzy rules. The fuzzy ADRC control method can adjust control parameters online according to different working states and obtain the most suitable control parameters within the set parameter variation range. Fuzzy control is integrated into the ADRC controller, and the control parameters of ADRC are adaptively adjusted based on the deviation and deviation rate of change.

NLSEF is added to the error integration link Equation (19), and the fuzzy control method is used to achieve self-tuning of the NLSEF parameters in ADRC. Fuzzy rules are used for fuzzy inference based on the input deviation e_1 and the change rate e_2 of the deviation to achieve online adjustment of the NLSEF coefficients and achieve adaptive ability:

$$\begin{cases} e_1(k) = z_{11}(k) - z_{21}(k) \\ e_2(k) = z_{12}(k) - z_{22}(k) \\ e_0 = \int e_1 dt \\ u_0(k) = \beta_0 fal(e_0(k), \alpha_0, \delta) + \beta_1 fal(e_1(k), \alpha_1, \delta) + \beta_2 fal(e_2(k), \alpha_2, \delta) \\ u(k) = u_0(k) - \frac{z_{23}(k)}{b_0} \end{cases} \tag{19}$$

The inputs of fuzzy controllers in fuzzy ADRC are e_1 and e_2, and the outputs are $\Delta\beta_0$, $\Delta\beta_1$, and $\Delta\beta_2$. In fuzzy PID control, based on the variation of e_1 and e_2, fuzzy subsets of five language variables, namely {"Negative Big (NB)," "Negative Small (NS)," "Zero (ZO)," "Positive Small (PS)," and "Positive Big (PB)"} are often used, or fuzzy subsets of seven language variables, namely {"Negative Big (NB)," "Negative Medium (NM)," "Negative Small (NS)," "Zero (ZO)," "Positive Small (PS)," "Positive Medium (PM)," and "Positive Big (PB)"} are often used. The control accuracy of seven fuzzy subsets is better than that of five fuzzy subsets. Here, β_0, β_1, and β_2 have the same control effect as k_i, k_p, and k_d, so seven subsets are selected here. Common membership functions include triangle membership, Z/S membership, trapezoid membership, and Gaussian membership, and in order to reduce the workload of operations, triangular membership functions are used for each fuzzy variable. The established fuzzy rules are presented in Table 1.

Table 1. $\Delta\beta_0$, $\Delta\beta_1$, and $\Delta\beta_2$ fuzzy rules.

e_1	e_2						
	NB	**NM**	**NS**	**ZO**	**PS**	**PM**	**PB**
NB	NB/PB/PS	NB/PB/NS	NM/PM/NB	NM/PM/NB	NS/PS/NB	ZO/ZO/NM	ZO/ZO/PS
NM	NB/PB/PS	NB/PB/NS	NM/PM/NB	NS/PS/NM	NS/PS/NM	ZO/ZO/NS	ZO/NS/ZO
NS	NB/PM/ZO	NM/PM/NS	NS/PM/NM	NS/PS/NM	ZO/ZO/NS	PS/NS/NS	PS/NS/ZO
ZO	NM/PM/ZO	NM/PM/NS	NS/PS/NS	ZO/ZO/NS	PS/NS/NS	PM/NM/NS	PM/NM/ZO
PS	NM/PS/ZO	NS/PS/ZO	ZO/ZO/ZO	PS/NS/ZO	PS/NS/ZO	PM/NM/ZO	PB/NM/ZO
PM	ZO/PS/PB	ZO/ZO/NS	PS/NS/PS	PS/NM/PS	PM/NM/PS	PB/NM/PS	PB/NB/PB
PB	ZO/ZO/PB	ZO/ZO/PM	PS/NM/PM	PM/NM/PM	PM/NM/PS	PB/NB/PS	PB/NB/PB

The variation surfaces of β_0, β_1, and β_2 obtained from the domain of each variable and fuzzy reasoning are shown in Figure 5.

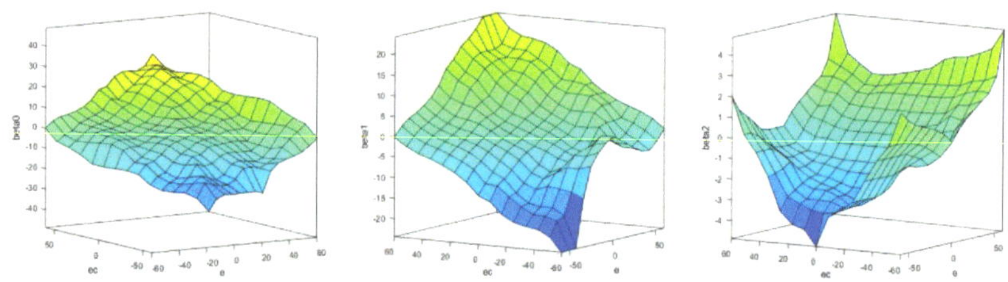

(**a**) β_0 variation surface (**b**) β_1 variation surface (**c**) β_2 variation surface

Figure 5. β_0, β_1, and β_2 variation surfaces.

The modified parameters $\Delta\beta_0$, $\Delta\beta_1$, and $\Delta\beta_2$ are obtained using the fuzzy rule table and the deblurring algorithm. The control parameters in NLSEF are obtained after correction by using Equation (20). Thus, ADRC parameter self-tuning is realized, and the adaptive ability of the system can be improved by adjusting and controlling the control parameters in NLSEF in real time. β_{00}, β_{10}, and β_{20} are the initial values; select the initial value according to the empirical method:

$$\begin{cases} \beta_0 = \beta_{00} + \Delta\beta_0 \\ \beta_1 = \beta_{10} + \Delta\beta_1 \\ \beta_2 = \beta_{20} + \Delta\beta_2 \end{cases} \tag{20}$$

The structural diagram of fuzzy ADRC is shown in Figure 6.

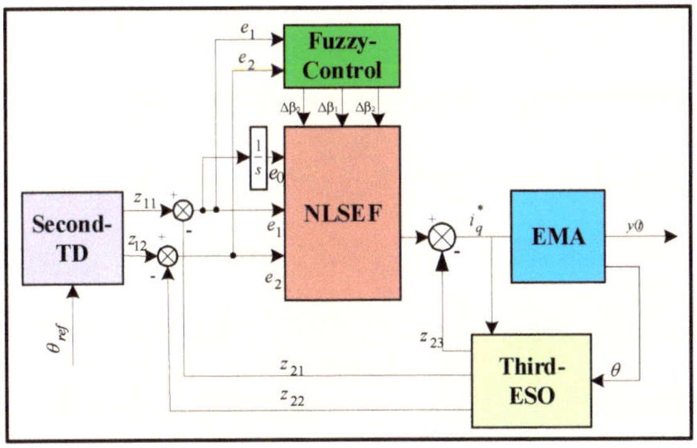

Figure 6. Fuzzy ADRC control block diagram.

5. EMA Control System Design

The three-dimensional cross-sectional and physical views of the EMA developed with an integrated hollow shaft harmonic reducer are depicted in Figure 7. The incremental encoder disk is fixed on the hollow shaft of the spindle by using an adhesive that has high aging resistance, impact resistance, and shear strength. To ensure the reliability of bonding, the viscosity is 750–1750 cps, and the shear strength is greater than 19 MPa. The main parameters of the harmonic reducer in Table 2. The main parameters of the PMSM in Table 3.

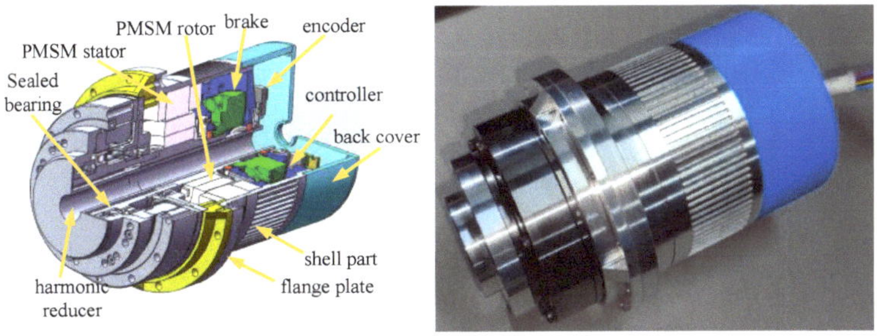

Figure 7. Structure diagram and photograph of the EMA.

Table 2. Parameters of the harmonic reducer.

Reduction Ratio	Transmission Direct Efficiency at Rated Load	Max Torque (N.m)	Max Input Speed (rev/min)	Theoretical Lifespan (h)	Weight (kg)
100	0.69	49	7000	15,000	0.8

Table 3. PMSM parameters.

Resistance (Ω)	Inductance (mH)	Rated Torque (N.m)	Peak Torque (N.m)	Max Speed (rev/min)	Peak Current (A)	Inertia
100	0.65	0.72	3.8	3100	27	$3.04 * 10^{-5}$ kgm

For the proposed IADRC control algorithm, STM32F4 is used as the main control chip for verification. The controller possesses abundant built-in resources, supports floating-point operations, and encompasses various communication interfaces, including two advanced timers, TIM1 and TIM8, dedicated to motor control. Functions such as position detection, current detection, USART, CAN, and RS485 can be performed using this chip. The hardware circuit structure of the EMA is illustrated in Figure 8.

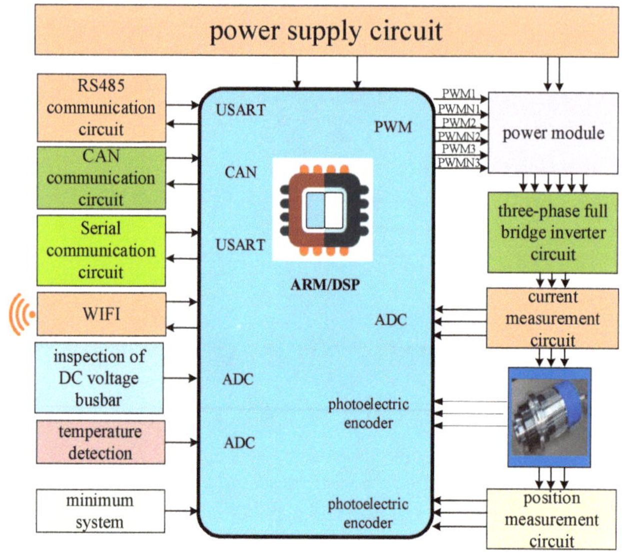

Figure 8. Hardware circuit structure of the EMA.

The N-type IRFS3607 MOSFET is used as the power device in the inverter circuit, and IR2101S is used as the power driver chip. The driving circuits for the V and W phases in a three-phase system are consistent. Using the U-phase as an example, the inverter circuit is briefly explained. The U-phase drive circuit for three-phase current is illustrated in Figure 9.

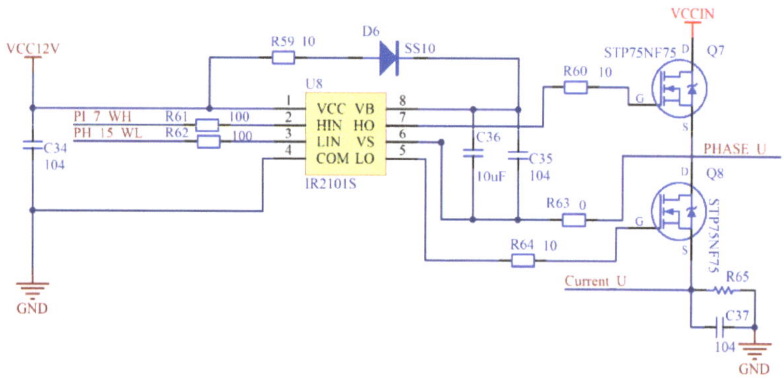

Figure 9. U-phase drive circuit.

The IO ports corresponding to advanced timer 1 and advanced timer 8 in STM32F4 can output six complementary and symmetrical PWM waves. The working voltage of the IR2101S power driver chip is 12V, and IR2101S receives PWM signals from the MCU to drive IRFS3607. IRFS3607 is an N-type MOSFET.

Rotor position data constitutes crucial information in the FOC process. Current resistance sampling methods encompass single, double, and triple resistance sampling. The single resistance sampling method, while structurally simple, complicates software processing. Conversely, double resistance sampling may induce three-phase asymmetry. Triple resistance sampling requires an operational amplifier, which is costly; however, it offers the advantages of accurate sampling and relatively simple program processing. For the convenience of software processing, the triple resistance sampling method is adopted in the control system. The U, V, and W three-phase control circuits are the same. Here, the U-phase is taken as an example; the U-phase sampling circuit is shown in Figure 10.

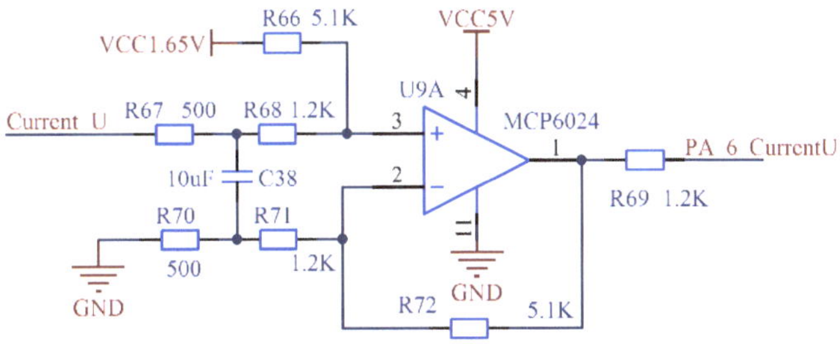

Figure 10. U-phase current sampling circuit.

MCP6024 has a large magnification. According to the virtual shorting of the amplifier, there is no current flowing through both ends of the operational amplifier. The current flowing through R2 and R7 is equal, and the current flowing through R10 and R14 is equal:

$$\frac{V_+ - V_{cc}}{R_2} = \frac{V_{in} - V_+}{R_6 + R_7}, \frac{V_{out} - V_-}{R_{14}} = \frac{V_-}{R_9 + R_{10}} \tag{21}$$

Let a = R6 + R7 = R9 + R10 and b = R2 = R14, Substituting these into Equation (21), we obtain

$$\frac{V_+ - V_{cc}}{b} = \frac{V_{in} - V_+}{a}, \frac{V_{out} - V_-}{b} = \frac{V_-}{a} \tag{22}$$

Solving Equation (22) yields

$$V_+ = \frac{bV_{in} + aV_{cc}}{a+b}, V_- = \frac{aV_{out}}{a+b} \tag{23}$$

Under virtual shorting, V+ = V−, Can be obtained

$$V_{out} = \frac{b}{a}V_i + V_{cc} = 1.65 + 5.1V_{in} \tag{24}$$

As can be seen from Equation (24), the voltage at both ends of the sampling resistor is biased by 1.65 V and amplified by 5.1 times. The sampling resistor is selected as a high-precision resistor of 10 m Ω and 2 W, with a theoretical maximum sampling current of 14.14 A. If the maximum amplitude of the sinusoidal current of the motor is 10 A, the voltage range input to the amplifier end is −0.1–0.1 V. According to Equation (24), the output voltage of the amplifier is calculated as 1.14–2.16 V, which can be directly inputted into the ADC sampling pin of the motor, providing a large safety margin.

6. Experimental Analysis

According to Equation (13), friction torque can be determined by measuring the torque current i_q at a constant speed without load. This article performed experimental analysis on frictional forces in the counter-clockwise rotation direction. The inertia of the reducer was disregarded, and it was assumed that the torque during no-load operation equals the friction torque during uniform motion. A friction model was developed by measuring torque values at various speeds and fitting the data. This model was incorporated into the control system through feedforward compensation, effectively eliminating friction disturbances. Friction torque testing was performed on the RT-Cube platform, which is capable of achieving a minimum control cycle for the motor within 100 µs. Moreover, this platform allows for the online modification of any control parameter and the online monitoring of any system variable during the control process. The tests were made at a room temperature of approximately 25 °C and a relative humidity ranging from 40% to

70%RH. The experimental platform and the test results obtained using the Gaussian fitting method are shown in Figures 11 and 12, respectively.

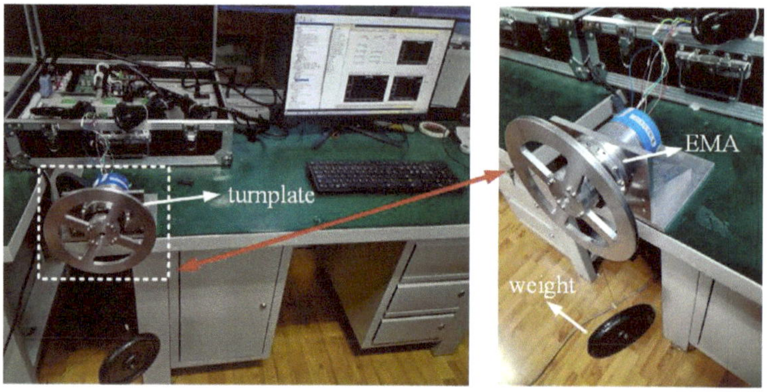

Figure 11. EMA friction torque test bench.

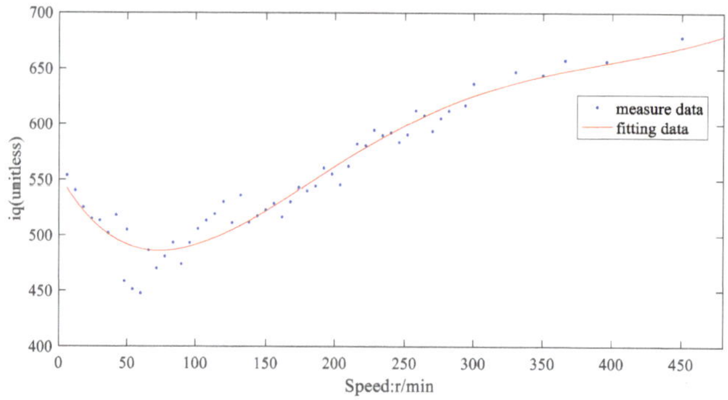

Figure 12. EMA friction moment fitting experiment.

According to the fitting equation, the friction force at different speeds was obtained. The friction force, corresponding to the torque current, was compensated for and attenuated by adjusting the torque current at various speeds. The IADRC controller was constructed by integrating the frictional torque current into the second-order fuzzy ADRC control model through feedforward compensation, as shown in Figure 13.

In the EMA speed mode, the current loop of all three control methods adopts PI control mode, and the speed loop adopts PI, ADRC, and IADRC, respectively. An IADRC controller with a step speed of 6 rev/min was used. The control parameters for the three controllers were empirically set. The main parameters to be adjusted in the TD are the r and h_1. The r affects the tracking effect. A larger r corresponds to a shorter transition time and thus a faster tracking response. However, very large r leads to overshoot and oscillation. When the r is constant and the h_1 is large, the tracking signal error is large; when the h_1 is small, the noise suppression is more prominent. However, when the h_1 is too small, the ability of the TD to suppress noise will be weakened. The disturbance compensation factor b_0 mainly affects the disturbance compensation capacity. If the system disturbance is significant, b_0 should be slightly larger; if the system disturbance is small, b_0 should be marginally lower. Directional adjustment is adopted. When we set $\alpha_{01} = \alpha_{02} = 1$, $fal(e,\alpha,\delta)$ can be linearized to $fal(e,\alpha,\delta) = e$. The values for the parameters β_{01}, β_{02}, and β_{03} need to be adjusted in practical applications according to the system output. The tuning rules for these parameters are listed in Table 4. Notably, when one parameter is tuned, the other two remain constant.

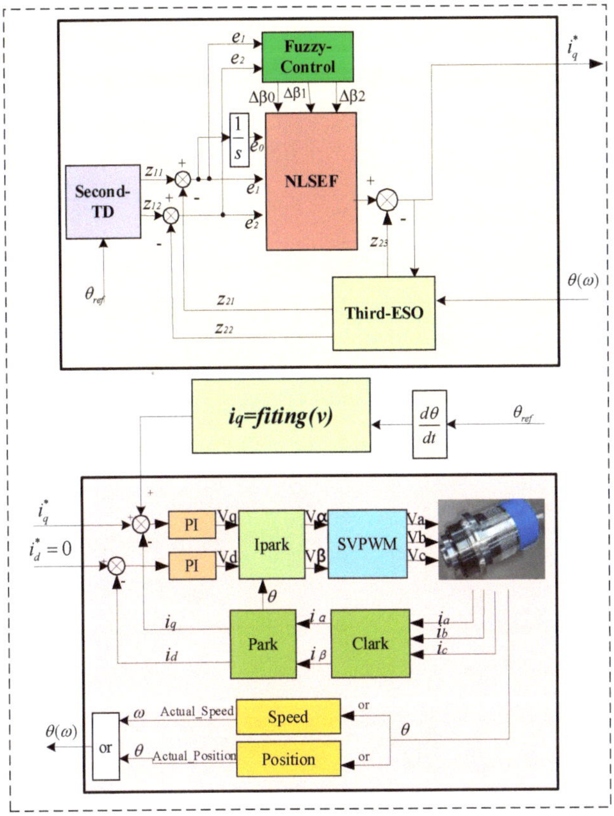

Figure 13. IADRC control block diagram.

Table 4. Tuning rules for β_{01}, β_{02}, and β_{03}.

Constant Parameters	System Response Phenomena	Tuning Rules
β_{02}, β_{03}	Oscillation occurs	Decrease β_{01}
	Divergence occurs	Decrease β_{01}
	Steady-state high-frequency oscillation occurs	Increase β_{01}
β_{01}, β_{03}	High-frequency oscillation occurs	Decrease β_{02}
	Disturbance rejection performance decrease	Increase β_{02}
	Oscillation amplitude increase	Increase β_{02}
β_{01}, β_{02}	Overshoot occurs	Increase β_{02}
	Response time is long	Increase β_{03}
	Large oscillation occurs	Decrease β_{03}

The results obtained using PI control method and the enlarged image of the step response are shown in Figure 14. The results obtained using ADRC control method and the enlarged image of the step response are shown in Figure 15. The results obtained using IADRC control method and the enlarged image of the step response are shown in Figure 16. As can be seen in the locally enlarged image A, the PI control method, ADRC control method, and feedforward compensation fuzzy IADRC reached a steady state in 0.65, 0.25, and 0.20 s, respectively. The PI control method experienced an overshoot before reaching the steady state, with a maximum speed of 6.2 rev/min and an overshoot of 3.33%. The other two control methods quickly achieved the target speed without overshooting. The steady-state speed error of all three control methods was 0.1 rev/min. By comparing the locally enlarged image B of ADRC and IADRC, it can be concluded that the IADRC control method has a lower speed oscillation frequency in the steady state.

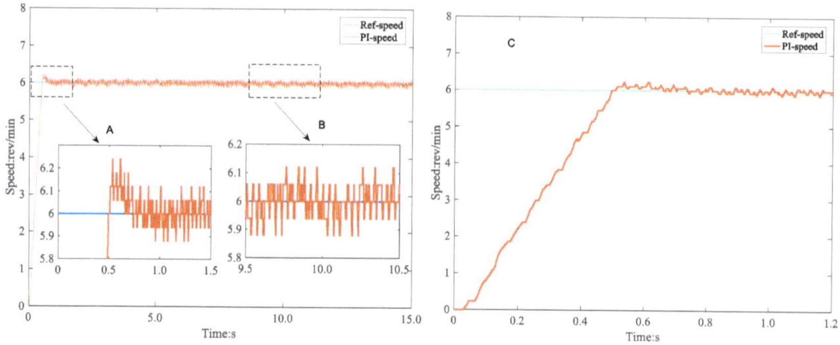

Figure 14. PI test results of no-load step signal.

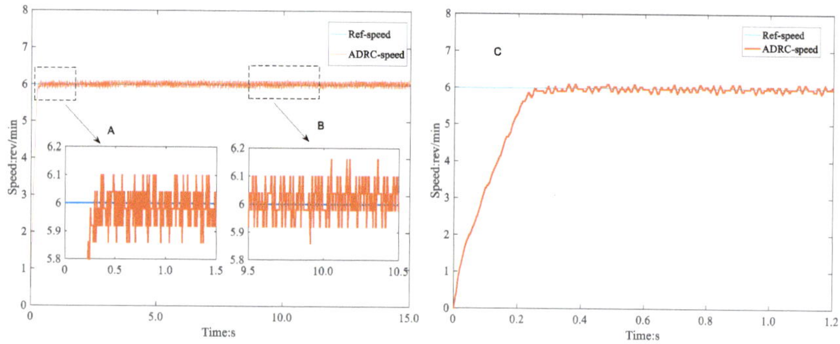

Figure 15. ADRC test results of no-load step signal.

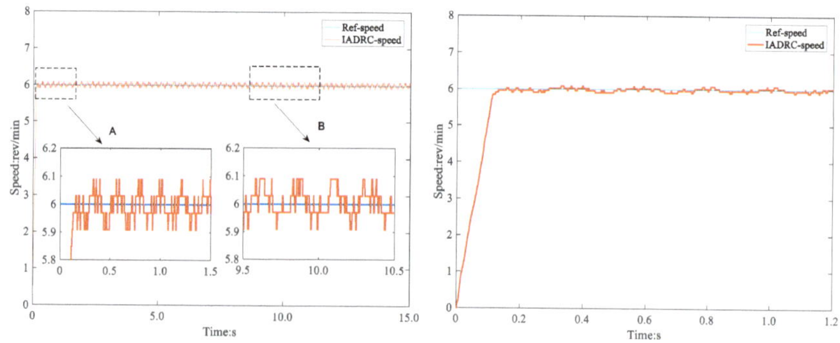

Figure 16. IADRC test results of no-load step signal.

Common performance indicators of the control system include integrated square error (ISE), integrated time square error (ITSE), integrated absolute error (IAE), and integrated time absolute error (ITAE). Different performance indicators have different priorities.

The *ITAE* criterion can better reflect the system's response speed, oscillation characteristics, and steady-state errors, and has good selectivity for different controllers:

$$ITAE = \int_{0}^{\infty} t|e(t)|dt \tag{25}$$

The *ITAE* calculation results for the three control methods within 0–1s are presented in Table 5.

Table 5. ITAE calculation results of three methods.

Control Mode	PI	ADRC	IADRC
ITAE	47.714 (rev/min)*s^2	11.559 (rev/min)*s^2	5.727 (rev/min)*s^2

The unit of ITAE is "(rev/min)*s^2". The ITAE calculation result within 0–1 s of IADRC was 15.445 (rev/min)*s^2, thus indicating the optimal control performance of IADRC. The number of encoder lines is 2880, and after fourfold frequency, it is 11,520. The position input signal is y = 115,200*sin(0.05*pi*t), and the unit of y is the carving line number of the encoder (LNE). The main parameters in the experimental are shown in Table 6.

Table 6. Main parameters in the experiment.

Name	Parameter Value
Encoder	2880 PPR
Reduction ratio	50
Counter weight	25 N
Disc radius	0.1 m
Load	2.5 N·m

In the EMA position mode, three control configurations were implemented: PID (position loop) + PI (speed loop) + PI (current loop); ADRC (position loop) + PI (current loop); and IADRC (position loop) + PI (current loop). Data were recorded after the system stabilized. The position tracking results under no-load conditions for the three control methods are illustrated in Figure 17. The corresponding position tracking error results are presented in Figure 18.

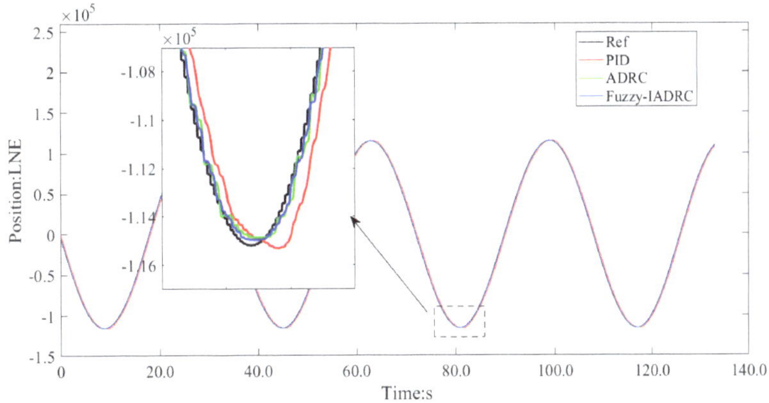

Figure 17. No−Load position tracking results.

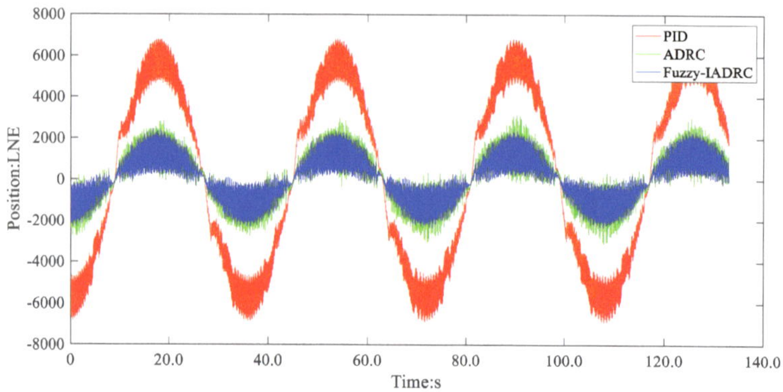

Figure 18. No−Load position error tracking results.

The RMSE and peak-to-peak calculation results of tracking error are presented in Table 7.

Table 7. No−load test results.

Control Mode	RMSE	Peak-to-Peak Values
PID	4165.1	13685 LNE
ADRC	1201.4	6218 LNE
IADRC	1040.8	4780 LNE

As can be seen from Figure 18 and Table 7, the ADRC control method yielded higher accuracy than the PID control when under load. After adding friction feedforward compensation, the RMSE and peak-to-peak values of position error improved. The peak-to-peak value of IADRC was 1438 less than that of ADRC. The RMSE of IADRC reduced by approximately 160.6 compared with ADRC.

Record data after the system stabilizes. In the position-mode test under 2.5 Nm load conditions, the position tracking results for the three control methods are depicted in Figure 19. Correspondingly, the position tracking error results for the three control methods are illustrated in Figure 20. It is noteworthy that this load (2.5 Nm) represents 6.25% of the rated torque.

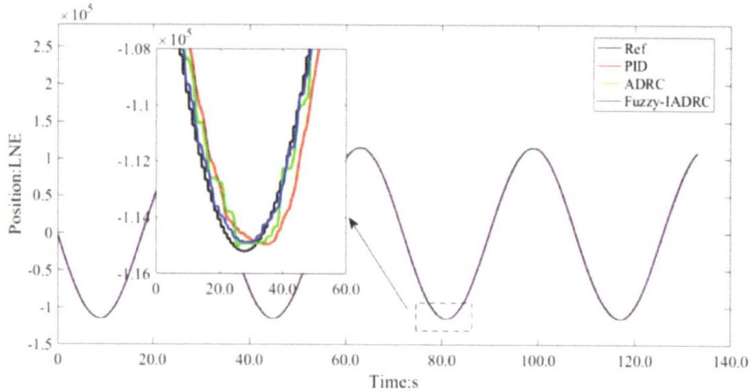

Figure 19. Load position tracking results.

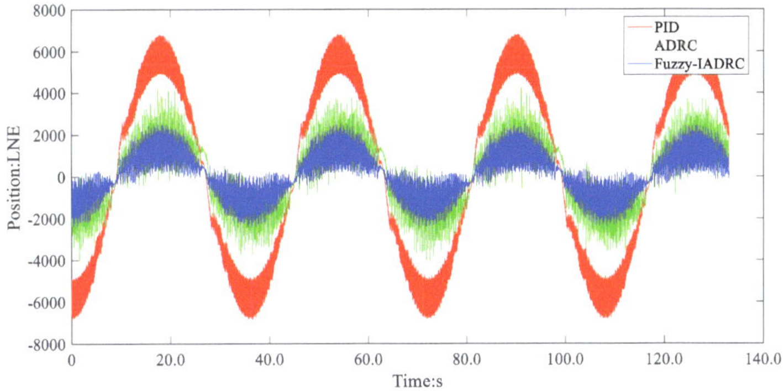

Figure 20. Load position error tracking results.

The RMSE and peak-to-peak calculation results of PID, ADRC, and friction feedforward compensation fuzzy IADRC control methods under load are presented in Table 8.

Table 8. Load test results.

Control Mode	RMSE	Peak-to-Peak Values
PID	4183.6	13,773 LNE
ADRC	1636.4	8242 LNE
IADRC	1046.8	4869 LNE

As can be seen from Figure 20 and Table 8, the ADRC control method yielded higher accuracy than the PID control when under load. After the addition of friction feedforward compensation, the root mean square and peak-to-peak values of position error improved. The peak-to-peak value of IADRC was 3373 less than that of ADRC. The RMSE of IADRC was reduced by approximately 410.4 compared to ADRC.

7. Conclusions

Aiming at the problem of EMA control accuracy, this paper adopts high-performance IADRC and friction feedforward compensation methods. The PMSM mathematical model was established, and a second-order composite ADRC control strategy was developed for the PMSM speed loop and position loop based on the FOC model. The ADRC controller demonstrates several superior characteristics not present in the PI controller. To address the issue of ADRC controller parameter adaptation, fuzzy control was integrated into the nonlinear state error feedback link, facilitating self-tuning of ADRC parameters. Furthermore, a model for EMA transmission was developed, and the relationship between friction torque and torque current i_q was analyzed. Furthermore, on the RT-Cube platform, the torque current i_q at different speeds was measured and then added to the current loop control through feedforward compensation, determining controller parameters through empirical methodologies. In addition, speed-mode and position-mode experiments were conducted in the PI control mode, ADRC control mode, and IADRC control mode. Moreover, the experimental results of the speed step response were analyzed using the IATAE criteria. The IADRC control mode yielded the smallest calculation result and the best control performance. Neglecting the inertia of the reducer, assuming that the no-load running torque is equal to the friction torque during uniform motion, the experimental results of sinusoidal position tracking were analyzed, and the results were evaluated using RMSE and peak-to-peak values. Under conditions of pure inertial load, the integration of friction feedforward compensation combined with the implementation of the IADRC control method enhances the accuracy of EMA transmission.

Author Contributions: Conceptualization, P.Z. and Z.S.; methodology, P.Z. and B.Y.; software, P.Z.; validation, P.Z. and B.Y.; formal analysis, P.Z.; investigation, P.Z.; resources, Z.S.; data curation, P.Z.; writing—original draft preparation, P.Z.; writing—review and editing, P.Z. and B.Y.; visualization, P.Z.; supervision, B.Y.; project administration, Z.S.; funding acquisition, Z.S. All authors have read and agreed to the published version of the manuscript.

Funding: This research was funded by the National Natural Science Foundation of China Youth Fund (Granted No. 52305046).

Data Availability Statement: Data are contained within the article.

Conflicts of Interest: The authors declare no potential conflicts of interest with respect to the research, authorship, and publication of this article.

References

1. Zhang, Y.; Zhao, C.; Dai, B.; Li, Z. Dynamic simulation of permanent magnet synchronous motor (PMSM) electric vehicle based on Simulink. *Energies* **2022**, *15*, 1134. [CrossRef]
2. Zhou, D.; Luo, K.; Shen, Z.; Zou, J. Vector-Space-Decomposition-Based Power Flow Control of Single-Stage-Multiport-Inverter-Fed PMSM Drive for Hybrid Electric Vehicles. *IEEE Trans. Ind. Electron.* **2023**, 1–11. [CrossRef]
3. Chuyen, T.D.; Van Hoa, R.; Co, H.D.; Huong, T.T.; Ha, P.T.T.; Linh, B.T.H.; Nguyen, T.L. Improving control quality of PMSM drive systems based on adaptive fuzzy sliding control method. *Int. J. Power Electron. Drive Syst.* **2022**, *13*, 835–845. [CrossRef]
4. Assoun, I.; Idkha, L.; Nahid-Mobarakeh, B.; Meibody-Tabar, F.; Monmasson, E.; Pacault, N. Wide-Speed Range Sensorless Control of Five-Phase PMSM Drive under Healthy and Open Phase Fault Conditions for Aerospace Applications. *Energies* **2022**, *16*, 279. [CrossRef]
5. Li, P.; Xu, X.; Yang, S.; Jiang, X. Open circuit fault diagnosis strategy of PMSM drive system based on grey prediction theory for industrial robot. *Energy Rep.* **2023**, *9*, 313–320. [CrossRef]
6. Spong, M.; Khorasani, K.; Kokotovic, P. An integral manifold approach to the feedback control of flexible joint robots. *IEEE J. Robot. Autom.* **1987**, *3*, 291–300. [CrossRef]
7. Gandhi, P.S.; Ghorbel, F.H.; Dabney, J. (Eds.) Modeling, Identification, and Compensation of Friction in Harmonic Drives. In Proceedings of the 41st IEEE Conference on Decision and Control, Las Vegas, NV, USA, 10–13 December 2002; pp. 160–166.
8. Taghirad, H.D.; Belanger, P.R. Modeling and parameter identification of harmonic drive systems. *ASME J. Dyn. Syst. Meas. Control* **1998**, *120*, 439–444. [CrossRef]
9. Maré, J.C. Friction modelling and simulation at system level: Considerations to load and temperature effects. *Proc. Inst. Mech. Eng. Part I J. Syst. Control. Eng.* **2015**, *229*, 27–48. [CrossRef]
10. Martineau, J.P.; Chedmail, P. Caractérisation Expérimentale et Modélisation du Comportement des Réducteurs Harmonic Drive. In Proceedings of the Congrès Mondial des Engrenages et des Transmissions de Puissance, Paris, France, 16–17 March 1999; pp. 1089–1100.
11. Martineau J, P. Modelisation Experimentale des Reducteurs Harmonic Drive, Application a la Determination des Parametres Minimaux d'un Robot Souple Trois Axes[D]. Ph.D. Thesis, Nantes University, Nantes, France, 1996.
12. Marc, M.J.C.B. *Conception Préliminaire des Actionneurs Électromagnétiques Basée sur les Modèles: Lois d'estimations et Règles de Conception pour la Transmission de Puissance Mécanique*; INSA: Toulouse, France, 2012.
13. Yadav, D.; Verma, A. Comperative performance analysis of PMSM drive using MPSO and ACO techniques. *Int. J. Power Electron. Drive Syst.* **2018**, *9*, 1510–1522. [CrossRef]
14. Salem, W.A.A.; Osman, G.F.; Arfa, S.H. (Eds.) Adaptive Neuro-Fuzzy Inference System Based Field Oriented Control of PMSM & Speed Estimation. In Proceedings of the 2018 Twentieth International Middle East Power Systems Conference (MEPCON), Cairo, Egypt, 18–20 December 2018; pp. 626–631.
15. Liu, Z.; Wei, H.; Zhong, Q.-C.; Liu, K.; Xiao, X.-S.; Wu, L.-H. Parameter estimation for VSI-fed PMSM based on a dynamic PSO with learning strategies. *IEEE Trans. Power Electron.* **2016**, *32*, 3154–3165. [CrossRef]
16. Barkat, S.; Tlemçani, A.; Nouri, H. Noninteracting adaptive control of PMSM using interval type-2 fuzzy logic systems. *IEEE Trans. Fuzzy Syst.* **2011**, *19*, 925–936. [CrossRef]
17. Han, J. From PID to active disturbance rejection control. *IEEE Trans. Ind. Electron.* **2009**, *56*, 900–906. [CrossRef]
18. Gao, S.; Wei, Y.; Zhang, D.; Qi, H.; Wei, Y. A modified model predictive torque control with parameters robustness improvement for PMSM of electric vehicles. *Actuators* **2021**, *10*, 132. [CrossRef]
19. Wang, X.; Zhu, H. Active disturbance rejection control of bearingless permanent magnet synchronous motor based on genetic algorithm and neural network parameters dynamic adjustment method. *Electronics* **2023**, *12*, 1455. [CrossRef]
20. Bao, H.; He, D.; Zhang, B.; Zhong, Q.; Hong, H.; Yang, H. Research on dynamic performance of independent metering valves controlling concrete-placing booms based on fuzzy-LADRC controller. *Actuators* **2023**, *12*, 139. [CrossRef]
21. Jin, K.; Song, J.; Li, Y.; Zhang, Z.; Zhou, H.; Chang, X. Linear active disturbance rejection control for the electro-hydraulic position servo system. *Sci. Prog.* **2021**, *104*, 00368504211000907. [CrossRef]

22. Hu, X.; Han, S.; Liu, Y.; Wang, H. Two-axis optoelectronic stabilized platform based on active disturbance rejection controller with LuGre friction model. *Electronics* **2023**, *12*, 1261. [CrossRef]
23. Sira-Ramírez, H.; Linares-Flores, J.; García-Rodríguez, C.; Contreras-Ordaz, M.A. On the control of the permanent magnet synchronous motor: An active disturbance rejection control approach. *IEEE Trans. Control. Syst. Technol.* **2014**, *22*, 2056–2063. [CrossRef]
24. Li, S.; Li, J. Output predictor-based active disturbance rejection control for a wind energy conversion system with PMSG. *IEEE Access* **2017**, *5*, 5205–5214. [CrossRef]
25. Du, B.; Wu, S.; Han, S.; Cui, S. Application of linear active disturbance rejection controller for sensorless control of internal permanent-magnet synchronous motor. *IEEE Trans. Ind. Electron.* **2016**, *63*, 3019–3027. [CrossRef]
26. Lu, Y.S.; Hwang, C.S. Tracking Control of a Harmonic Drive Actuator with Sliding-Mode Disturbance Observers. In Proceedings of the 2009 IEEE/ASME International Conference on Advanced Intelligent Mechatronics, Singapore, 14–17 July 2009; pp. 1798–1803.
27. Huang, Y.; Xue, W. Active disturbance rejection control: Methodology and theoretical analysis. *ISA Trans.* **2014**, *53*, 963–976. [CrossRef] [PubMed]
28. Jingqing, H. Active Disturbance Rejection Control Technique—The Technique for Estimating and Compensating the Uncertainties. Master's Thesis, National Defense Industry Press, Beijing, China, 2008.
29. Wang, Y.; Fang, S.; Hu, J. Active disturbance rejection control based on deep reinforcement learning of PMSM for more electric aircraft. *IEEE Trans. Power Electron.* **2022**, *38*, 406–416. [CrossRef]
30. Rong, Z.; Huang, Q. A new PMSM speed modulation system with sliding mode based on active-disturbance-rejection control. *J. Cent. South Univ.* **2016**, *23*, 1406–1415. [CrossRef]

Disclaimer/Publisher's Note: The statements, opinions and data contained in all publications are solely those of the individual author(s) and contributor(s) and not of MDPI and/or the editor(s). MDPI and/or the editor(s) disclaim responsibility for any injury to people or property resulting from any ideas, methods, instructions or products referred to in the content.

actuators

MDPI

Article

Enhancing Vibration Control in Cable–Tip–Mass Systems Using Asymmetric Peak Detector Boundary Control

Leonardo Acho † and Gisela Pujol-Vázquez *,†

Department of Mathematics, Universitat Politècnica de Catalunya-BarcelonaTech (ESEIAAT), 08222 Terrassa, Spain; leonardo.acho@upc.edu
* Correspondence: gisela.pujol@upc.edu; Tel.: +34-9373-98159
† These authors contributed equally to this work.

Abstract: In this study, a boundary controller based on a peak detector system has been designed to reduce vibrations in the cable–tip–mass system. The control procedure is built upon a recent modification of the controller, incorporating a non-symmetric peak detector mechanism to enhance the robustness of the control design. The crucial element lies in the identification of peaks within the boundary input signal, which are then utilized to formulate the control law. Its mathematical representation relies on just two tunable parameters. Numerical experiments have been conducted to assess the performance of this novel approach, demonstrating superior efficacy compared to the boundary damper control, which has been included for comparative purposes.

Keywords: boundary control; flexible cable; partial differential equation; control design; peak detector model

Citation: Acho, L.; Pujol-Vázquez, G. Enhancing Vibration Control in Cable–Tip–Mass Systems Using Asymmetric Peak Detector Boundary Control. *Actuators* **2023**, *12*, 463. https://doi.org/10.3390/act12120463

Academic Editor: Guang-Hong Yang

Received: 16 November 2023
Revised: 29 November 2023
Accepted: 6 December 2023
Published: 11 December 2023

Copyright: © 2023 by the authors. Licensee MDPI, Basel, Switzerland. This article is an open access article distributed under the terms and conditions of the Creative Commons Attribution (CC BY) license (https://creativecommons.org/licenses/by/4.0/).

1. Introduction

In numerous industrial applications, systems are often represented by partial differential equations (PDEs), where the targeted physical quantity relies on both position and time, as noted by Morris [1]. There are two primary PDE control settings depending on the nature of control actuation; it can either be distributed throughout the system's domain, or the actuation and sensing are confined solely to the boundary conditions [2–5]. Boundary control is regarded as more physically realistic due to the non-intrusive nature of the actuation and sensing, as emphasized by Kao and Stark [6]. In fact, the design of boundary control for cable-based systems presents a significant challenge and finds relevance in various control engineering applications, such as floating platforms for offshore wind turbines [7,8], overhead cranes equipped with flexible cable mechanisms [9–11], conveyor belt devices [12], oil-drilling actuators [13], and so on [14–22]. See Figures 1 and 2 for examples of these applications.

These applications are prominent due to the favorable attributes of cables, such as their relatively low weight, flexibility, strength, and ease of storage, as noted by de Oliveira and Cajueiro [23]. However, if the induced vibrations are not effectively filtered out in the cable system, they can significantly deteriorate system performance and eventually lead to critical failures. In the existing literature, various control strategies have been proposed to address this issue. Therefore, boundary vibration control remains a crucial area of focus in these applications. For an in-depth review on this topic, see Zhao et al. [21] and the references therein. To achieve a state-of-the-art understanding of vibration control applied to cable mechanical systems, the cable–tip–mass model serves as a reference challenge. Additionally, in [24], the author reduce riser vibration through stochastic control methods, while Zou et al. [25] present an adaptive control system with backlash. Furthermore, Koshal et al. [26] and Zhang et al. [27] propose an observer-based boundary control approach, as in [28]. In [29], the authors present an active disturbance rejection controller, where the energy system converges to equilibrium with an exponential manner. Adaptive control methods, as described in [30], are applied, but despite not relying on the measurement or estimation of system states, the energy consumption is found to be excessive.

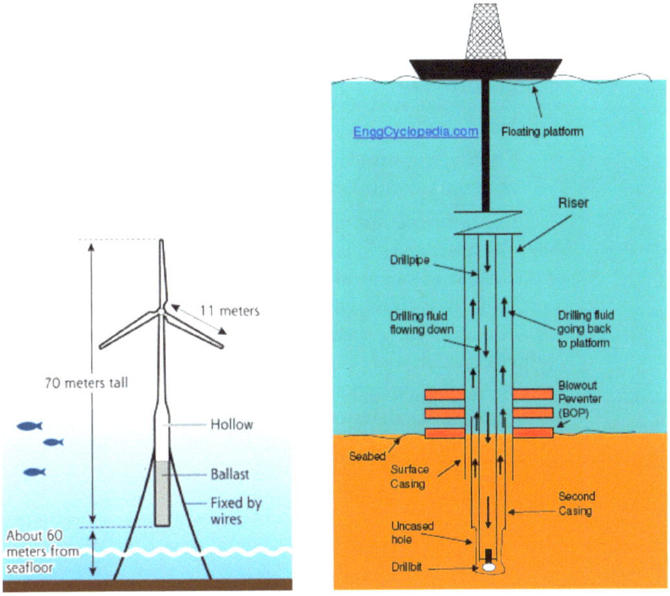

Figure 1. Floating platform for offshore wind turbines (cleantechnica.com); oil-drilling actuators (EnggCyclopedia.com).

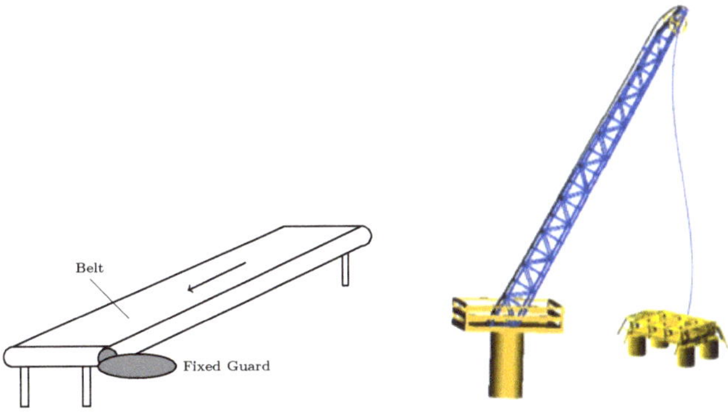

Figure 2. Conveyor belt device (scheme); overhead crane with flexible cable mechanisms.

Moreover, the application of the asymmetric peak detector mechanism has demonstrated its utility in reducing vibrations in flexible structures, as highlighted in Pujol's work [31]. The conventional peak detector system identifies peak values within the input signal, as mentioned by Meng [32]. The current modification introduces two parameters to regulate the behavior of the input signal, resulting in the development of a non-symmetric controller configuration. The primary goal is to derive a modified control input that enhances vibration attenuation. We implement this approach for a cable–tip–mass system, considering a modification of a standard boundary controller for comparative analysis. According to numerical experiments, the performance of this modified approach surpasses that of the damper controller case [23,33,34]. Specifically, we effectively adapt the peak detector algorithm to our boundary control design for attenuating vibrations along the cable or string in our mechanical cable–tip–mass device. Additionally, we provide a formal proof of this assertion using Lyapunov theory. This paper represents an enhanced version of our prior work presented in [5]. The primary contribution of this paper lies in validating a seemingly unrelated technique, such as the peak detector model, when applied to the cable–tip–mass system. This paper offers several key contributions:

- Introduction of a boundary strategy centered on detecting peak vibration values and subsequent proof of its bounded-input bounded-output (BIBO) stability.
- Proposal of two design parameters aimed at enhancing the flexibility of peak detection. Their values can be determined using specific performance indices.
- Conducting simulations that compare the performance of this approach with a classic boundary controller, demonstrating its efficacy.

Furthermore, the practical implementation of the control algorithm presents a considerable challenge in technological progress. Reference [35] provides a detailed account of particular mechanical configurations within the mechatronic stiffness concept, offering insights into its traits, behavior, and the achieved control outcomes.

The remaining sections of the paper are organized as follows. Section 2 outlines the mathematical model of the cable–tip–mass system employed for our purposes, along with the pertinent assumptions relevant to real-world applications. It also introduces the novel asymmetric peak detector model, with its stability established in terms of Lyapunov theory. Section 3 comprises several numerical simulations that demonstrate the effectiveness of the proposed control design, discussed in Section 4. Finally, Section 5 presents the key conclusions derived from this study.

2. Materials and Methods

2.1. Cable–Tip–Mass System

The system illustrated in Figure 3 depicts the boundary-actuated cable–tip–mass system, which can be mathematically represented by [23]:

$$\rho u_{tt}(x,t) - T_o u_{xx}(x,t) = 0, \quad (1)$$

$$u(0,t) = 0, \quad t \geq 0, \quad (2)$$

$$m u_{tt}(L,t) + T_o u_x(L,t) = f(t), \quad t \geq 0, \quad (3)$$

where ρ denotes the mass per unit length of the cable, m is the mass of the actuator located at the free boundary space, T_o is the applied tension to the cable, L is the cable length, and $x \in [0, L]$ represents the independent position variable. Variable $u(x,t)$ represents the transverse position at the x-position for a t-time, and $f(t)$ denotes the boundary control force. Finally, with regards to notation, the provided subscripts denote the corresponding partial derivatives, as is customary [23]. To establish a practical control framework, two assumptions must be introduced [21]:

(i) The amplitude of $u(x,t)$ is very small.
(ii) T_o is constant all along the cable.

In summary, the aim of the boundary control is to determine a controller $f(t)$ that decreases the intensity of cable vibrations. It can be stated in terms of bounded-input bounded-output (BIBO) stability; $|u(x,t)| \in L_\infty$ if $f(t)$ is bounded too. To conclude this section, we observe that the open-loop response (with $f(t) = 0$) of this system is undamped.

The implementation of the control algorithm, schematically represented in Figure 3, is a challenging issue of technological development. Reference [35] describes selected mechanical arrangement of the mechatronic stiffness concept, its features, behavior, and control results.

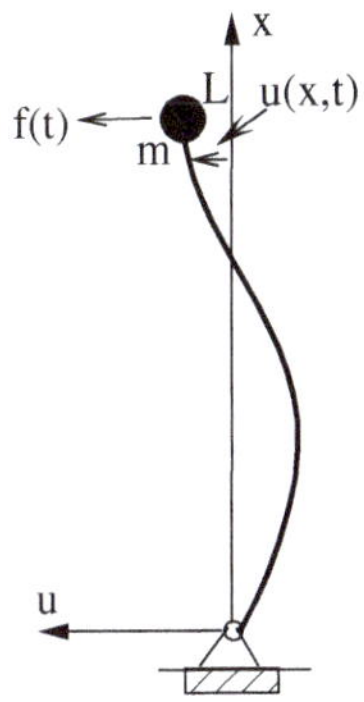

Figure 3. A simplified cable–tip–mass system, with local coordinates and boundary conditions. One end of the string is pinned while the other end is linked to an actuator $f(t)$. Our assumption accounts for uniform tension along the entire length of the string.

2.2. Asymmetric Peak Detector Model

In electronics [32], a peak detector system is applied to estimate the peak voltage value of a given signal. In this section, we present the mathematical model of the asymmetric peak detector system, proving its BIBO stability.

2.2.1. Definition and Characterization

Essentially, a standard peak detector system is implemented by utilizing a diode (D), a capacitor (C), and a resistor (R), as depicted in Figure 4. In this configuration, the input signal $v_i(t)$ is fed into the peak detector system, and subsequently, the output signal $y(t)$ provides an estimate of the peak voltage value of $v_i(t)$. The retention duration for storing the peak value of $v_i(t)$ in the capacitor is regulated by the specific values of R and C.

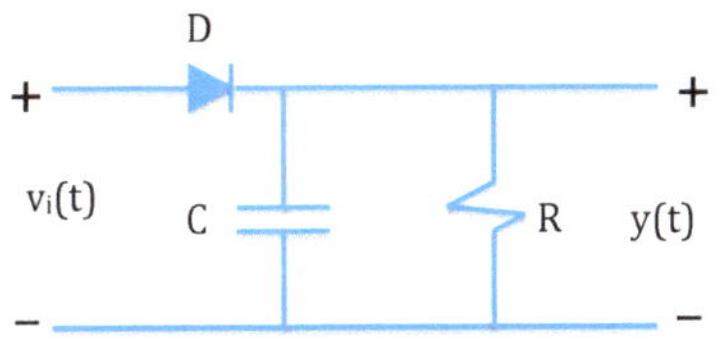

Figure 4. Simplified electronic circuit of the peak detector system. $y(t)$ displays the peak information on the input signal $v_i(t)$. The memory time to keep the peak value of $v_i(t)$ stored in the capacitor is controlled through the values of R and C.

By applying Kirchhoff laws to the electronic circuit, we derive the differential equation governing this system:

$$\dot{y}(t) = \frac{\alpha}{2}\Big((v_i(t) - y(t))(\text{sign}(v_i(t) - y(t)) + 1) + y(t)(\text{sign}(v_i(t) - y(t)) - 1)\Big), \tag{4}$$

where $\alpha = \frac{1}{RC}$.

To introduce a degree of adaptability into the system, we propose a modification of the peak detector model (4), outlined as follows:

$$\dot{y}(t) = \frac{\alpha_1}{2}\,(v_i(t) - y(t))\Big(\text{sign}(v_i(t) - y(t)) + 1\Big) + \frac{\alpha_2}{2}\,y(t)\Big(\text{sign}(v_i(t) - y(t)) - 1\Big), \tag{5}$$

where the parameter α in (4) is decomposed into two design parameters α_1 and α_2. Numerical simulations are conducted to validate our proposed peak detector system. To demonstrate the robustness of our approach, we subject the system to various classes of external disturbances. For instance, refer to Figure 5, which illustrates the input signal $v_i(t)$ and the peak detector signal $y(t)$, as described in (5). Based on these simulations, it can be inferred that the parameters α_1 and α_2 govern the memory dynamics of $y(t)$. When α_1 is significantly smaller than α_2, as observed in Figure 5a,d,g, the peak value is not reached. Conversely, in Figure 5b,e,h, the peak value is achieved, but the response behaves slowly and does not precisely align the input but the dynamics of $v(t)$ is remembered. Finally, in Figure 5c,f,i, we obtain the non-modified system. The main idea of the asymmetric peak detector system is to activate the control when a peak value is detected, or almost detected, as desired.

Hence, by tuning adequately the values of parameters α_1 and α_2, we can modulate different outputs, capturing the input desired values. The asymmetry comes from the decoupled parameter, allowing us to detect a desired value.

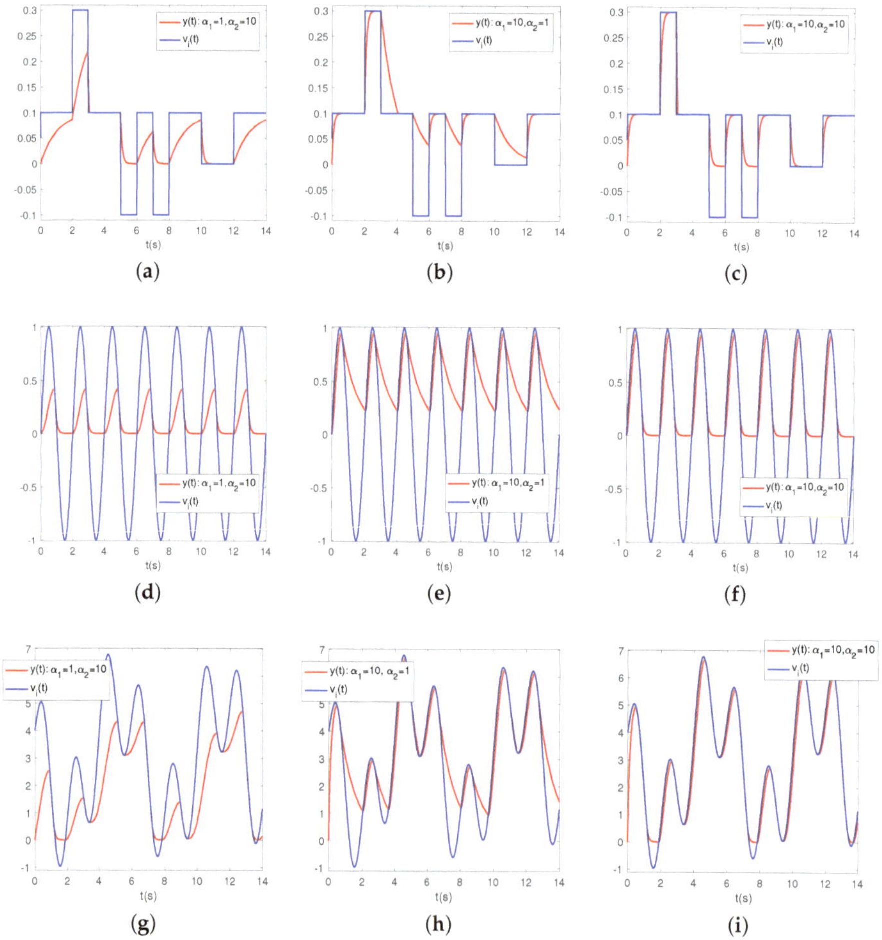

Figure 5. Simulation results of the asymmetric peak detector system (5), when different classes of input $v_i(t)$ are considered; in blue: input $v_i(t)$; in red: output $y(t)$. Three cases of parameters α_1 and α_2 values in (5) are considered for each input function, to expose the performance of the proposal. We obtain a positive response, with different input values detected. The designer needs to determine the preferred scenario for their system.

2.2.2. Bounded-Input Bounded-Output Analysis

According to Figure 5, it appears that our system is Bounded-Input Bounded-Output (BIBO)-stable; if the system input is bounded, that is, if there exits $K > 0$ such that $|v_i(t)| \leq K$, then $y(t)$ is bounded [36]. Let us prove this mathematical hypothesis. Observe that system (5) is equivalent to:

$$\dot{y}(t) = \begin{cases} \alpha_1(v_i(t) - y(t)) & \text{if } v_i(t) \geq y(t) \\ -\alpha_2 y(t) & \text{if } v_i(t) < y(t) \end{cases}, \tag{6}$$

By solving the corresponding ordinary differential Equation (6), we obtain:

- If $v_i(t) \geq y(t)$, then $|y(t)| \leq |v_i(t)| \leq K$. In fact, we can obtain:

$$|y(t)| \leq Ke^{-\alpha_1 t_0}.$$

- If $v_i(t) < y(t)$, then $|y(t)| \leq |y_0|e^{-\alpha_2(t-t_0)}$, for $t \geq t_0$ with $y(t_0) = y_0$.

Therefore, the system is exponentially stable, and a Lyapunov function $V_a(t)$ can be stated.

Note that considering $\alpha_1 >> \alpha_2$, there exists t_0 such that $|y(t)| \leq K|y_0|e^{-\alpha_2(t-t_0)}$, for all $t \geq t_0$.

2.3. *Asymmetric Peak Detector Boundary Controller*

First, let us introduce the boundary damper controller, previously employed for the vibration control of a cable–tip–mass cable mechanism (refer, for instance, to [23,33,34]):

$$f_d(t) = -k_d u_t(L,t), \tag{7}$$

where $k_d > 0$ is the control gain defined by the designer, and $u_t(L,t)$ is the corresponding feedback signal. The performance of the previous boundary damper controller is comparable, for instance, to the one based on model reference technique design [23]. So, the standard boundary control input in (3) is $f(t) = f_d(t)$.

In our design, the controller $f(t)$ in (3) is constructed, modifying this standard controller $f_d(t)$, as follows. The signal $f_d(t)$ generated by (7) is supplied to our peak detector algorithm (6), i.e., $v_i(t) = f_d(t)$. Then, we define the asymmetric peak detector controller as $f(t) = y(t)$, to be supplied to our cable–tip–mass system (1)–(3). The mathematical model of the asymmetric peak detector controller is then:

$$\dot{y}(t) = \begin{cases} \alpha_1(f_d(t) - y(t)) & \text{if } f_d(t) \geq y(t) \\ -\alpha_2 y(t) & \text{if } f_d(t) < y(t) \end{cases}, \tag{8}$$

where α_1 and α_2 are positive constant parameters, and $f_d(t)$ is its input signal. The response $y(t)$ is piecewise continuous and it is available for a given boundary damper controller. Moreover, as said in Section 2.2.2, for any piecewise continuous and bounded signal $f_d(t)$, $y(t)$ is bounded, with $\alpha_1 >> \alpha_2$. Indeed, there exists a Lyapunov function $V_a(t) = V_a(y(t))$ such that $\dot{V}_a(t) \leq 0$. Hence, we can state the following stability statement:

The closed-loop system (1)–(3) with control input $f(t) = y(t)$ defined in (7) and (8), and with $\alpha_1 >> \alpha_2$, is BIBO-stable.

The proof is as follows. Consider the following energy-kinetic-like Lyapunov function ([23,33] but the last term):

$$V(t) = \frac{1}{2}\rho \int_0^L u_t^2(x,t)dx + \frac{1}{2}T_o \int_0^L u_x^2(x,t)dx + \frac{1}{2}mu_t^2(L,t) + V_a(t). \tag{9}$$

Then, it is straightforward to obtain (for simplicity, in some functions, their arguments are intentionally omitted):

$$\dot{V}(t) = \rho \int_0^L u_t u_{tt} dx + T_o \int_0^L u_x u_{xt} dx + m u_t(L,t) u_{tt}(L,t) + \dot{V}_a(t). \quad (10)$$

Taking into account that $\dot{V}_a \leq 0$, by invoking (1) and (3), we have:

$$\dot{V}(t) \leq T_o \int_0^L u_t u_{xx} dx + T_o \int_0^L u_x du_t + u_t(L,t)[-T_o u_x(L,t) + f(t)]. \quad (11)$$

Then, after employing integration by parts and some algebraic simplifications, we arrive to

$$\dot{V}(t) \leq y(t) u_t(L,t). \quad (12)$$

From (8), we have to consider two cases:

(a) If $f_d(t) \geq y(t)$, then $\dot{V}(t) \leq -k_d u_t(L,t) u_t(L,t) = -k_d u_t^2(L,t) \leq 0$.

(b) If $f_d(t) < y(t)$, then $-y(t) < -f_d(t)$. Moreover, from (7) we induce that $u_t(L,t) = -\frac{1}{k_d} f_d(t)$. So, $\dot{V}(t) \leq y(t) u_t(L,t) = -y(t) \frac{1}{k_d} f_d(t) < -\frac{1}{k_d} f_d^2(t) \leq 0$.

We can affirm that $\dot{V}(t) \leq 0$, which means that $V(t)$ is bounded, thus concluding our main proof.

We are using the fact that, in real physical systems, if the energy of the systems is bounded, then all surrounding dynamic signals of the closed-loop system are bounded too [23].

3. Results

To analyze the proposed controller design of $f(t)$ as applied to the cable–tip–mass (1)–(3), we prepare two control cases for comparison:

(i) Standard boundary controller: $f(t) = f_d(t)$ (7);

(ii) Asymmetric peak detector controller: $f(t) = y(t)$ (7)-(8).

We set the following data: $m = 1$ Kg, $T_o = 1$ N, $\rho = 0.25$ kg/m, and $L = 1$ m. As initial conditions, we impose $u(x,0) = 0.01 \sin(\pi x)$ and $u_t(x,0) = 0$. The time interval is $[0, 120]$s. The value of the control gain k_d will be discussed in the next section. In programming, we use the numerical difference method with $dx = 0.1$ m and $dt = 0.005$ s. This discretization time was also employed for our peak detector system. To evaluate the total system energy employed by the controller, we consider the following functional energy index:

$$E(f) = \frac{1}{T} \int_0^T |f(t)| dt. \quad (13)$$

The current objective in control design is to determine the design parameters α_1 and α_2 with the aim of reducing, if possible, this performance index (13). To conclude the study, we consider the non-perturbed and externally perturbed systems in the following two sections.

3.1. Unperturbed Case Experiments

Consider the system defined in (1)–(3). The asymmetric peak detector parameters are set as $\alpha_1 = 1000$ and $\alpha_2 = 100$, verifying the BIBO constraint. First, we will discuss the controller behavior in the function of the value of the control gain k_d. We consider the case when $k_d = 100$ in (7). In this scenario, the conventional boundary controller exhibits instability. Conversely, the asymmetric peak detector boundary controller demonstrates strong performance, showcasing BIBO stability. Figure 6 depicts the performance of our proposed approach, displaying notable vibration attenuation, in contrast to the instability exhibited by the standard controller, as illustrated in Figure 7. For this experiment, we obtain that the energy index (13) for the proposed controller is $E(y) = 74.9$, in front of the boundary control energy value of $E(f_d) = 3432.6$.

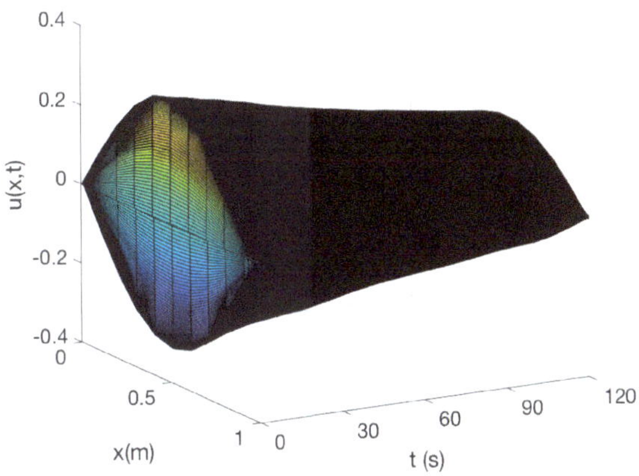

Figure 6. Simulation results for time interval $[0, 120]$, by using the asymmetric peak detector controller (7) and (8), with $k_d = 100$, $\alpha_1 = 1000$ and $\alpha_2 = 100$. The control objective is reached; the vibration decreases.

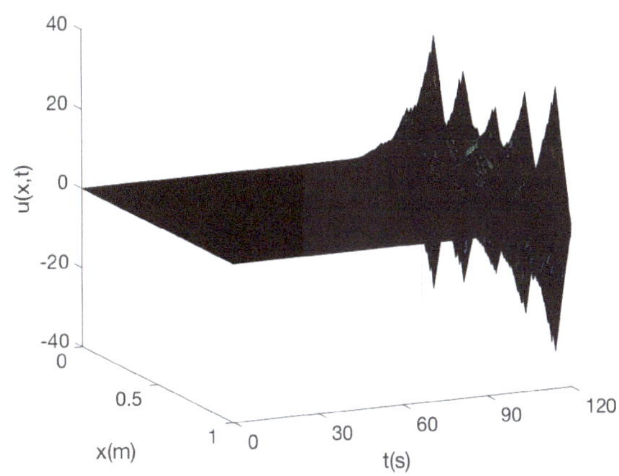

Figure 7. Simulation results for time interval $[0, 120]$, by using the standard boundary damper controller (7), with $k_d = 100$. In contrast to Figure 6, the vibration here does not exhibit bounded behavior.

One alternative for enhancing our performance is to raise the value of the control gain. We set now $k_d = 1000$ (7). The obtained results are shown in Figures 8–11. Using the asymmetric peak detector boundary controller, we induce from Figures 8 and 9 that in twenty seconds, the cable vibration is almost dissipated. On the other hand, under the standard boundary controller, we need fifty seconds to reach similar performance, as illustrated in Figures 10 and 11. We can observe that both controllers spend similar functional energy: $E(y) = 49.0$ and $E(f_d) = 43.3$. Finally, to exemplify the advantage of our asymmetric model, we now consider the symmetric case by setting $\alpha_1 = \alpha_2 = 100$ in (8). These values were tuned online, and the simulation result is presented in Figure 12. It can be inferred that the asymmetric model should be considered.

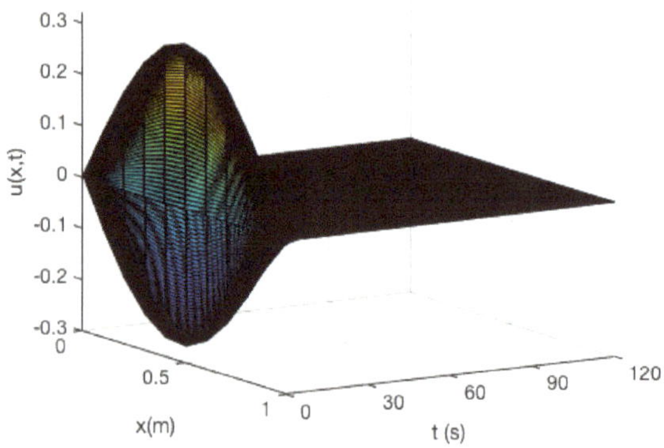

Figure 8. Simulation results of $u(x,t)$ by using the asymmetric peak detector controller (7) and (8), with $k_d = 1000$, $\alpha_1 = 1000$ and $\alpha_2 = 100$.

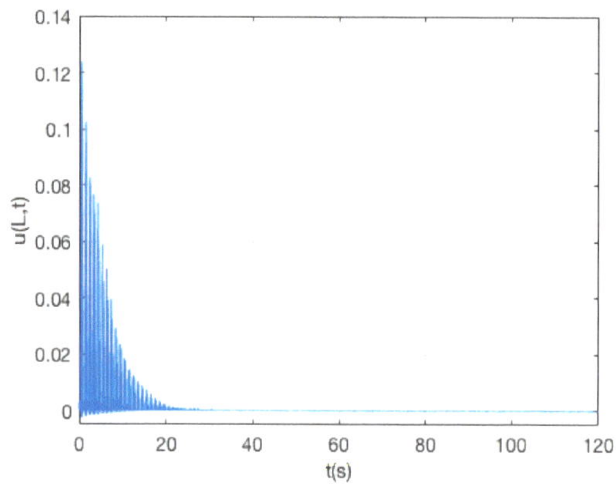

Figure 9. Simulation of $u(L,t)$ versus t, to appreciate the vibration attenuation time of twenty seconds, when the asymmetric peak detector controller (7) and (8) is considered. The rapid response of our controller is attributed to its consideration of peak values.

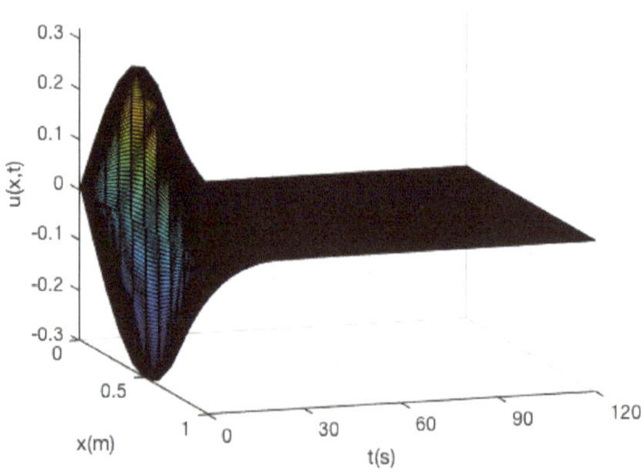

Figure 10. Simulation results $u(x,t)$ by using the standard boundary damper (7), with $k_d = 1000$.

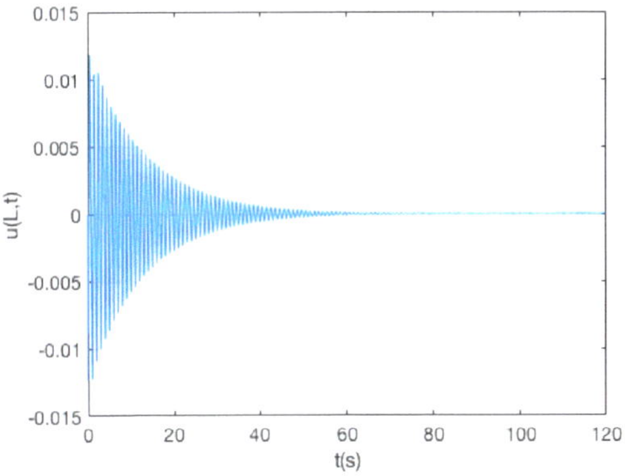

Figure 11. Numerical result of $u(L,t)$ versus t, by using the standard boundary damper (7). The vibration attenuation is reached in fifty seconds.

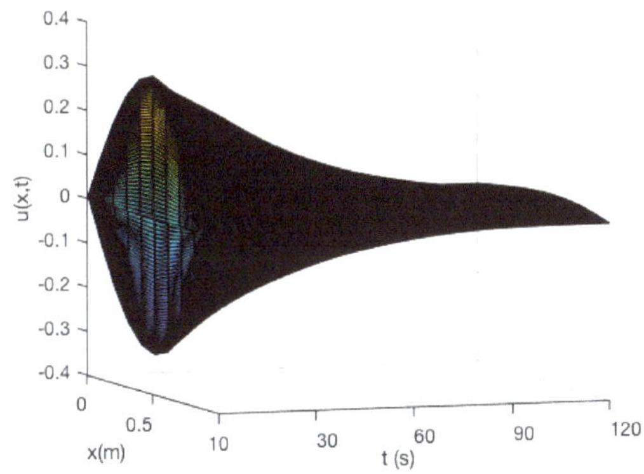

Figure 12. Numerical experiment using the standard peak detector controller (7) and (8), with $\alpha_1 = \alpha_2 = 100$, and control gain $k_d = 1000$.

3.2. External Disturbance Case Experiments

To evaluate the performance of the asymmetric peak detector controller, we examine the realistic case by considering external disturbances affecting the cable. These perturbations are represented by $g(x,t)$ as the disturbance along the cable, and $d(t)$ as the perturbation on the related boundary condition. The system equations are then:

$$\rho u_{tt}(x,t) - T_o u_{xx}(x,t) = g(x,t), \tag{14}$$

$$u(0,t) = 0, \quad t \geq 0, \tag{15}$$

$$m u_{tt}(L,t) + T_o u_x(L,t) = f(t) + d(t), \quad t \geq 0. \tag{16}$$

The corresponding results are presented in Figures 13–16, from which we deduce that both controllers exhibit similar performance, yet our proposal notably reduces the transient behavior time. The following specific cases are examined. The first case solely considers boundary disturbances:

$$g(x,t) = 0\,, \quad d(t) = 0.1 + 0.001\sin(0.2t). \tag{17}$$

Figures 13–15 illustrate the results. We observe that the asymmetric peak detector controller diminishes the time response, even though the total energy (13) is higher for the peak detector control: $E(y) = 64.51$ versus $E(f_d) = 58.15$. The thickness band of 0.1 shown in Figure 14 is due to the definition of $d(t)$ (17), where the term 0.1 is considered as a boundary condition. The cable inclination is illustrated in Figures 13 and 15.

The second case involves considering only external perturbations along the cable:

$$g(x,t) = 0.001\cos(\pi x t), \quad d(t) = 0. \tag{18}$$

This disturbance $g(x,t)$ is taken as a reference input, so the cable tries to maintain this shape, as can be observed in Figure 16. In this case, the compared behavior is very similar; the total energy is greater for our proposal ($E(y) = 132.16$, $E(f_d) = 99.73$), but again the time response is reduced.

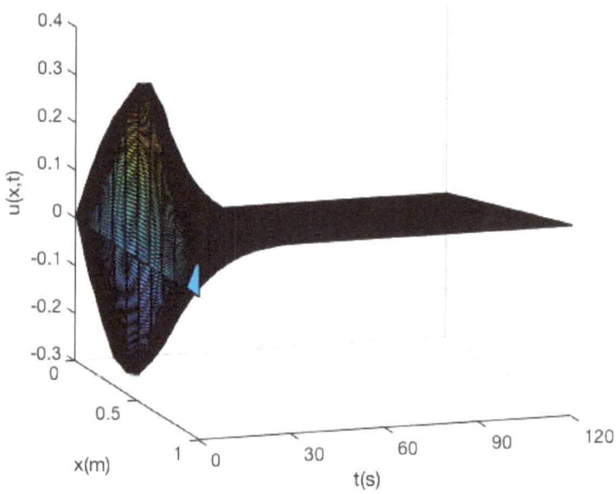

Figure 13. Simulation results by using the standard boundary controller (7), with $k_d = 1000$, of the boundary perturbed system: $g(x,t) = 0$ and $d(t)$ in (17).

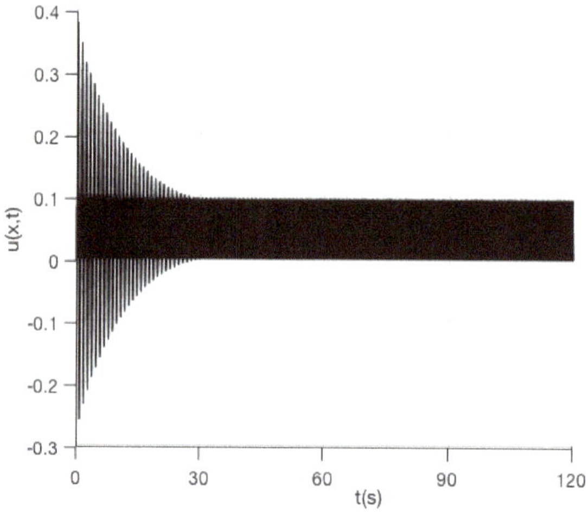

Figure 14. View of the plane $u(x,t)$ versus t (perspective of Figure 13). The thickness band of 0.1 is due to $d(t)$ (17).

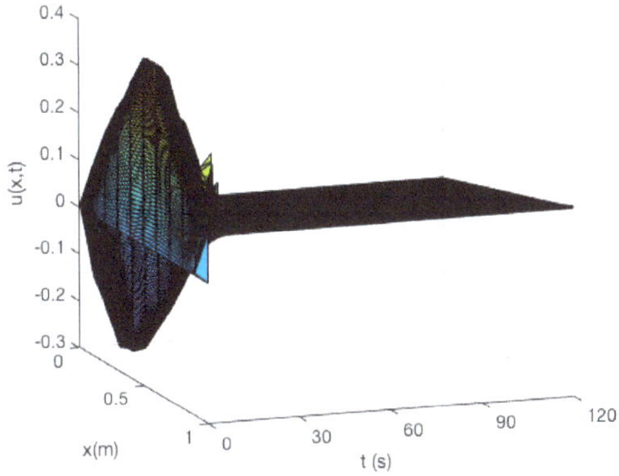

Figure 15. Simulation results by using our design (7) and (8), with $k_d = 1000$, when $g(x,t) = 0$ and $d(t)$ in (17).

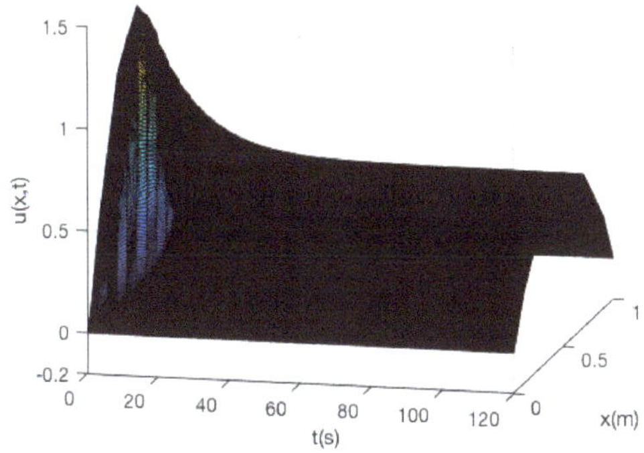

Figure 16. Simulation results by using (8), with $k_d = 1000$, when only disturbance along the cable is considered: $g(x,t)$ in (18) and $d(t) = 0$. This external perturbation $g(x,t)$ is taken as a reference input, so the cable tries to maintain this shape, as can be observed.

The inclusion of the asymmetric peak detector modification in the controller generally enhances the overall performance of the control design, which might specifically manifest as an improvement in the system's response time.

4. Discussion

The present study unveils a modified adaptation of a conventional boundary controller, devised to facilitate a more prompt response while minimizing energy usage in the management of vibrations in a cable–tip–mass system. This innovative approach involves a strategic modification of the peak detector system, incorporating the decoupling of a key design parameter to introduce what we have termed the asymmetric peak detector boundary control. The comparative analysis of its efficacy with the standard boundary damper forms a pivotal aspect of this investigation. Based on these simulations, it can be inferred that the parameters α_1 and α_2 govern the memory dynamics of the system. Hence, by tuning adequately the values of peak detector parameters, we can modulate different outputs, capturing the desired input values. The asymmetry comes from the decoupled parameter, allowing us to detect a desired value.

The results of numerical experiments highlight the effectiveness of the asymmetric peak detector controller in effectively reducing cable vibration, while operating with significantly reduced energy consumption. This successful mitigation not only prevent potential mechanical damage, a common concern associated with the use of unmodified controllers, but also underscores the substantial efficiency improvements brought about by this novel control strategy. When external perturbations are present, our approach may not reduce the total energy, but it significantly reduces the rise time, indicating a noticeable improvement.

Furthermore, the study highlights the moderate increase in total energy consumption when aiming for a swifter system response. These findings serve to emphasize the notable potential of integrating the asymmetric peak detector modification into the controller, thus offering a promising avenue for augmenting the overall efficacy of the control design.

5. Conclusions

In this paper, we introduce a novel modification of a standard boundary controller aimed at achieving a faster response with reduced energy consumption for controlling the vibration of a cable–tip–mass system. To achieve this, we alter the peak detector system by decoupling a design parameter, defining it as the asymmetric peak detector boundary control. Its performance is compared with the standard boundary damper. Numerical experiments indicate that the asymmetric peak detector controller effectively mitigates cable vibration with lower energy consumption, preventing mechanical damage that may arise with the unmodified controller. Additionally, when a faster response is desired, the total energy increase is moderate. The simulations presented in this study suggest that incorporating the asymmetric peak detector modification into the controller can significantly enhance the performance of the control design.

Author Contributions: Conceptualization, L.A. and G.P.-V.; methodology, L.A. and G.P.-V.; software, L.A. and G.P.-V.; validation, G.P.-V.; formal analysis, L.A. and G.P.-V.; investigation, L.A. and G.P.-V.; writing—original draft preparation, L.A. and G.P.-V.; writing—review and editing, L.A. and G.P.-V.; funding acquisition, G.P.-V. All authors have read and agreed to the published version of the manuscript.

Funding: This work has been funded by the Generalitat de Catalunya through the research projects 2021-SGR-01044.

Informed Consent Statement: Not applicable.

Data Availability Statement: No new data were created.

Conflicts of Interest: The authors declare no conflict of interest.

Abbreviations

The following abbreviations are used in this manuscript:

PDE	Partial Differenial Equation
BIBO	Bounded-Input Bounded-Output
D	Diode
C	Capacitor
R	Resistor

References

1. Morris, K. *Control of Systems Governed by Partial Differential Equations*; The IEEE Control Theory Handbook; CRC Press: Boca Raton, FL, USA, 2010.
2. Lasieka, I.; Triggiani, R. *Control Theory for Partial Differential Equations: Continuous and Approximation Theories*; Cambridge University Press: Cambridge, UK, 2000.
3. Lagnese, J. *Boundary Stabilization of Thin Plates*; SIAM: Philadelphia, PA, USA, 1989.
4. Bensoussan, A.; Prato, G.D.; Delfour, M.; Mitter, S. *Representation and Control of Infinite-Dimensional Systems*; Birkhauser: Basel, Switzerland, 2006.

5. Acho, L.; Pujol-Vazquez, G. A boundary control technique to the string-tip-mass system based on a non-symmetric peak-detector model. In Proceedings of the International Conference on System Theory, Control and Computing, Sinaia, Romania, 19–21 October 2017; pp. 759–762.
6. Krstic, M.; Smyslyaev, A. *Boundary Control of PDEs: A Course on Backstepping Desings*; SIAM: Philadelphia, PA, USA, 2008.
7. Luo, N.; Vidal, Y.; Acho, L. *Wind Turbine Control and Monitoring*; Springer: Berlin/Heidelberg, Germany, 2014.
8. Dibaj, A.; Gao, Z.; Nejad, A.R. Fault detection of offshore wind turbine drivetrains in different environmental conditions through optimal selection of vibration measurements. *Renew. Energy* **2023**, *203*, 161–176. [CrossRef]
9. d'Andréa Novel, B.; Coron, J.M. Exponential stabilization of an overhead crane with flexible cable via a back-stepping approach. *Automatica* **2000**, *36*, 587–593. [CrossRef]
10. Ballaben, J.S.; Sampaio, R.; Rosales, M.B. Stochastic dynamics of a non-linear cable-beam system. *J. Braz. Soc. Mech. Sci. Eng.* **2016**, *38*, 307–316. [CrossRef]
11. Caverly, R.J.; Forbes, J.R.; Mohammadshahi, D. Dynamic modeling and passivity-based control of a single degree of freedom cable-actuated system. *IEEE Trans. Control. Syst. Technol.* **2015**, *23*, 898–909. [CrossRef]
12. Qu, Z. An iterative learning algorithm for boundary control of a stretched string on a moving transporter. In Proceedings of the 3rd Asian Control Conference, Shanghai, China, 4–7 July 2000.
13. Bekiaris-Liberis, N.; Krstic, M. Compensation of wave actuator dynamics for nonlinear systems. *IEEE Trans. Autom. Control* **2014**, *59*, 1555–1570. [CrossRef]
14. Morgul, O.; Rao, B.P.; Conrad, F. On the stabilization of a cable with a tip mass. *IEEE Trans. Autom. Control* **1994**, *39*, 2140–2145. [CrossRef]
15. Guo, B.Z. Riesz basis approach to the stabilization of a flexible beam with a tip mass. *SIAM J. Control Optim.* **2001**, *39*, 1736–1747. [CrossRef]
16. Baicu, C.; Rahn, C.; Nibali, B. Active boundary control of elastic cables: Theory and experiment. *J. Sound Vib.* **1996**, *198*, 17–26. [CrossRef]
17. Zhang, F.; Dawson, D.; Nagarkatti, S.; Rahn, C. Boundary control for a general class of non-linear string–actuator systems. *J. Sound Vib.* **2000**, *229*, 113–132. [CrossRef]
18. He, W.; Zhang, S.; Ge, S.S. Adaptive control of a flexible crane system with the boundary output constraint. *IEEE Trans. Ind. Electron.* **2014**, *61*, 4126–4133. [CrossRef]
19. Li, Y.; Aron, D.; Rahn, C.D. Adaptive vibration isolation for axially moving strings: Theory and experiment. *Automatica* **2002**, *38*, 379–390. [CrossRef]
20. Shahruz, S. Boundary control of a nonlinear axially moving string. *Int. J. Robust Nonlinear Control* **2000**, *10*, 17–25. [CrossRef]
21. Zhao, Z.; Liu, Y.; Luo, F. Boundary Control for a Vibrating String System with Bounded Input. *Asian J. Control* **2018**, *20*, 323–331. [CrossRef]
22. Neusser, Z.; Martin, N.; Pelikán, J.; Pawlik, V.; Beneš, P.; Jan, Z.; Volech, J.; Halamka, V.; Machálka, M.; Valasek, M.; et al. Active vibration damping for manufacturing machines using additional cable mechanisms: Conceptual design. *Int. J. Adv. Manuf. Technol.* **2022**, *122*, 3769–3787. [CrossRef]
23. De Queiroz, M.S.; Dawson, D.M.; Nagarkatti, S.P.; Zhang, F. *Lyapunov-Based Control of Mechanical Systems*; Springer Science & Business Media: Berlin/Heidelberg, Germany, 2012.
24. Do, K. Stochastic boundary control design for extensible marine risers in three dimentional space. *Automatica* **2017**, *77*, 184–197. [CrossRef]
25. He, W.; He, X.; Zou, M.; Li, H. PDE Model-based boundary control design for a flexible robotic manipulator with input backlash. *IEEE Trans. Control Syst. Technol.* **2018**, *27*, 790–797. [CrossRef]
26. Krstic, M.; Guo, B.Z.; Balogh, A.; Smyshlyaev, A. Control of a tip-force destabilized shear beam by observer-based boundary feedback. *J. Control Optim.* **2008**, *47*, 553–574. [CrossRef]
27. Zhang, S.; He, W.; Li, G. Boundary Control Design for a Flexible System with Input Backlash. In Proceedings of the American Control Conference, Boston, MA, USA, 6–8 July 2016; pp. 6698–6702.
28. Zhang, H.W.; Wang, J.M. Boundary stabilization of an unstable parabolic PDE with a time-varying domain and the external disturbance. *Int. J. Control* **2023**, 1. [CrossRef]
29. Kelleche, A.; Saedpanah, F.; Abdallaoui, A. On stabilization of an axially moving string with a tip mass subject to an unbounded disturbance. *Math. Methods Appl. Sci.* **2023**, *46*, 15564–15580. [CrossRef]
30. Homayounzade, M. Adaptive robust output-feedback boundary control of an unstable parabolic PDE subjected to unknown input disturbance. *Int. J. Syst. Sci.* **2021**, *52*, 2324–2337. [CrossRef]
31. Pujol-Vázquez, G.; Acho, L. Semi-active control of flexible structures using acceleration information via peak-detector systems. In Proceedings of the 6th World Conference on Structural Control and Monitoring: 6WCSCM. Centre Internacional de Mètodes Numèrics en Enginyeria (CIMNE), Barcelona, Spain, 15–17 July 2014; pp. 1893–1901.
32. Floyd, T. *Electronic Devices*; Prentice-Hall Electronics: Hoboken, NJ, USA, 2005.
33. Cheng, M.B.; Su, W.C. Boundary stabilization and matched disturbance rejection of hyperbolic PDE systems: A sliding-mode approach. In Proceedings of the American Control Conference (ACC), Montreal, QC, Canada, 27–29 June 2012; IEEE: Piscataway, NJ, USA, 2012; pp. 5360–5365.

34. Chen, G. Energy decay estimates and exact boundary-value controllability for the wave-equation in a bounded domain. *J. Mathématiques Pures Appliquées* **1979**, *58*, 249–273.
35. Neusser, Z.; Necas, M.; Pelikán, J.; Karlícek, J.; Pawlik, V.; Benes, P.; Machálka, M.; Sika, Z.; Valásek, M. Mechatronic stiffness of cable?driven mechanisms: A study on production machine model. *Int. J. Adv. Manuf. Technol.* **2022**, *123*, 431?446. [CrossRef]
36. Ozbay, H. *Introduction to Feedback Control Theory-Chapter 4*; CRC Press: Boca Raton, FL, USA, 2000.

Disclaimer/Publisher's Note: The statements, opinions and data contained in all publications are solely those of the individual author(s) and contributor(s) and not of MDPI and/or the editor(s). MDPI and/or the editor(s) disclaim responsibility for any injury to people or property resulting from any ideas, methods, instructions or products referred to in the content.

MDPI AG
Grosspeteranlage 5
4052 Basel
Switzerland
Tel.: +41 61 683 77 34

Actuators Editorial Office
E-mail: actuators@mdpi.com
www.mdpi.com/journal/actuators

Disclaimer/Publisher's Note: The title and front matter of this reprint are at the discretion of the Guest Editors. The publisher is not responsible for their content or any associated concerns. The statements, opinions and data contained in all individual articles are solely those of the individual Editors and contributors and not of MDPI. MDPI disclaims responsibility for any injury to people or property resulting from any ideas, methods, instructions or products referred to in the content.

www.ingramcontent.com/pod-product-compliance
Lightning Source LLC
Chambersburg PA
CBHW041359210926
51850CB00013B/3129